Contemporary Studies and Theories in Tourism

İrfan Yazıcıoğlu / Özgür Yayla / Alper Işın /
Can Aktuna / Eren Yalçın (eds.)

Contemporary Studies and Theories in Tourism

Berlin · Bruxelles · Chennai · Lausanne · New York · Oxford

Library of Congress Cataloging-in-Publication Data
A CIP catalog record for this book has been applied for at the
Library of Congress.

**Bibliographic Information published by the
Deutsche Nationalbibliothek**
The Deutsche Nationalbibliothek lists this publication in the Deutsche
Nationalbibliografie; detailed bibliographic data is available online at
http://dnb.d-nb.de.

ISBN 978-3-631-92205-7 (Print)
E-ISBN 978-3-631-92828-8 (E-PDF)
E-ISBN 978-3-631-92829-5 (E-PUB)
10.3726/b22454

© 2024 Peter Lang Group AG, Lausanne
Published by Peter Lang GmbH, Berlin, Germany

info@peterlang.com - www.peterlang.com

All rights reserved.

All parts of this publication are protected by copyright. Any
utilisation outside the strict limits of the copyright law, without
the permission of the publisher, is forbidden and liable to
prosecution. This applies in particular to reproductions,
translations, microfilming, and storage and processing in
electronic retrieval systems.

Table of Contents

Hakan Oğuz Ari
Chapter 1 Health Tourism in Türkiye 9

Filiz Özlem Çetinkaya and Nurcan Çetiner
Chapter 2 Using SWOT Analysis in Creating Destination Image: The Case of Karaman Province 23

Ahmet Aydin
Chapter 3 A Theoretical Evaluation of the Use of Sociocultural Values for Branding in Tourism 33

Ayşe Topaloğlu and Sabri Arici
Chapter 4 Cultural Tourism Potential of Anamur 45

Havva-Gözgeç Mutlu and Volkan Akgül
Chapter 5 Wellness Hotel Practices: Flexibility and Mind 57

Didem Kutlu
Chapter 6 A Conceptual Overview of Regenerative Tourism 69

Münevver Çiçekdaği, Ayşe Cabi Bilge and Seda Özdemir Akgül
Chapter 7 What Is Türkiye's Tourism Industry's Place in the Digital Divide? 83

Savaş Yildiz and Zafer Yildiz
Chapter 8 Possible Trend in Tourism of Türkiye: Treasure Hunting Tourism 95

Cevat Ercik
Chapter 9 Touristic Motivation in the World of Flavors 109

Begüm Ilbay Vatan
Chapter 10 Volunteer Tourism Proposal for the Post-Kahramanmaraş Earthquakes Recovery 123

Uğur Can Aykanat and Gamze Şanli Ak
Chapter 11 The Effects of Zero Waste and Composting Method on the Costs of Fertilizer Production from Food Waste of Hotels in Seferihisar District of Izmir Province 135

Gözde Oğuzbalaban
Chapter 12 Green Marketing Practices in Tourism Businesses 145

Neslihan Onur and Ayşen Ertaş Sabanci
Chapter 13 Gastrotourists' Approaches towards Destinations 155

Olca Sezen Dogancili and Ramazan Guzel
Chapter 14 Artificial Intelligence and Post-Luddism in Tourism Industry 165

Duygu Doğan and Hasan Köşker
Chapter 15 The Role of Effective Communication in Crisis Management within the Tourism Sector 175

Haldun Demirel, Muhammed Demiralp and Evren Güçer
Chapter 16 Tourism in Turkey after 1980 187

Çağri Erdoğan and Zeynep Yamaç Erdoğan
Chapter 17 Tourism Product Generation Function of Wars and the Reflections on Tourism 199

Yasin Ozaslan
Chapter 18 Service Robots and Artificial Intelligence in the Hospitality Industry: A Literature Review 211

Nihat Çeşmeci
Chapter 19 Affiliate Marketing in Tourism and Its Implementation by Travel Bloggers 225

Yasemin Ersoy and Fuat Bayram
Chapter 20 Cultural Structure of Turkish Cuisine and Kitchen Equipment Used in Turkish Cuisine 239

Yağmur Kaplan
Chapter 21 Green Marketing in Tourism 247

Betül Buladi Çubukcu
Chapter 22 Green Marketing Practices in Tourism 257

Yusuf Bayraktar
Chapter 23 Hidden Treasure Street Foods Valued in Tourism: From
Cultural Interaction to Marketing Opportunities 269

Erkan Denk and Furkan Zirzakiran
Chapter 24 Analyzing Communicative Strategies in Online Hotel
Reviews: A Case Study of Winter Tourism Corridor 279

Mehmet Necati Cizrelioğulları and Tuğrul Günay
Chapter 25 The Impact and Strategic Implications of Digital Marketing
on the Tourism Industry 299

Hakki Çilginoğlu and Kaan Berk Dalahmetoğlu
Chapter 26 The Conceptual Framework of Dark Tourism 313

Ozan Esen
Chapter 27 Employee Theft in Tourism Businesses 321

Diğdem Eskiyörük
Chapter 28 Holistic Review of Organizational Mindfulness and Mindful
Leadership in the Perspective of Sustainable Tourism
Management 329

Engin Tengilimoğlu
Chapter 29 The Role of Virtual Reality in Tourism: Pre-travel Decision
Making and On-Visit Experience Enhancement 341

Gulsun Yildirim and Sena Bakir
Chapter 30 Developing Gastronomy Tourism in Turkey through Food
Geographical Indication 355

Sabahat Deniz, Günay Ahmadli and Hakan Koç
Chapter 31 Green Management Practices in Tourism Businesses 363

Erdem Şimşek
Chapter 32 The Future of Travel Assistance: SWOT Insights into Chatbot
Usage in Tourism Industry 377

Kürşat Başkan
Chapter 33 Integration of Technology in Small Tourism Enterprises 387

Murat Hacimurtazaoğlu and Gülsün Yildirim
Chapter 34 Evaluation of the Use of Fuzzy Logic in Decision-Making
              Processes in Tourism 399

Seval Kurt and Cansu Uzun Güripek
Chapter 35 Ecological Approach to Tourism System 411

Hakan Oğuz Ari[1]

Chapter 1 Health Tourism in Türkiye

Introduction

The advancement of science and technology, the acceleration of globalization and the increase in expectations from health systems have caused people to seek international access to health services. As individuals' desire to access fast and quality health services gained an international dimension, the concept of health tourism emerged, and a new economic market was created for countries. Cross-border mobility of people with the desire to access faster, cost-effective, and reliable health services is defined as health tourism (Rai, 2019). Health tourism; it is recognized as a dynamic, rapidly growing, multidisciplinary field of economic activity and knowledge. Many factors such as quality of health services, cost, culture, social norms, and treatment duration are effective in country preferences (Zhong et al., 2021).

Since health tourism occurs as a result of international mobility, cross-border travels for different purposes need to be categorized. Cohen (2014); evaluated health tourists in three subcategories: (a) Tourists for holiday and treatment purposes: Those who prefer countries whose travel reasons are not only for treatment purposes and have sufficient facilities for the treatment of their existing illnesses. (b) Vacationing patient: These are people who use their preferred region for a holiday after treatment. After the treatment, they continue holiday. They take the availability of these opportunities into account when choosing a country. (c) Mere patients: The main purpose of these people's travel destination preferences is the use of healthcare services.

According to the generally accepted classification of health tourism worldwide, its three main sub-components are it is seen that there are tourism services including medical tourism, thermal tourism and spa-wellness tourism, care and rehabilitation services for the elderly and the disabled (Tengilimoğlu, 2017). Increasing costs of health institutions affect social security institutions financially. To solve such difficulties, social security institutions and private health insurance organizations focus on providing health services to their citizens at

---

1 Ph.D., University of Health Sciences Ankara, Türkiye, Faculty of Gülhane Health Sciences, Department of Healthcare Management, hakanoguz.ari@sbu.edu.tr

a lower cost by making lower-cost contracts with countries with quality health facilities. Thanks to technological developments, especially in medical services, a globalization effort has begun in healthcare services around the world, and with the development of transportation facilities between countries, such medical journeys have tended to increase.

Health tourism, which was previously considered an alternative type of tourism, has developed as a sector over time, its economic power has rapidly increased, and it has become an attractive field. In an increasingly globalizing world, people's health needs and the ways they meet them have also differed. Motivated by factors such as inadequate insurance coverage in some countries, long waiting times, high healthcare fees, receiving better quality service and benefiting from high technology, health tourism has become a growing market and competitive area. In this context, Türkiye is on its way to becoming a strong actor in the field of health tourism with its tourism potential and health sector capacity. In this study, the general situation of health tourism in Türkiye will be examined in detail under the headings of medical, thermal, elderly, and disabled health tourism categories. In addition, the studies carried out by public authorities in Türkiye to regulate the health tourism sector will be mentioned and its potential will be evaluated.

Health Tourism at a Glance in Türkiye

Health tourism is a developing field all over the world due to basic factors such as people's inability to access health services produced by qualified workforce in their own countries, their inability to access high technology, and the provision of better-quality health services at lower costs in other countries (Kaymaz, 2018). Globalization around the world, long waiting times in some countries, awareness of consumers, the search for new markets by the sectors, the ability to perform illegal procedures (such as abortion) in different countries, and the desire of the tourist mass, who also have a holiday habit, to find a cure for their chronic diseases while traveling are other factors that positively affect the development of health tourism.

Health tourism creates significant added value to economies due to its structure that affects many sectors such as hotels, agencies, and transportation, especially health facilities. For this reason, health tourism has the power to be a source of growth for the economies of the countries. While health tourism enables people to access quality health services, it has become a rapidly growing sector that supports economies by increasing the employment rate. For this reason, health tourism is considered as a priority sector by many governments, and various targets are set and supported by investments and incentives (Kılavuz, 2018).

Countries such as the United States, India, Germany, South Africa, Thailand, Brazil, Mexico, and Singapore are the leading countries in health tourism in the world. In a study conducted by Global Healthcare Resources (GHR) in 2016, the top seven most preferred countries for health tourism were stated as the USA, Germany, Türkiye, India, England, Thailand, and the United Arab Emirates (Tarınç, 2019).

Türkiye is one of the most ideal regions for medical tourism and alternative health tourism thanks to its climate, sea, beaches, thermal springs, thalassotherapy facilities, forests, plateaus, as well as the ease of transportation it offers with an international airline brand that flies to the most destinations in the world (Turkish Airlines). At the same time, Türkiye has become a rising value in the field of health tourism with the worldwide recognition of cities such as Istanbul, Ankara, Izmir and Antalya, tourism infrastructure that has become a state policy and its share from the world tourism market (Tontus, 2019a).

The progress that the government has made in the field of health, especially with the Health Transformation Program implemented since 2003, the policies it has developed in the field of health tourism, especially in the period covering recent years, and the projects it has carried out, have given Türkiye the image of a great health country. In this context, in addition to the development of health services, Türkiye is taking firm steps towards becoming one of the leading countries in the field of health tourism, thanks to the development of health services and building a health system that exports health services and sets an example to the world, developing bilateral agreements with other countries, regulating this field through legal regulations, and developing its institutional structure.

Table 1.1 shows Türkiye's number of tourist and health tourist, revenues from tourism and health tourism by years.

Table 1.1. Total Number of Tourists and Health Tourists and Tourism and Health Tourism Income by Years, 2003–2023, Türkiye

Year	Tourism Income (Thousand $)	Health Tourism Income (Thousand $)	Share of Health Tourism Income in Tourism Income (%)	Number of Tourists	Number of Health Tourists	Share of the Number of Health Tourists in the Total Number of Tourists (%)
2003	13 854 868	203 703	1.5	16 302 050	153 223	0.9
2004	17 076 609	283 789	1.7	20 262 642	204 790	1.0
2005	20 322 110	343 181	1.7	24 124 502	269 801	1.1
2006	18 593 950	382 412	2.1	23 148 669	230 171	1.0
2007	20 942 500	441 677	2.1	27 214 988	223 882	0.8
2008	25 415 067	486 342	1.9	30 979 979	261 240	0.8
2009	25 064 481	447 296	1.8	31 972 377	222 597	0.7
2010	24 930 996	433 398	1.7	33 027 944	183 070	0.6
2011	28 115 693	488 443	1.7	36 151 328	208 524	0.6
2012	29 689 249	627 862	2.1	36 463 921	240 682	0.7
2013	33 073 502	772 901	2.3	39 226 226	300 102	0.8
2014	35 137 949	837 796	2.4	41 415 070	473 896	1.1
2015	32 492 212	638 622	2.0	41 617 530	395 019	0.9
2016	22 839 468	715 438	3.1	31 365 330	400 699	1.3
2017	27 044 542	827 331	3.1	38 620 346	467 302	1.2
2018	30 545 924	863 307	2.8	45 628 673	594 851	1.3
2019	38 930 474	1 492 438	3.8	51 860 042	701 046	1.4
2020	14 817 273	1 164 779	7.9	15 826 266	407 423	2.6
2021	30 173 587	1 726 973	5.7	29 357 463	670 730	2.3
2022	46 477 871	2 119 059	4.6	51 369 026	1 258 382	2.4
2023	54 315 542	2 307 130	4.2	57 077 440	1 398 504	2.5

Source: TurkStat.

In 2003, Türkiye earned a total of $13.8 billion in tourism revenue, of which only 1.5 % was health tourism revenues ($203.7 million). By 2010, this rate had increased to 1.7 % and $433.4 million. In the following years, both Türkiye's tourism revenues and health tourism revenues gradually increased. The Covid-19 epidemic, which emerged in December 2019, has deeply affected the health and health tourism sector as well as many other sectors. However, despite the pandemic conditions in Türkiye, the income from health tourism and its share in tourism revenues continued to increase. Türkiye, which earned $1.1 billion in health tourism revenue (7.9 % of total tourism revenue) in 2020, exceeded $2 billion in revenue for the first time in 2022 and earned $2.3 billion in health tourism revenue in 2023, earning 4.2 % of its total tourism revenue from health tourism. The number of foreign patients coming to Türkiye fluctuates over the years. The number of health tourists, which approached the limit of five hundred thousand in 2017, exceeded seven hundred thousand in 2019. It decreased in 2020 due to the pandemic, and in 2022, it exceeded the one million. In 2023, it is seen that 1.4 million health tourists come to Türkiye (TurkStat, 2024).

Medical Tourism in Türkiye

Although Türkiye entered the health tourism sector later compared to its Far Eastern country competitors, which entered the health tourism sector in the 1970s, it has become competitive in the sector thanks to its geographical location, qualified health manpower, price advantage, infrastructure regulations, quality tourism management and famous Turkish hospitality.

Medical tourism in Türkiye refers to the visit of foreign patients to the country for health services. Türkiye has significant potential in this field and is among the top 10 medical tourism destinations in the world (MedicalNews, 2024). When we look at the general situation of medical tourism in Türkiye, it is seen that it has strengths such as its geographical location, rich historical, cultural, and touristic texture, specialist physicians, experienced healthcare personnel and quality healthcare services, advanced medical device park, competitive price advantage, state support and infrastructure opportunities. The low number of employees who speak foreign languages and the slow functioning of the bureaucracy in public hospitals are the weaknesses. The ease of transportation to Türkiye and the higher risk of epidemics in Southeast Asia can be seen as an opportunity for Türkiye. Political instability and its possible repercussions in neighboring countries and the low-price policies of other countries in the sector are seen as a threat (Akbolat and Gülçindeniz, 2017; USTTAK, 2024).

Türkiye has become a very popular destination in terms of health tourism in recent years. The high-quality health services, modern medical technology, and affordable prices offered have made Türkiye a world-class health tourism center. Some of the prominent areas in health tourism in Türkiye are listed as follows (Turizm Günlüğü, 2022):

Hair Transplant: Türkiye is a destination that has gained a worldwide reputation for hair transplantation. Hair transplantation procedures are successfully performed with specialist doctors and modern facilities.

Eye Diseases Treatment: In the field of eye health, Türkiye is known for its quality services and specialist doctors. It is preferred for cataract surgeries, laser eye treatments and other eye diseases.

Dentistry: Türkiye is also popular for cosmetic dentistry, implants, and other dental treatments. Service is provided by specialist dentists and modern clinics.

Aesthetic Surgery: Türkiye is preferred for plastic surgery and body contouring procedures. Specialist plastic surgeons and reliable health institutions provide services in this field.

IVF Treatment: Türkiye is also an important center for IVF treatment. With high success rates and experienced specialists, couples prefer Türkiye in this field.

In addition to these services, health services such as cardiac surgery, orthopedics, rheumatology, and check-up are also popular in Türkiye. Health tourism services in our country are at world standards and can compete with European Union countries. Therefore, a wide range is offered for patients who are considering Türkiye for health tourism.

Accreditation is one of the most important factors that give health institutions a reputation on an international scale and increase their brand value. In particular, the Joint Commission International (JCI), which is the first accreditation institution in health services, gives a roadmap for health institutions to reach an excellent structure and ensures that the success of the health institution is recognized by the public (Avcıl and Uslu, 2022). As of May 2024, Türkiye has 40 internationally accredited healthcare providers by JCI (JCI, 2024). Hospital accreditation, which is believed to be effective in the selection of health facilities in the countries planned to be visited, can be seen as an indicator of how ready the countries are for international medical tourism. From the point of view of medical tourism, one of the most striking advantages of Türkiye is the abundance of accredited health facilities.

As of 2021, Türkiye has been ranked 30th in the Medical Tourism Index (MTI), which ranks countries according to their potential in the field of health tourism, by the Medical Tourism Association, one of the leading authorities in the global medical tourism sector. Among the European countries where Spain ranks first, Türkiye has found itself in the eighth place (MTA, 2024).

Another important advantage that Türkiye has in medical tourism compared to other countries is the affordability of the prices of medical procedures. The prices of medical tourism services in Türkiye are 50 % to 80 % cheaper than in the United States and the United Kingdom. For example, while the cost of a dental implant in the United States is 3,000 dollars, it is 540 dollars in Türkiye. It is possible to have five implants in Türkiye for the same price as one implant in the United States. The prices of dental, orthopedic, and aesthetic procedures in Türkiye, the UK and the USA are compared in the Table 1.2 below (Medical Center, 2023):

Table 1.2. Health Tourism Services Prices by Country

Procedure	United Kingdom	USA	Türkiye
Titanium implant	$1,960	$3,000	$540
Teeth whitening	$520	$600	$250
Root canal	$1,700	$700	$125
Knee replacement	$15,000	$20,000	$2,300
Hip replacement	$15,000	$21,500	$2,370
Face lift	$13,000	$7,500	$2,200
Breast Augmentation	$4,500	$5,000	$2,600
Rhinoplasty	$5,800	$5,400	$1,800
Hair transplant	$4,000	$4,000	$1,250

Source: Medical Center.

In general, the difference between costs and service quality are effective and decisive in people's departures abroad. In this context, it is seen that Türkiye has a lower cost in many medical procedures, especially in developed countries, and it is thought that it can gain a place in the medical tourism market with this advantage (Binler, 2015). Although price is not the only determining factor in the destination selection of health tourists, an affordable price policy that meets quality criteria has the potential to offer very attractive opportunities, as in the case of Türkiye.

Thermal Health Tourism in Türkiye

The concept of "Thermal Tourism", which is a type of health tourism, dates to 500 BC according to written sources in Europe. Even thousands of years ago, it is known that there were facilities like today's modern SPA facilities. The therapeutic effects of thermal water are well known even in ancient times, and thermal bath culture was widely used in Ancient Greece (Szabó et al., 2013). Therefore, it can be said that thermal health tourism is one of the oldest types of tourism in terms of history (Boroviç and Marković, 2015). Türkiye is among the top ten countries in the world in terms of geothermal resource richness. Thermal waters in Türkiye; flow rate and temperatures, quantities, various physical and chemical properties, namely, it is superior to the thermal waters in Europe in terms of quality and quantity and provides a wide treatment area. It is widely located in every region of the country and is intertwined with other types of tourism. (Özbek, 1991).

Türkiye is a country located in the Alpine-Himalayan belt and therefore has a high geothermal potential. Geothermal resources in Türkiye are the product of volcanism, which is associated with the northern, eastern, and western Anatolian fault zones. Türkiye has more than 1,500 thermal springs with a temperature above 20 °C and flow rates ranging from 2 to 500 lt/sec (Şengül and Bulut, 2019).

Türkiye, which is one of the richest countries in the world in terms of water resources suitable for thermal health tourism, has made significant developments in recent years, especially in order to make effective use of these opportunities. It is seen that modern thermal touristic businesses have been opened in many regions, especially in locations such as Afyon, Pamukkale, Bursa, and Kızılcahamam (Tontus, 2019b). There are around 190 hot springs in 46 provinces in Türkiye. Facilities with thermal tourism operation certificates and many facilities with spa operation licenses serve in the field of thermal tourism (GEKA, 2009). However, despite all these developments, inadequacies in business, human resources and promotional problems cause these resources to remain idle. Only around 10 % of the existing thermal resources in Türkiye are used and it seems that the potential cannot be used effectively (Okumuş, 2023).

There are a total of 53 5-star thermal hotels in Türkiye. These thermal facilities serve thermal tourists with a total of 12,304 rooms and 25,794 beds. There are 43 thermal hotels with 4 stars and 21 thermal hotels with 3 stars. In addition to these, there is one 5-star thermal resorts with tourism operation certificates and 4 thermal detached apart hotels with tourism operation certificates in Türkiye. Thermal facilities with investment and operation certificates in Türkiye have a total capacity of 19,244 rooms and 40,866 beds (Ministry of Culture and Tourism, 2023).

Due to high healthcare costs in developed countries, there is an increasing search to reduce treatment services by providing healthcare services at lower costs from countries that provide quality services. After Sweden, Norway and Denmark, some private insurance companies in the Netherlands and Germany have decided to cover some or all of the treatment costs of their citizens who will receive health services for thermal treatment in Türkiye. This issue creates a significant market share in terms of the Turkish thermal health tourism market and is described as motivating developments for increasing demand and making the necessary investments (Aydemir Atay, 2019). In order to better utilize its thermal resources and activate its potential in this field, Türkiye has made developing thermal health tourism a state policy by preparing strategy documents and action plans within the scope of the 10th Development Plan, Thermal Tourism Master Plan and Türkiye Tourism Strategy.

Elderly and Disabled Health Tourism in Türkiye

It is seen that the concept of elderly health tourism, which is called geriatric tourism, third age tourism or old age tourism, is still in the development stage in the general health tourism market. Elderly tourism is a type of health tourism that has emerged with the increase in the care and treatment needs of individuals aged 65 and over in parallel with their average age in the last 20–30 years (Aydın et al., 2011). Elderly care tourism; it is a common working area of tourism, health and care sectors. In addition to the target group's demands for old-age tourism for healthy aging, there may be demands for care and health tourism to meet long-term care and, when necessary, health service needs. Demands related to healthy aging can be met mainly through elderly care tourism strategies. Tourism activities aimed at improving health are within the scope of health tourism, regardless of age group (Ökem and Çelik, 2019).

Disabled tourism, on the other hand, is a tourism activity carried out by disabled individuals. It is also called barrier-free tourism and accessible tourism. Accessible tourism is defined as a tourism activity that provides access to tourism opportunities for all disabled and non-disabled individuals, especially disabled and elderly individuals (Denizli, 2023). Especially in European countries, there is an increase in the rate of disabled people as well as in the elderly population.

The Ministry of Health of the Republic of Türkiye states that elderly and disabled tourism includes elderly and disabled tourism (sightseeing tours and occupational therapies), elderly care services (nursing homes or rehabilitation services), rehabilitation services in clinical hotels and special care and sightseeing tours for the disabled, and these services are provided in areas such as clinical hotels, holiday villages and nursing homes (İzmir Ticaret Odası (İTO), 2024).

Due to its geographical location, Türkiye is in a location that is more easily accessible to both elderly and disabled individuals. Factors such as the provision of rehabilitation services in thermal facilities in Türkiye, its natural resources, the richness of historical and touristic places, the existence of alternative treatment opportunities (hippotherapy, etc.), accredited hospitals, specialist physicians and health professionals, less waiting times, lower treatment and care costs, the existence of rehabilitation centers and holiday villages built for elderly-disabled tourism in Türkiye are important advantages for the tourism and also health tourism (Denizli, 2023). As of 2023, 14,428 elderly people are served by the Ministry of Family and Social Services in Türkiye with a capacity of 169 nursing homes and 17,648 beds. In addition, 21 nursing homes belonging to other public institutions and 268 nursing homes belonging to the private sector have a capacity of 20,583 beds. In addition, elderly living homes and elderly day care centers are also institutions that provide services in this context (Aile, Çalışma ve Sosyal Hizmetler Bakanlığı, 2023).

Türkiye ratified the Convention on the Rights of Persons with Disabilities, which was adopted by the United Nations in 2006, in 2008 and supports the active participation of people with disabilities in tourism activities like other people around the world. In this way, Türkiye, like other developed countries, has put the protection and defense of the rights of the disabled on its agenda. In addition, with this convention, people with disabilities have the opportunity to seek their rights on the international platform (Türk et al., 2022) Türkiye has implement special arrangements for people with disabilities in order to enable people with disabilities to participate in tourism activities. The Association of Turkish Travel Agencies (TÜRSAB) established the "Barrier-Free Tourism Committee for All" in 2006. This committee carries out studies in order to enable people with disabilities to participate in tourism activities. TÜRSAB and this committee are among the 400 registered members of the European Barrier-Free Tourism Network (ENAT).

Although Türkiye has not yet achieved the success it has gained in medical tourism in advanced age and disabled health tourism, it is making legal arrangements in order to use its potential more effectively and is also expanding its investments in order to have a say in this field.

Institutional Structuring and Legal Regulations

In order to become an important actor in the globally growing health tourism sector, Türkiye has carried out a series of institutional structures and implemented some legal regulations. In 2010, with the "Health Tourism Coordinatorship" established within the Ministry of Health, health tourism policies were officially determined. The following year, this unit was structured as the Department. In the following years, the concept of health tourism found its place in Development Plans. In order to ensure inter-sectoral integration related to health tourism, the "Health Tourism Coordination Board (SATURK)" was established in 2011 with the Prime Ministry Circular. One of the important building blocks for regulating the field of health tourism, such as the International Health Tourism and

Tourist Health Regulation, entered into force in mid-2017 within the scope of this initiative.

USHAŞ (International Health Services Inc.), which is the one of the relevant institution of the Ministry of Health, started its operations in early 2019 in order to promote the services offered in Türkiye in the field of international health services, to support and coordinate the activities of the public and private sectors for health tourism, to make recommendations to the Ministry on policies and strategies regarding international health services, service delivery standards and accreditation criteria.

Currently, tasks such as planning services related to Health Tourism and Tourist's Health, granting the necessary permissions, ensuring coordination with all relevant institutions, establishing the admission criteria of patients coming from abroad and examining the requests and complaints of these patients, providing assistance and consultancy services and keeping records are carried out by the Health Tourism Department operating under the Ministry of Health, General Directorate of Health Services.

"Compulsory Professional Liability Insurance", known as complication insurance, which provides an important assurance to patients in case healthcare professionals harm third parties while providing services, has been made compulsory in Türkiye. In this way, it is aimed to protect both service providers and service recipients. With the legal regulations, the opening of treatment centers that can provide elderly care and rehabilitation services, the provision of dialysis services for tourists and health tourism in accommodation facilities and clinical guest houses, and the opening of physical therapy and rehabilitation centers with accommodation to serve health tourism. Additionally, in order to control the prices of services offered within the scope of health tourism, minimum wages are determined by the Ministry of Health and are monitored periodically. Thanks to these regulations, Türkiye plans to achieve the targets for thermal, medical and elderly-disabled health tourism that it has determined regarding health tourism within the scope of the 10th Development Plan and implements this as a state policy.

Results

Türkiye is one of the rare countries that has the ability to provide services in all types of health tourism with its resources and potential. It is making a name for itself, especially thanks to the investments and legal regulations made in recent years. However, in order to achieve the goals, it planned to achieve in previous years and its aim of becoming one of the leading countries in health tourism,

Türkiye should make investments and studies not only in medical tourism but also in thermal and elderly-disabled health tourism (Arı, 2022).

Millions of people around the world prefer health tourism to undergo eye, dental, orthopedic, and especially aesthetic surgery operations at lower costs and with shorter waiting times, and to perform various treatments in a quality and effective manner. This population group also wants to visit historical, cultural, and natural beauties and have a better time after treatment.

In this context, in Türkiye, treating the health tourism sector as a state policy and controlling it by determining the authorization criteria of health service providers and agencies through legal regulations makes it easier for world citizens to receive health services safely. In addition, important incentive mechanisms provided by the state accelerate the development of all categories of health tourism and the making of additional investments.

Türkiye is known not only for its geographical location and health institutions with developed infrastructure, but also for its trained and educated manpower in the health and tourism sectors, advanced technological infrastructure in the field of health, having the most important geothermal resources in Europe, historical and natural beauties, rich cultural heritage and It is one of the important destinations that can be preferred in the field of health tourism due to its low costs and quality health services.

References

Aile, Çalışma ve Sosyal Hizmetler Bakanlığı. (2023). Engelli ve Yaşlı İstatistik Bülteni, Engelli ve Yaşlı Hizmetleri Genel Müdürlüğü, Ankara.

Akbolat, M., Gülçin Deniz, N. (2017). Türkiye'de Medikal Turizmin Gelişimi ve Bazı Ülkelerle Karşılaştırılması. *Uluslararası Global Turizm Araştırmaları Dergisi, 1(2)*, 123–139.

Arı, H.O. (2022). Türkiye'deki Sağlık Turizmi Politikalarının, Sektörel Hedefler Bağlamında Mevcut Durumunun Değerlendirilmesi (Evaluation of the Current Situation of Health Tourism Policies in Turkey in the Context of Sectoral Targets). *Journal of Tourism and Gastronomy Studies*. 10.21325/jotags.2022.1005.

Avcıl, S., Uslu, K. (2022). JCI Sağlık Standartlarının Akreditasyonunun Türkiye'de Uygulanmasında Üniversite ve Devlet Hastanelerinin Hizmet Kalitesinin İncelenmesi. *Doğuş Üniversitesi Dergisi, 23(1)*, 279–297.

Aydemir Atay, A. (2019). Uluslararası Termal Sağlık Turizmi Potansiyelinin Arttırılmasında Stratejik Pazarlama Planlaması: Pamukkale Karahayıt Termal

Turizm Bölgesi İçin Bir Araştırma. Yüksek Lisans Tezi, Pamukkale Üniversitesi Sosyal Bilimleri Enstitüsü, Denizli.

Aydın, D., Aktepe, C., Sahbaz, R. P., Arslan, S. (2011). Türkiye'de Medikal Turizmin Geleceği. Sağlık Bakanlığı Yayınları, 1–22.

Binler, A. (2015). Türkiye'nin Medikal Turizm Açısından Değerlendirilmesi ve Politika Önerileri. Kalkınma Bakanlığı, Sosyal Sektörler ve Koordinasyon Genel Müdürlüğü, Uzmanlık Tezi.

Boroviç, S., Marković, I. (2015). Utilization And Tourism Valorisation of Geothermal Waters İn Croatia. *Renewable And Sustainable Energy Reviews,* 44:52–63.

Cohen, E. C. (2014). MedicalTourism in Thailand. *AU-GSB E-JOURNAL,* 1(1).

Denizli, F. (2023). Yaşlı ve Engelli Bakım Turizm Potansiyelinin Değerlendirilmesi "Fırsatlar ve Tehditler": Kayseri İli Örneği. *Süleyman Demirel Üniversitesi Sosyal Bilimler Enstitüsü Dergisi (46),* 59–75.

Ministry of Culture and Tourism. (2023). Statistics of Tourism Certified Facility.

Global Healthcare Resources (GHR). (2016). Global Buyers Survey.

Güney Ege Kalkınma Ajansı (GEKA). (2009). Güney Ege Bölgesinde Termal Turizm.

İzmir Ticaret Odası (İTO). (2024). İzmir'de Sağlık Turizmi. https://www.izto.org.tr/tr/tg/izmirde-saglik-turizmi. (Access Date: 2.05.2024)

Joint Comission International (JCI). (2024). JCI-Accredited Organizations. https://www.jointcommissioninternational.org/who-we-are/accredited-organizations/#sort=%40aoname%20ascending&f:@aocountry=[Turkey] (Access Date: 28.04.2024)

Kaymaz, Ç. (2018). 2010 Sonrasında Türkiye'de Sağlık Turizmi'nin Gelişimi, Yüksek Lisans Tezi, Namık Kemal Üniversitesi Sosyal Bilimler Enstitüsü, Tekirdağ.

Kılavuz, E. (2018). Medical tourism competition: The case of Turkey. *International Journal of Health Management and Tourism.* 42–58.

Medical Center. (2023). Turkey: Health Tourism in 2023. https://www.medicalcenterturkey.com/turkey-health-tourism-in-2023/. (Access Date: 1.05.2024)

Medical Tourism Association. (2024). Medical Tourism Index 2020–2021.

MedikalNews. (2024). Türkiye'de Sağlık Turizmi, https://www.medikalnews.com/turkiyede-saglik-turizmi/ (Access Date: 28.04.2024)

Okumuş, N. (2023). TR33 Bölgesinde Turizm Potansiyelinin Değerlendirilmesi, Termal Sağlık Turizmi. Regions 2030 Projesi ve TR33 Bölgesi'nde Sürdürülebilir Kalkınma Hedefleri Bölgesel Toplantısı, Afyonkarahisar, Türkiye.

Ökem, G.Z., Çelik, H. (2019). Türkiye Hizmet İhracatında Yeni Hedefler: Yaşlı Bakım Turizmi. TÜSİAD Yayınları, Yayın No: TÜSİAD-T/2019-11/610.

Özbek, T. (1991). Dünya'da ve Türkiye'de Termal Turizmin Önemi. *Anatolia: Turizm Araştırmaları Dergisi, 2(3),* 15–29.

Rai, A. (2019). Medical Tourism in Kolkata, Eastern India, Global Perspectives on Health Geography, Springer International Publishing AG, part of Springer Nature, ISBN 978-3-319-73272-5 (eBook).

Szabó, Z., Koscondi, J., Lakne, Z. (2013). Role Of Thermal Tourism in Regional Development-A Case Study from Hungarian Side of The Hungarian–Croatian Border. *Podravina*, 2013;23:70–76.

Şengül, H., Bulut, A. (2019). Sağlık Turizmi Çerçevesinde Türkiye'de Termal Turizm; Bir Swot Analizi Çalışması. *ESTÜDAM Halk Sağlığı Dergisi.* 4(1):55–70.

Tarınç, A. (2019). Alternatif Turizm Çeşitliliği Kazandırmak Amacı ile Gönüllü Turizm: Türkiye Uygulaması, Doktora Tezi, Selçuk Üniversitesi Sosyal Bilimler Enstitüsü, Konya.

Tengilimoğlu, D. (2017). Sağlık Turizmi (2. Baskı). Siyasal Kitabevi: Ankara.

Tontus, O. (2019a). Türkiye'nin Sağlık Turizmindeki Önemi. SATURK Yayınları, Ankara.

Tontus, O. (2019b). Türkiye'de Termal Sağlık Turizmi. SATURK Yayınları, Ankara.

Turizm Günlüğü. (2022). Sağlık turizminde veriler açıklandı, https://www.turizmgunlugu.com/2022/04/11/turkiye-saglik-turizmi/. (Access Date: 29.04.2024)

Türk, E., Erdoğan, Y., Çalışkan, G. (2022). Türkiye'nin Engelli Turizmi Politikaları ve Mevcut Durumun Değerlendirilmesi. *Yönetim, Ekonomi ve Pazarlama Araştırmaları Dergisi, 6(6),* 318–337.

TurkStat. (2024). Tourism Statistics, Tourism Income by Expenditure Types.

Uluslararası Sağlık Turizmi Tesisleri ve Aracı Kuruluşlar Derneği (USTTAK). (2024). Türkiye'de Medikal Turizm, https://usttak.org/medikal-turizm.aspx. (Access Date: 26.04.2024)

Zhong, L., Deng, B., Morrison, A.M., Coca-Stefaniak, J.A., and Yang, L. (2021). Medical, Health and Wellness Tourism Research—A Review of the Literature (1970–2020) and Research Agenda. *Int. J. Environ. Res. Public Health, 18,* 10875.

Filiz Özlem Çetinkaya[1] and Nurcan Çetiner[2]

Chapter 2 Using SWOT Analysis in Creating Destination Image: The Case of Karaman Province*

Introduction

In today's intensely competitive environment, there are many different ways for organizations to achieve their goals. One of the most important of these ways is to have an effective image. Corporate image is an expression of how an organization is perceived by people both inside and outside the organization. Corporate image is an important element that helps institutions to continue their course of life and gain competitive advantage in the constantly changing and developing external environment. Just like institutions, destinations need to have a positive image in terms of the sustainability of tourism activities and the destinations' ability to gain competitive advantage.

The issue of image, which has become an important element for the tourism sector, is effective in creating a certain perception of the target audience of destinations and in shaping attitudes and behaviors towards the destination (such as choosing the destination and revisiting the destination) (Artuğer and Çetinsöz, 2014: 367). There are many destinations in our country. Karaman's destination, which is the subject of this study, is one of them. Karaman destination has different types of tourism potential. However, Karaman's destination is not well known because it is not promoted enough. For this reason, most local tourists show interest. It is a destination that contains many beauties that are not often visited by foreign tourists or from other provinces. When we look at the history of Karaman, it is seen that it has hosted many civilizations and that there are many buildings, some of which, although not all, have survived to the present day and have carried their cultures and lifestyles to the present day.

---

1   Assistant Professor, Kırşehir Ahi Evran Üniversitesi, Çiçekdağı Meslek Yüksekokulu, Büro Hizmetleri ve Sekreterlik Bölümü, f.cetinkaya@ahievran.edu.tr
2   Associate Professor, Karamanoğlu Mehmetbey Üniversitesi, Kazım Karabekir Meslek Yüksekokulu, Büro Hizmetleri ve Sekreterlik Bölümü, nurcancetiner@kmu.edu.tr
*   It was published as a summary paper in the proceedings book of the 3rd International Travel and Tourism Dynamics Congress.

When the literature is examined, SWOT analysis is one of the methods used in evaluating the tourism potential of destinations, in other words, in evaluating the tourism potential of a tourist destination (Acar et al., 2017: 1334). This study aims to reveal the destination image of Karaman province by using SWOT analysis, which is widely used in evaluating tourism potential.

Destination Image

Image, which is one of the marketing concepts of great importance in the tourism sector (Şahbaz and Kılıçlar, 2009: 34), is used in conveying the desired message to the target audience about a place or products specific to that region, and in redefining and positioning products, services and places (Göker, 2011: 10–11) is an effective communication tool (Şimşek, 1999: 353).

Destination image is the way the differences between the destinations are perceived by tourists (Ceylan, 2011: 92) and can be expressed as the sum of tourists' perspectives, impressions, and thoughts about the destination (Suhartanto et al., 2018: 171).

Features that make a destination successful include ease of transportation, harmony between the price paid and the quality of the service received, climate, event presentation, diversity of facilities, and proximity to cultural and/or natural heritage. Even if all the features are not together, it will be easier to increase and maintain the attraction power of the destination with good promotion of the natural and cultural heritage of the destination and efforts to create a strong image (Özdoğan and Öter, 2005: 129).

Creating the Destination Image

With the growth in the tourism sector, the destination alternatives offered to consumers have begun to expand. It has become easier for today's consumers, whose income levels and free time have increased in parallel with the developments in transportation networks, to reach these destinations. The fact that consumers have the opportunity to choose among a wide variety of destinations in the tourism sector, which has become a complex and competitive global market, has revealed the necessity for tourism sector managers to influence the decision-making processes of consumers. One of the most important ways to meet this need is to create and manage a distinctive, attractive, and positive image of the destination, which allows it to differentiate a destination from its competitors and position it positively in the minds of the target audience (Echtner and Ritchie, 2003: 37).

Positive destination image has an increasing effect on tourists' preferences to visit the destination. In addition to the physical characteristics of the destination in question, factors such as the security situation and the awareness level of the local people are also effective in creating a positive destination image and in the destination selection process of tourists (Kutlu and Ünal, 2022: 71). In creating the destination image, first of all, the boundaries of the destination must be determined, and in the next stage, the parties that affect the creation of the image of the destination in question and those who will be affected by this image must be determined. On the other hand, the target image of the destination is tried to be created by clearly stating the current image of the destination in question and the image perceived by the target audience (Özdoğan and Öter, 2005: 129). In this study, Karaman's current tourism potential was primarily discussed to reveal the destination image of Karaman.

Karaman's Tourism Potential

As a result of archaeological excavations, it was revealed that there was a settled life in Karaman and its surroundings in 8000 BC, with documents indicating that it was an important settlement, culture, and trade center. In the city, which hosted the Hittites, Roman, and Byzantine civilizations, after the collapse of the Anatolian Seljuk State, independence was declared and the Karamanoğlu State was established (Karaman Belediyesi, 2022). Karaman, which was connected to Konya after the declaration of the Republic, became a province in 1989.

Karaman, which went down in history as the capital of Turkish as Turkish was accepted as the official language, is also accepted as the Capital of Turkish Culture by the Turkish Republics in Central Asia (MEVKA, 2011:117). Karaman province; has attraction elements where faith tourism, history and culture tourism, eco-tourism, cave tourism, and air sports tourism types can be done together (Karaman İl Kültür ve Turizm Müdürlüğü, 2022).

In terms of *religious tourism*, there are Aktekke Mosque, Yunus Emre Mosque, İmaret Mosque, Taşmescit, Hacıbeyler Mosque, Dikbasan (Fasih) Mosque, Karabaş Veli Complex-Siyahser Mosques related to Islam in Karaman. There are Karadağ Binbir Church, Çeşmeli Church, Taşkale Manazan Caves, Fisandon Mosque Church, Derbe Mound, Derbe, Madenşehir Ruins, Değle Ruins, Mahalaç Church related to Christianity. In terms of *cultural tourism*, there are Karaman Museum, Hatuniye Madrasa, Tol Madrasa, Tartanlar Mansion, Hürrem Dayı House, Başdağ Pool and Military Castles, Karaman Castle, Manazan Caves, Grain Stores, Gökçeseki Ruins, Hartapus Monument, Zeyve Market, Gödet Hidden Paradise, Rock Church Caves and the Crater Pit. In terms of *eco-tourism*, there are some photo safari areas in Karaman such as Karadağ Mahallaç Hill, Taşkale Sinat Region and Granaries, Ermenek Firan Castle, Zeyve Market, Gödet Hidden Paradise and Nunu Valley Safari area.Karaman also presents opportunities for nature walks in Mennan Castle, Değle-Krater Pit-Çeşme-TRT Tower, Hakkı Teke Forest, Mahallaç Hill-Halisgümü-Başdağ Roman Pool, Gödet Hidden Paradise, Ermenek-Çatak locality, Ermenek Damlaçatı, Narlıdere Avgan-Bucakkışla, Manazan-Gürlük and Zeyve Market. In terms of *plateau tourism*, there are Ayrancı Berendi Plateau, Başyayla District Plateau, Taşkale Village Avdan Plateau, Sarıveliler Barçın Plateau, and Ermenek Balkusan Plateau. In terms of *farm (agro) tourism*, there are Bucakkışla Village, Ermenek Bağları, Mandenşehri, and Üçkuyu Villages and rural areas of Karaman. In terms of *camping and caravan tourism*, there are areas such as Gökçe Çamlığı, Hacıbaba Mountain Kızılyaka Village, Taşeli Plateau, Göksu Valley, Karadağ Crater, Yeşildere Akköprü, Taşkale Village Gürlük locality, Gödet Valley, Berendi Plateau, Barçın Plateau, Avdan Plateau.Ermenek Dam and Göksu Valley are suitable for river *sports tourism* (canoe-boat-water bike-sailing). In terms of *bicycle tourism*, Karaman: (1) Karaman-Yollarbaşı-Pınarbaşı-Karaman, (2) Karaman-Paşabağı roundtrip, (3) Karaman-Sertavul road-Dereköy-Baraj-Karaman, (4) Karaman-Kılbasan-Karadağ TRT transmitter roundtrip. It has trails between return, (5) Karaman-Değirmenbaşı-Tarlaören-Seyitasan-Karaman, (6) Yeşildere Akköprü-Taşkale-İncesu Cave.Dereköy Gödet Dam, Yeşildere Dam, Ermenek Turquoise Lake, and Göksu River are suitable for sport *angling tourism*. Areas such as the Taurus Mountains, Ermenek Dam, Ereğli Marshes, Karadağ, and Göksu Valley are used for *bird-watching tourism* (ornithology).For *botanical tourism*, Karaman has the Taurus Mountains, Karadağ, Barçın, and Avdan plateaus. In terms of *wildlife observation tourism*, there are wild animals such as deer, mountain goats, bears, wild boars, foxes, partridges, quails, grouse, rabbits, hornbills, and wild ducks in the foothills of the Taurus Mountains, as well as wild horses and Anatolian wild sheep in the Karadağ region. For *cave tourism*, there are İncesu Cave, New World Cave, Yeşildirek Cave, and Maraspoli Cave. In terms of *air sports tourism*, paragliding on Mahalaç hill in Karadağ; In terms of *ski tourism*, Bozdağ Ski Center facility is one of the tourism attraction centers of Karaman (Baycan, 2022).

Karaman's SWOT Analysis

SWOT analysis is one of the analysis methods used to evaluate the tourism potential of a tourist destination (Acar et al., 2017: 1334). It is an analysis technique that allows scientific situation analysis and has strategic importance for organizations to identify the opportunities and threats offered/to be presented by the external environment by determining the strong and weak sides of the organization, institution, region, technique, or process under examination (Çoban and Karakaya, 2010: 347).

As it is known, in terms of strategic management, the harmony between the environmental conditions of the organization or the region and its existing capabilities and capacities is important. The process of examining the current situation and capabilities of the organization, company, or region as a whole, identifying their superior and weak sides, and harmonizing them with the environmental conditions as a whole is called SWOT analysis (Dinçer, 2007:142). SWOT analysis, one of the situation determination matrices, takes its name from the first letters of the words "Strengths, Weaknesses, Opportunities, Threats". While the opportunities and threats mentioned in the SWOT analysis include elements related to the external environment of the organization, technique, or region, the advantages and weaknesses are the characteristics of the organization, technique, or region. It offers an analysis of the elements of the internal environment (Ülgen and Mirze, 2004: 65–66). As a result of the SWOT analysis of Karaman province, which is the subject of this study;

Strengths: Karaman, which has a very deep-rooted historical past, has structures known to belong to the Hittite, Roman, Byzantine, and Karamanid periods. Some of these structures are important for Muslims and Christians in terms of faith. Some caves attract attention with their natural beauty. Taşkale Granaries and Manazan Cave, which keep the temperature and humidity constant with their clayey and calcareous structure, are an important center of attraction for the region. It has great tourism potential with its many naturally beautiful caves, recreation areas, and dams. It is quite diverse in terms of tourism activities (cultural tourism, nature tourism, sports tourism, cave tourism, …). As a geographical location, Karaman serves as a bridge between the Central Anatolia Region and the Mediterranean Region. It is a destination where many types of tourism (faith, culture, water sports,…) can be done together. It has a variety of touristic activities that can be done in every season. There is Zeyve Market where you can access the products produced by the local people. Some structures give an idea about ancient lifestyles and architecture. There are accommodation facilities with tourism operation certificates and municipality certificates in the center of Karaman and in Ermenek, where many tourist attractions are located. Karaman can be easily reached by road and rail. It is considered the capital of Turkish.

Weaknesses: Some of the tourist destinations of Karaman are not included on Turkey's official promotion page (GoTürkiye, 2022). The level of competition is low compared to surrounding provinces. The historical texture of the city, which dates back to 8000 BC, has not been adequately preserved. Today, many buildings are in ruins. Some shooting locations (İnsuyu cave, Manazan caves, Taşkale granaries, etc.) can only be reached by private vehicles, which limits the possibilities of visiting. Local people do not act consciously to reflect the image of the destination. The people living in the touristic area are not sensitive about protecting the touristic buildings. The areas opened to tourism are not adequately maintained and controlled. The lack of an airport in Karaman poses a problem for those who want to travel by plane. Tourism demand is not at the desired level.

Opportunities: Since Karaman is a crossing point connecting the Ankara-Mersin road, it is a frequent destination for visitors going in both directions. Thanks to its proximity to the Mediterranean, it attracts daily tourists. It has the potential to bring its diversity in terms of tourism activities (cultural tourism, nature tourism, sports tourism, cave tourism, etc.) to more tourists. Binbirkilise buildings, which were once a bishopric center for Christians, are located in Karaman. There is a high diversity of endemic plant species in rural areas. Due to its flat and mountainous land structure, Karaman; has great potential for the development of touristic activities such as mountaineering, transhumance, hiking, and paragliding. There are tourism activities that can be done throughout the four seasons. Alternative tourism potential is high. It has various elements to create the touristic image. The existence of many tourist attractions close to each other makes it possible to create alternative travel routes.

Threats: Karaman's natural beauties (caves, ruins, etc.) and structures are negatively affected by neglect, human destruction and lack of due care. Touristic activities that can be done unconsciously (such as trekking, bicycle tours, camping-caravan tourism, etc.) are considered negative for the region.

Conclusion

One of the important factors in creating the destination image is; to determine the strengths and weaknesses of the destination in question regarding its current situation. Based on the data obtained by analyzing the opportunities and threats in the environment, supporting the weaknesses, eliminating the threats, preserving the strengths, and evaluating the opportunities to make it preferable among the competitors. For this purpose, the promotion and tourism potential of Karaman province, which is the subject of the study, was prioritized. Based on the data obtained, the current image of the destination in question was tried to be revealed through SWOT analysis.

It can be seen that Karaman has many elements that make it a strong destination. These elements; the cultural riches it has accumulated due to having hosted different civilizations that provide information about its historical texture, lifestyle, and architecture are human-shaped structures of natural formations originating from its geographical features. In addition, the natural beauties of its flora and fauna, the ease of transportation due to its location, the ability to realize tourism types specific to every season, and the fact that it is the capital of the Turkish language are among these factors. With these aspects, Karaman draws a strong image with its historical and natural structures.

The weak points of the destination are; Insufficient promotion of the destination and low level of competition; Failure to pay due attention to cultural structures; and limited access to some attractions; local people do not have the perception of creating a destination image and do not pay enough attention to protecting these structures; Insufficient maintenance and control of touristic places by the relevant authorities; The destination is only accessible by road and rail and the number of visitors is not at the expected level.

The elements that create opportunities for the destination are as follows. Karaman province's location connecting the Ankara-Mersin road has the potential to attract daily visitors; while visitors turn to sea tourism, it is a destination that combines many types of tourism such as culture, faith, nature, and sports; being one of the faith centers for Christians; it hosts different endemic plant species due to its climate diversity and geographical structure; it has tourism potential in all seasons and alternative travel routes can be created for different tourism centers located close to each other. With these opportunities, Karaman draws the image of a destination that can become one of the alternative tourism centers in the future in terms of increasing religious tourism and tourism diversity.

The destruction of natural beauties and historical buildings due to lack of due care and the uncontrolled and unconscious carrying out of many alternative tourism activities of the destination are seen as factors that may pose a threat to Karaman.

Some suggestions are presented based on the findings obtained as a result of the SWOT analysis. By effectively promoting the natural and cultural heritage of the destination and thus creating a positive and strong image, it will be easier to increase and protect the attraction power of the destination in question. It is recommended that all stakeholders of the destination organization (local people, public and private sector, non-governmental organizations, etc.) act jointly in the promotion of this destination, where many types of tourism come together. For the protection and development of the destination image and ensuring the sustainability of touristic activities; It is recommended to carry out studies (training, meetings, seminars, etc.) to raise awareness of local people and tourist facility employees. To increase tourist arrivals to these attraction centers, it is recommended that municipalities and tourism stakeholders act together to facilitate access. It would be appropriate to advertise and promote the festivals and festivities in a way that will attract participation from different regions. It is known that people's comments about the experiences of previous tourists about the destination in question are effective in their decision to visit tourist destinations. For this reason, it is recommended to draw touristic routes where participants in the festivals and festivities held in Karaman can experience the touristic products of the destination. It is recommended that authorized units and tourism organizations organize the different tourism activities of the destination in a way that does not disrupt the existing structure and facilitates their control.

In future studies, it is recommended that researchers conduct studies on creating alternative tourist routes by including Karaman's gastronomy.

References

Acar, V., Avcı, E. & Erat, B. (2017). Birgi'nin turistik bir destinasyon olarak SWOT analizi kapsamında değerlendirilmesi. *Ulakbilge*, 5(14), 1334–1372.

Artuğer, S. & Çetinsöz, B. C. (2014). Destinasyon imajı ile destinasyon kişiliği arasındaki ilişkiyi belirlemeye yönelik bir araştırma. *İşletme Araştırmaları Dergisi*, 6 (1), 366–384.

Baycan, S. (2022). Karaman ilinin turizm değerleri. URL: http://www.rotamizkaraman.com/Karaman_Tr.aspx (Erişim Tarihi: 26.07.2022).

Ceylan, S. (2011). Destinasyon marka imajı ve Pamukkale yöresinde bir uygulama. *Uluslararası İktisadi ve İdari İncelemeler Dergisi*, (7), 89–102.

Çoban, B. & Karakaya, Y. E., (2010). Geleceği planlamada stratejik yönetim ve Swot analizi: kavramsal yaklaşımlar, *e-Journal of New World Sciences Academy Social Sciences*, 5(4), 342–352.

Dinçer, Ö. (2007). *Stratejik Yönetim ve İşletme Politikası*, 8. Baskı, İstanbul: Alfa Basım Yayım.

Echtner, C. M. & Ritchie, J. R. B. (2003). The meaning and measurement of destination image. *The Journal Of Tourism Studies*, 14 (1), 37–48.

GoTürkiye (2022). *Karaman rotaları*. URL: https://gokaramanturkiye.com/tr/karaman-rotalari (Erişim Tarihi: 01.08.2022).

Göker, G. (2011). Destinasyon Çekicilik Unsuru Olarak Gastronomi Turizmi (Balıkesir İli Örneği). Balıkesir Üniversitesi Sosyal Bilimler Enstitüsü Turizm İşletmeciliği ve Otelcilik Anabilim Dalı. Yayınlanmamış Yüksek Lisans Tezi.

Karaman Belediyesi (2022). *Karaman tarihçe*. URL: https://www.karaman.bel.tr/tarihce (Erişim Tarihi: 01.08.2022).

Kutlu, D., & Ünal, E. A. (2022). "Turizmde güvenlik algısının destinasyon imajı açısından önemi: Yerli turistler üzerine bir araştırma". *İktisadi ve idari bilimlerde güncel araştırmalar* (Ed. Mustafa METE, Aytaç TOPTAŞ, Yahya KATI) 61–92, Ankara: Gece Kitaplığı.

Karaman İl Kültür ve Turizm Müdürlüğü (2022). *Turizm.* URL: https://karaman.ktb.gov.tr/TR-95849/turizm.html (Erişim Tarihi: 01.08.2022).

MEVKA (Ağustos-2011). *Karaman bölgesi turizm sektörü alansal varlık, uluslararası rekabetçilik ve makro düzey kümelenme çalışması.* Vezir Araştırma ve Danışmanlık Ltd.

Özdoğan, O. N. & Öter, Z. (2005). Kültür amaçlı seyahat eden turistlerde destinasyon imajı: Selçuk-Efes Örneği, *Anatolia: Turizm Araştırmaları Dergisi,* 16 (2), 127–138.

Suhartanto, D., Lu, C. Y., Hussein, A. S. & Chen, B. T. (2018). Scrutinizing shopper and retailer perception on shopping destination image. *Advances in Hospitality and Tourism Research (AHTR),* 6 (2), 169–187.

Şahbaz, P.R. & Kılıçlar, A. (2009). Filmlerin ve televizyon dizilerinin destinasyon imajına etkileri. *İşletme Araştırmaları Dergisi.* 1 (1), 31–52.

Şimşek, M. Ş. (1999). Yönetim ve Organizasyon. Ankara: Nobel.

Ülgen, H. & Mirze, S. K. (2004). İşletmelerde Stratejik Yönetim, 2. Baskı, İstanbul: Literatür Yayınları.

Ahmet Aydin[1]

Chapter 3 A Theoretical Evaluation of the Use of Sociocultural Values for Branding in Tourism

Introduction

To understand the concept of sociocultural in detail and correctly, it is necessary to mention the origin and historical significance of the words that make up the concept. It can be said that the origin of the concept of sociocultural, which 'expresses elements related to the combination or interaction of social and cultural elements' (https://www.dictionary.com), is based on the concepts of 'social' and 'culture'. It is possible to evaluate the facts and values related to the daily lives, working lives, communication styles, hobbies, habits, etc., activities of individuals, families, and masses that make up the society within the scope of the concept of social. On the other hand, it can be stated that culture is the existence of societies, regions and countries consisting of living conditions such as language, religion, writing, folklore, greetings, food types and eating styles, clothing and dressing preferences, architectural structure, art, settlement style. According to Najib et al. (2021: 4), "cultural factors are a set of values, norms, beliefs and shared behavioural patterns that take place in the minds of society and enable them to understand each other". When the concepts of social and culture are analysed in terms of meaning and in the context of social life, it is seen that they cannot be separated from each other. Accordingly, it is understood that many material and immaterial elements expressed as sociocultural values have an important place in the life of societies.

Scott and Palincsar (2013: 1) stated that "the focus of the sociocultural perspective is on the roles that participation in social interactions and culturally organised activities play in influencing psychological development". It Looking events-phenomena and developments from a sociocultural perspective is also important for enterprises in all areas of life, apart from the science of psychology. Because it is seen by every business operator that daily, medium and long-term needs will arise in the environment where individuals and masses are present

---

1 Assistant of Professor Dr., İstanbul Nişantaşı University, Faculty of Economics, Administrative and Social Sciences, Department of Business Administration, ahmet.aydin@nisantasi.edu.tr

and it is known that they act in accordance with enterprise strategies in this direction. In line with marketing and brand strategies, enterprises can analyse the sociocultural values of the society such as language, religion, clothing, lifestyle, habits, hobbies, etc. in detail and accurately, and thus develop appropriate products and services by understanding the needs of the masses correctly. This is because individuals and institutions may prioritize purchasing products and services with content appropriate to their sociocultural values. In this context, designing and launching brands with products-services developed by giving importance to sociocultural values can make significant contributions to the survival of businesses operating in different sectors as a sustainable enterprise as well as market success.

Accommodation, recreation, spa, sightseeing-tour, food and beverage, agricultural activities, etc. Hotels, motels, spa facilities, agencies, restaurants, touristic agricultural farms, etc., where products-services are offered, are the leading enterprises where employees compatible with sociocultural values take place and products-services are offered. Considering the interest of individuals and societies of different countries in the sociocultural values of other countries and their desire to interact, it is seen that there are very important opportunities for tourism enterprises and it is understood that these opportunities are tried to be utilised. In tourism enterprises that want to be permanent in the market and want to be branded; it may be necessary to design products-services in accordance with the sociocultural values of the country, to make physical conditions suitable for sociocultural values, to dress employees in accordance with local-local culture and to be qualified or trained for sociocultural values, and ultimately to provide appropriate-effective service to consumers-customers. Providing these requirements may not only contribute to the branding of tourism enterprises but also have a positive impact on their sustainability. In this study, which is produced in the light of these considerations, the concept of sociocultural value, its development and its importance in terms of social aspects have been mentioned and a theoretical framework has been tried to be determined about the advantages related to branding and brand strategies that may arise in tourism enterprises where sociocultural values are prioritised and what kind of contributions they can make to the marketing strategies of tourism enterprises.

Place of Sociocultural Values in Societies and Sociocultural Theory

Culture is defined as the order of values in society. Sociocultural values are closely related to human life. Some sociocultural values-elements are formed from human life and for the continuation of human life (Muhammad & Hariyati, 2021: 29). Although the characteristics of the elements related to the values order and life of each society are different, they are the same or similar in phenomena such as the transfer of cultural values, communication, socialisation and sharing. Despite the difference in content and characteristics, as can be understood from the similarities in basic characteristics, sociocultural values are very important for societies. The determining effect of sociocultural values can be noticed in many environments related to human life, from the formation and change of behavioural patterns in families, neighbours, relatives, villagers, groups-associations, educational institutions, workplaces, etc.

Individuals continue their lives in both individual and social areas from birth to death. In addition to the cultural values created by the individual, the cultural values transmitted by the society from generation to generation are of great importance. These cultural values created by the individual and society turn into a cultural identity over time. These identities, which reflect the characteristics of societies, include familial, religious, social, political and many human factors. Sociocultural values-elements formed by human factors have an important place in the self-expression of society (Söğütlü, 2019: 1). Experts state that "a prosperous person has a good education, a good job, access to modern information technologies, the ability to travel the world, buy quality products and clothes, and increase their income" (Gritskikh et al., 2019: 145). Considering this view mentioned by Gritskikh, it is understood that individuals continue to want to live in a happy and peaceful environment throughout their lives. It cannot be thought that this desire of individuals is limited to themselves because; the fact that interactions such as communication, socialisation, cultural activity, etc. occur in environments where the rights and values of the majority are protected and secured, such as families, groups, communities, villages, towns, cities and countries formed by individuals, further increases the importance of sociocultural values and it is understood that a theory or understanding based on sociocultural values should be given importance for economic and social development. Within this understanding, the position of young individuals and their compatibility with the environment they live in can be seen as a necessity in terms of the unity of sociocultural values with modern youth. In this context, the mobilization and more effective use of sociocultural values can provide benefits at different levels in different fields.

The mobilisation of sociocultural values has positive consequences for the theory of sustainable development and the management of existing natural protected areas and reveals the need to go beyond the humannature struggle (Liburd et al., 2023: 14). In contrast to the human-nature struggle, many enterprises, especially in the tourism sector, which can observe human-nature harmony, have actually taken an important step in terms of sustaining their own existence. Considering sociocultural values and adapting them to the field of activity as one of the most important success factors in being compatible with nature, society and technological developments may be an approach in line with the sociocultural value theory. It is possible that adopting this approach and carrying out activities accordingly will contribute to the branding process as an element that increases the satisfaction of consumerscustomers

Use of Sociocultural Values for Branding Purposes in Tourism Enterprises, Advantages and Disadvantages

The brand acts as a cue, allowing consumers to recall relevant information from memory, such as their past experiences, perceptions or associations with the brand. Branding acts as a signal that enables consumers to recognize a product as an element-value quickly they are familiar with or like (Chovanová et al., 2015: 616). When considered in this direction, it is seen that brand and branding constitute the basic step in the effort to create customer-consumer loyalty. The attempt to develop customer-consumer loyalty through brand and branding is one of the strategic objectives of enterprises in the tourism sector, as in all industries. Tourism enterprises aim to be preferred again and again by tourists through branding, and in order to achieve this, they offer products and services with appropriate content to tourists from different countries. Considering that tourists from different countries have different sociocultural values, the job of tourism enterprises becomes more difficult. Therefore, the focus can be on offering products-services with appropriate content for the individuals of the countries from which tourists come to the enterprise the most. When evaluated in terms of sociocultural values and considering that the majority of tourists from different countries change at different time intervals, companies may need to restructure. Prioritising content suitable for the countries where tourists come from in terms of number may have a positive effect on the recognition and branding of the tourism enterprise. In addition, the tourism enterprise can try to attract tourists who are interested in different cultures by emphasizing the sociocultural values in the country.

Prioritising national and local sociocultural values may have a positive impact on the tourism enterprise to become a brand that is recognised and recommended at home and abroad, as it may attract the curiosity and interest of tourists coming from different countries and enable them to have more social interaction and thus increase their satisfaction. Tourism enterprises can contribute to the branding of the placedestination (country-region-city-town-town-village) with a service understanding that prioritises sociocultural interaction and change. For this, it is necessary to determine and implement an effective communication strategy. As Vasavada and Kour (2016: 2) point out, "the main objective of destination branding is to produce a coherent, focused communication strategy based on the selection of a collection of core intangible values that exist in the minds of consumers". Accordingly, the use of effective communication strategies in the presentation of places and sociocultural values that may be attractive to tourists can be an important factor that will ensure the branding of the tourism enterprise.

The image of a city depends on the physical, subjective and objective spheres, i.e. how people objectively and subjectively perceive city artefacts, architecture, amenities, etc. Therefore, an important focus of city branding as innovation is innovation in the process of finding and combining the meanings of images and city characteristics in a unique blend of elements that reflect the nature and culture of a city, with the socioeconomic prospects of appealing to more investors, visitors and audiences (AlShaalan and Durugbo, 2024: 4). The presence of sociocultural values that are open to innovations and interactions can offer some advantages to businesses. "Sociocultural values and norms are important but less recognised sources of international competitiveness. Many studies in the literature have shown that sociocultural factors positively influence various aspects of international competitiveness, entrepreneurship, innovation, productivity and international cooperation. Sociocultural factors provide an opportunity to develop competitiveness strategies based on unique advantages" (Apsalone and Šumilo, 2015: 276). Accordingly, tourism enterprises aiming to compete better with companies engaged in similar activities in the market and to appeal to more customers can take the opportunity to increase the satisfaction level and satisfaction rate of customers-consumers and thus gain customer loyalty by including more sociocultural values in the content of their products and services.

In addition to the effective inclusion of sociocultural values, it is clear that following the economic conditions in the market and developing solutions to the problems that may arise, making adequate use of legal facilities-incentives, adapting to geographical-geopolitical conditionsdevelopments, adding new products-services that may attract the interest of consumers-customers to the product-service portfolio, adapting to technological innovations, and continuing these efforts will have a positive effect on the development and growth of tourism enterprises. The growth of tourism enterprises and providing more employment can also contribute to the activities related to 'sustainable development' in the country. According to Okech (2010: 341), "sustainable development became a common concept in the international development community in 1987". The concept of sustainable development, which has been in the literature for almost the last 40 years, continues to take place in many areas as sustainability. Especially in order to protect the environment and transfer natural resources to future generations, it has gained more place in the literature today. Although sustainability is mostly used in the field of natural resources and business administration, it can also be a concept used for the protection and transfer of sociocultural values that have an important place in the lives of societies from generation to generation. As a phenomenon and value, it is possible to transfer sociocultural elements from generation to generation through individuals-masses and authorities such as employees, operators, consumers-customers, governments, local governments and intermediary enterprises in different sectors. The sociocultural value exchange-transfer of sociocultural values transferred from generation to generation at national and regional level at international-global level between citizens of different countries and societies with other cultures is realised through tourism activities. Although the effects of this change-transfer on the individuals and institutions of various countries engaged in touristic activities-activities have not yet been fully measured, it can be said that it causes change and transformation in elements such as content, scope, price, service, behaviour, etc. in tourism enterprises and places visited for touristic purposes.

Although the effects of tourism have been extensively studied, researchers have rarely compared sociocultural transformations in destinations with and without a planned tourism intervention (Sebastian and Rajagopalan, 2009: 5). Tourism activities, which differ in terms of their scope and the way they are implemented, have led to the emergence of sociocultural interactions in tourism with the inclusion of individuals and families with different sociocultural structures in the sector. This interaction continues on the basis of transferring and sharing cultural values and social life-related characteristics and habits. Touristic activities and tourism activities that include sociocultural values can provide sociocultural change and transformation in the settlements and societies. It can be seen that touristic activities, which provide certain effects and transformations in individuals and institutions through the use of sociocultural values, contribute to national economies at the national level in terms of foreign exchange inflow and employment, and at the local level in terms of infrastructure, modern and environmentally sensitive construction, and greater recognition in different countries.

Sociocultural structuring and interaction in tourism enterprises, where social and cultural communication occurs intensively, may have some negative consequences as well as positive economic and social effects. As Uslu et al. (2020: 1) state, "excessive or unplanned development of the tourism sector, which has made significant contributions to local economies such as increasing employment, contributing to the economy and protecting cultural heritage, especially in the last two decades, may also have negative effects on local people, resources and sociocultural structure". In addition, it may result in the influence of dominant cultures on other cultures and the weakening or disappearance of local cultural values over time. Shahzalal (2016: 30) expresses this situation as follows; "acculturation is a process of adaptation and adjustment, and when local people welcome tourists, they adopt the needs, attitudes and values of tourists and eventually begin to follow them. This usually takes place in a less sophisticated society where the stronger culture dominates the weaker one. However, the diffusion of innovation of cultural components in a social framework depends on the compatibility, advantage and complexity of cultural objects". In terms of sharing and exchanging the components of sociocultural values, tourism enterprises and tourism activities have a very important position in terms of positive and negative effects.

Results

In its most basic sense, tourism is a socialising activity that aims to have fun and spend time with relatives, friends or other people during nonworking hours (Higgins-Desbiolles et al., 2021: 1). Based on this definition, it can be stated that tourism is a practical example of interacting with individuals with different sociocultural values. During the fulfilment of the need to have a holiday, which has an important place in the lives of individuals and societies, different sociocultural values and touristic places are visited-discovered and adopted, and some behaviours and habits with sociocultural values may be included in the lives of individuals. Thus, individuals and communities of different countries interact with each other. Knowing this, tourism enterprises should act in a conscious and planned manner and frequently remind their employees that they should be respectful and sensitive to tourists with different sociocultural values while doing their jobs. In addition to the scope and quality of products and services, customer-consumer satisfaction should be ensured by giving importance to the way of communicating with tourists in service delivery.

As in many sectors, ensuring customer-consumer satisfaction in the tourism sector is one of the most important requirements for branding. Achieving this necessity requires the importance of different sociocultural values more than other sectors. In this direction, priority should be given to the development and presentation of products-services with content-features appropriate to the traditional values and lifestyles of societies. However, at the same time, products-services that include local and national values that have become part of the society should also be prioritised. In this context, it would be more appropriate to include sociocultural values in products-services in a balanced manner for customer-consumer satisfaction-oriented branding purposes. As Zhuang et al. (2019: 14) state, "when the tourism industry reaches a certain level, traditional values and lifestyles will try to find a balance point. Ultimately, when tourism becomes a mature industry, a whole new socioculture will emerge".

Supporting the sociocultural values that are effective in the tourism sector becoming an important industry with different opportunities and technologies should be addressed. In particular, the inclusion of infrastructure and elements related to smart tourism destinations can ensure that settlements that contribute to the branding of tourism enterprises are more recognised and preferred. The existence of smart tourism destinations may have a positive impact on the establishment of cooperation between institutions and enterprises. As Cavalheiro et al. (2019: 243) state, "from a management and governance perspective, the infrastructure of a smart tourism destination can support close interaction and co-operation between multiple stakeholders in the tourism industry". Prioritising sociocultural values in the provision and continuation of this cooperation can contribute to the increase in the value and branding of tourism enterprises in the eyes of consumers and customers.

Today, culture-based tourism is recognized as one of the most attractive segments of the tourism industry, considering the volume of visitors or tourists interested in cultural activities (Amalu et al., 2021: 1692). By prioritising the sociocultural values of the different societies served and including the sociocultural values of the society in the settlements, the following suggestions can be made for the production-execution of culture-based tourism products-services and the branding of tourism enterprises in line with the current and future conditions that may arise:

- Tourism enterprises should notmaintaine fact that consumers-customers can come from both abroad and at home, and should optimally utilise sociocultural differences and similarities in product-service design and presentation.
- Employees who have the competence and qualifications to communicate with tourists from different cultures should be employed more, especially in enterprises with more foreign tourist customers.
- Physical structuring should be made by prioritising elements such as cleanliness, hygiene, order, respect and friendliness, which have an important place in many cultures, and the working order should be created accordingly.

- Care should be taken to include the culinary culture of different countries in a balanced way in the menu of restaurants or touristic restaurants in the tourism enterprise.
- Before each tourism season, employees should be given training on how to behave towards tourists with different sociocultural values and should be raised awareness.
- Socioeconomic, technological and cultural changes-developments should be followed and managers should have a positive approach to ensure that tourism enterprises adapt to these changesdevelopments.
- Attempts should be made to differentiate and brand in the market through geographical marking and for this, sociocultural values should be adequately included.
- As in other business activities, the understanding of sustainability should be followed in ensuring that sociocultural values can be used and consumed in tourism products and services.

References

AlShaalan, M. K., & Durugbo, C. M. (2024). City branding as innovation for tourism development: systematic review of literature from 2011 to 2023. *Management Review Quarterly*, 1–33. https://doi.org/10.1007/s11301-024-00431-2.

Amalu, T., Oko, U., Igwe, I., Ehugbo, U., Okeh, B., & Duluora, E. (2021). Tourism industry effects on sociocultural activities of host communities: Evidence from Cross River State, Nigeria. *GeoJournal*, 86, 1691–1703. https://doi.org/10.1007/s10708-020-10151-1.

Apsalone, M., & Šumilo, Ērika. (2015). Sociocultural factors and international competitiveness. *Business, Management and Economics Engineering*, 13(2), 276–291. https://doi.org/10.3846/bme.2015.302

Cavalheiro, M. B., Joia, L. A., & Cavalheiro, G. M. do C. (2019). Towards a Smart Tourism Destination Development Model: Promoting Environmental, Economic, Sociocultural and Political Values. *Tourism Planning & Development*, 17(3), 237–259. https://doi.org/10.1080/21568316.2019.1597763.

Chovanová, H. H., Korshunov, A. I., & Babčanová, D. (2015). Impact of brand on consumer behaviour. *Procedia Economics and Finance*, 34, 615–621. https://doi.org/10.1016/S2212-5671(15)01676-7.

Gritskikh, N., Reshetnikova, E., & Zagorodniy, V. (2019). Specifics of formation of comfortable sociocultural environment as factor in formation of well-being of modern students. In International Conference "Topical Problems of Philology and Didactics: Interdisciplinary Approach in Humanities and Social Sciences", (TPHD 2018) (pp. 145–149). Atlantis Press.

Higgins-Desbiolles, F., Doering, A., & Bigby, B. C. (2021). Socialising tourism: reimagining tourism's purpose. In "Socialising tourism-rethinking tourism for social and ecological justice". New York: Taylor & Francis.

Liburd, J., Menke, B., & Tomej, K. (2023). Activating sociocultural values for sustainable tourism development in natural protected areas. *Journal of Sustainable Tourism*, 32(6), 1182–1200. https://doi.org/10.1080/09669582.2023.2211245.

Muhammad, C.N., & Hariyati, Y. (2021). Prestigious perception of potato farming: An overview of the economy, socio-culture, and its existence. *Agricultural Socio-Economics Journal*, 21(1), 25–32. https://doi.org/10.21776/ub.agrise.2021.021.1.4.

Najib, M., Sumarwan, U., Septiani, S., Waibel, H., Suhartanto, D., & Fahma, F. (2021). Individual and Sociocultural Factors as Driving Forces of the Purchase Intention for Organic Food by Middle Class Consumers in Indonesia. *Journal of International Food & Agribusiness Marketing*, 34(3), 320–341. https://doi.org/10.1080/08974438.2021.1900015.

Okech, R. N. (2010). Sociocultural Impacts of Tourism on World Heritage Sites: Communities' Perspective of Lamu (Kenya) and Zanzibar Islands. *Asia Pacific Journal of Tourism Research*, 15(3), 339–351. https://doi.org/10.1080/10941665.2010.503624.

Scott, S., & Palincsar, A.S. (2013). The historical roots of sociocultural theory. https://dr-hatfield.com/theorists/resources/sociocultural_theory.pdf.

Sebastian, L. M., & Rajagopalan, P. (2009). Sociocultural transformations through tourism: a comparison of residents' perspectives at two destinations in Kerala, India. *Journal of Tourism and Cultural Change*, 7(1), 5–21. https://doi.org/10.1080/14766820902812037.

Shahzalal, M. (2016). Positive and negative impacts of tourism on culture: A critical review of examples from the contemporary literature. *Journal of Tourism, Hospitality and Sports*, 20(1), 30–34.

Söğütlü, G. (2019). Sosyo-kültürel etkenlerin tasarım ve tasarımcı üzerine etkileri. *Yüksek Lisans Tezi, Kütahya Dumlupınar Üniversitesi, Sosyal Bilimler Enstitüsü, Bileşik Sanatlar Ana Sanat Dalı*, Kütahya, Turkey.

Uslu, A., Alagöz, G., & Güneş, E. (2020). Sociocultural, economic, and environmental effects of tourism from the point of view of the local community. *Journal of Tourism and Services*, 11(21), 1–21.

Vasavada, F., & Kour, G. (2016). Heritage Tourism: How Advertising is Branding the Intangibles? *Journal of Heritage Management*, 1(1), 22–34. https://doi.org/10.1177/2455929616640688.

Zhuang, X., Yao, Y., & Li, J. (2019). Sociocultural impacts of tourism on residents of world cultural heritage sites in China. *Sustainability*, 11(3), 1–19. https://doi.org/10.3390/su11030840.

https://www.dictionary.com/browse/sociocultural. Access Date: 23.04.2024.

Ayşe Topaloğlu[1] and Sabri Arici[2]

Chapter 4 Cultural Tourism Potential of Anamur *

Introduction

The importance of cultural tourism, a subset of tourism, has been increasing annually. Various similar definitions have been proposed for cultural tourism. In one of these definitions, it is expressed as visiting historical or archaeological sites, participating in community events, watching traditional folk dances or ceremonies, or simply shopping for handicrafts (Besculides, Lee & McCormick, 2002). According to 2018 data, the proportion of cultural tourism in global tourism is 40 %. Public investments in cultural heritage assets in cultural tourism provide social benefits for both tourism stakeholders and visitors. These benefits extend beyond economic returns to include abstract but significant values and gains for the local community, such as a sense of identity acquisition, pride in the region, and the scientific appreciation of the region (Dragouni, Fouseki & Georgantzis, 2017). As the cultural tourists are typically the ones who are educated, who have high income, spend more during vacations and who are generally over the age of 50, who often travel in groups, and tend to travel during seasons other than summer (Aksu, 2004), their economic contribution to tourism becomes greater.

When the subject of the study is looked at from this perspective, it is important to highlight the potential of cultural tourism in Anamur. Anatolia has hosted over thirty civilizations due to its favorable climate, fertile and irrigable lands, its position as a bridge between the east and the west, and its rich surface and underground resources (Artal-Tur, 2018). The influences of these civilizations are evident in many areas of contemporary cultural life, particularly in archaeology. Traces of these civilizations can also be found in Anamur District of Mersin in Türkiye. However, the district does not currently occupy its deserved place in

---

1 Master's Degree, Mersin University, Tourism Faculty, Department of Tourism Guidance, aysekilincmeu@gmail.com Orcid ID: 0003-1925-1865
2 Assoc. Prof., Mersin University, Tourism Faculty, Department of Tourism Guidance, sabriarici@mersin.edu.tr Orcid ID: 0003-4925-6704
* This study is compiled from the Master's thesis titled "Evaluation of Anamur's Cultural Tourism Potential" prepared by Ayşe Topaloğlu under the supervision of Assoc. Prof. Dr. Sabri Arıcı in 2022.

cultural tourism. In a survey conducted, when the local people of Anamur were asked about the priority of feasible tourism types, 82 % of the local population prioritized sea-sand-sun tourism in Anamur while 6 % of the local population considered that cultural tourism was the most important. Additionally, 3 % of the local poplulation thought that highland tourism was the most important (Çetinsöz & Altınbıçak, 2008). In this context, the aim of the study is to reveal the cultural tourism potential of Anamur, evaluate the perspectives of stakeholders in the public, local government, and tourism-related civil society organizations to raise awareness, contribute to promotion, and contribute to the literature in the field. The significance of this study lies in the fact that direct interviews were conducted with key stakeholders of the sector, and it represents the latest comprehensive study on cultural tourism in the district.

Anemurium (Anamur) Ancient City

The history of Anamur begins with the ancient city of Anemurium (Anemur), situated within the boundaries of the Mountainous Cilicia in ancient geography, which is considered the most splendid in the region. Founded on a promontory, the city is located 10 km southwest of the district. Positioned on a cape, it gained significance as a port city due to its proximity to the Turkish Republic of Northern Cyprus and its proximity to Ermenek (Germanicopolis), providing overland access to Central Anatolia. Anemurium thrived from the time of Alexander the Great (333 BC), particularly experiencing its peak years within the Roman Empire until the Sassanian (Iranian) invasions around 260 AD. Due to frequent changes of ownership from this point until the 5th and 6th centuries AD, the city struggled to regain its former glory. With the assistance of Zenon, the Isaurian (Ermenek-Mut-Karaman Region) Roman Emperor (474–491 AD), the city experienced a resurgence, only to be devastated by a major earthquake around 580 AD. Subsequently, in the 12th and 13th centuries AD, a modest recovery occurred, with numerous remnants of baths and churches constructed. With the onset of Turkish rule in the region from the 13th century onwards, prominent structures that can be observed in the city include city walls, baths, theater, odeon (concert hall), agora (ancient marketplace), palaestra (sports hall), port street, mosaics, churches, aqueducts, and monumental vaulted tombs (Subaşı, 2018).

Mamure Castle

The castle, which located in 6 km southeast of Anamur on the coast, is one of the best-preserved Byzantine castles in Anatolia. The castle, which was demolished when it was captured by Seljuk Sultan Alaaddin Keykubat in 1221, was replaced by the current castle. Within the castle grounds, there is also a very sturdy mosque and a bathhouse structure outside the castle. The castle, which has 36 towers, is surrounded by a single row of moat for protection. The castle, consisting of two parts, has two inner walls and a road that surrounds the castle and provides passage between the towers is quite sturdy. The castle passed to the Karamanoğulları and later to the Ottomans over time. Due to being reacquired and restored, it acquired the name "Mamure" (Şahin, Özcan & Sol, 2021). In 2012, it was included in the UNESCO World Heritage temporary list (Kültür ve Turizm Bakanlığı, 2024).

The Ala Bridge

It is located on the Anamur (Dragon) River, at the 16th kilometer of the Anamur-Ermenek highway. Although it was suggested to have been built by the Karamanoğlu dynasty in the 14th century, recent research has determined that it was constructed by the Anatolian Seljuks in the 13th century. The Ala Bridge, boasting an monumental appearance, consists of a main arch and a relief arch (Göçmen, 2021).

The Anamur Archaeology Museum

The museum comprises archaeological and ethnographic artifacts. In addition to the findings from the ancient city of Anemurium, the museum also displays region-specific artifacts brought from the Silifke and Alanya museums. Among the prominent artifacts in the museum are the tomb findings from the ancient cities of Nagidos and Kelenderis, a floor mosaic dating back to the 5th century AD, votive offerings and funerary objects from the 4th to 3th centuries BC, terracotta oil lamps unearthed in the Anemurium settlement, as well as a bronze statue of Athena and bronze scale weights from the Roman era. The local kilims, clothing, as well as firearms and bladed weapons are exhibited in the ethnography section of the museum (Kültür Varlıkları ve Müzeler Genel Müdürlüğü, 2024).

The Yörük Cultural of Anamur

Originating from the Central Asian Turks and sustaining their livelihood through small-scale animal husbandry, these semi-nomadic groups are referred to as Yörüks (Eren, 1979; Sümer, 1992; Dulkadir, 1993: 481; Artun, 1996). Over the centuries, from the Seljuks to the Ottoman Empire, they have lived without significantly altering their cultural and lifestyle characteristics (Sümer, 1992). Various Yörük tribes such as Sarıkeçililer, Bahşiş, Tekmen, Abdal, and Tahtacılar are present throughout the district, albeit they do not strictly adhere to traditional ways of life in contemporary times, they still maintain their existence. Anamur Yörüks migrate to the Ermenek and Karaman plateaus during the summer months and settle in the coastal areas, which are not too far away during the winter months (Çetingöz & Temiz, 2018).

A Scientific Approach to the Cultural Tourism Potential of Anamur District

Qualitative research methodology is used in the study. Qualitative research methodology allows for an in-depth examination of data through commonly preferred methods such as observation, interviews, and document analysis. Qualitative research is utilized to richly portray participants' experiences, perceptions, and worldviews (Kozak, 2021; Yıldırım & Şimşek, 2021). This methodology enables a detailed and layered exploration of research questions, facilitating a multifaceted and thorough understanding of the subject matter. Given its capacity to provide a comprehensive and in-depth examination of stakeholders' views and recommendations regarding the current situation related to the topic in Anamur District, this method was preferred for its capability to facilitate induction (inductive reasoning). A semi-structured interview technique was employed as a data collection tool.

This research was conducted within the first six months of 2022, involving a total of 34 individuals through both online and face-to-face semi-structured interviews. The questionnaire prepared during the research process consists of 11 questions (Kozak, 2021; Yıldırım & Şimşek, 2021). The collected data were analyzed using the content analysis method. Furthermore, this study examining the cultural tourism potential of Anamur District was conducted by means of a comprehensive literature review which utilized various academic articles, web-based resources, as well as secondary data sources such as the websites of the Mersin Provincial Directorate of Culture and Tourism and the Governorship. In this process, numerous previous studies focusing on tourism types in Anamur District were particularly utilized (Çetinsöz & Altınbıçak, 2008; Çetinsöz & Temiz, 2018; Kayabaşı, 2008; Kuzu, 2010; Çetinsöz & Subaşı, 2014; Gölgeli, 2016; Çetinsöz & Atsan, 2018; Saygın, 2018; Onan, 2018; Nas, 2018; Karakeçili, 2018; Subaşı, 2018; Temiz, 2018). This literature review forms the foundation of the research and helps in better understanding the current status of the district in the context of cultural tourism.

In order to protect the confidentiality of the participants, most of them preferred not to disclose their names during the research process. Therefore, the principle of confidentiality was prioritized. The sampling for the research was conducted using the snowball sampling method, which is one of the purposive sampling techniques (Kozak, 2021; Yıldırım & Şimşek, 2021). This sampling method begins with a few initially identified participants and then aims to find new participants through these initial ones. In this way, researchers aim to reach a sample that possesses in-depth knowledge about the research topic and reflects the characteristics of the participants. This method allows researchers to maintain the participants' privacy while also accessing comprehensive and qualitative data.

The utilization of this method has enabled us to reach individuals who possess effective knowledge about cultural tourism in Anamur District and to reach the desired number of participants. It has been constructed from the viewpoints of stakeholders who could be related to cultural tourism in Anamur. In the last four questions of the research, the Grounded Zone of Feasibility Technique (GZFT) has been employed, which is one of the techniques allowing for a kind of situational analysis with its scientific aspect, enabling the evaluation of the organization's internal and external environments. GZFT helps in identifying the strengths and weaknesses of the examined subject and determining the opportunities and threats stemming from the external environment (Çoban & Karakaya, 2010).

To ensure the scientific quality of the research and to attain reliable results, careful attention has been paid to the concepts of "validity" and "reliability" (Kozak, 2021; Yıldırım & Şimşek, 2021). To establish the validity of the research, opinions of competent academics and research assistants in the field were got regarding the suitability and adequacy of the method used in the study. To enhance reliability, the research was conducted by obtaining opinions from participants who are specialized in the field of cultural tourism. In this way, data collection was facilitated.

The Findings of the Research

Significant findings regarding the cultural tourism potential of Anamur have emerged through the contributions of the participants. The obtained findings are as follows:

1. What can you say about Anamur's overall tourism potential?
    Discussions regarding Anamur's general tourism potential emphasize the city's significant natural and historical cultural assets. However, it is noted

that the focus is primarily on the sea. Historical structures such as the ancient city of Anemurium and Mamure Castle, as well as the coastline, stand out as factors that enhance Anamur's tourism potential. On the other hand, many participants emphasize that Anamur's tourism potential is not adequately utilized and that there are deficiencies in tourism infrastructure. It is suggested that common strategies for evaluating Anamur's tourism potential could be developed through collaboration between local governments, civil society organizations, and tourism associations.

2. How do you perceive the local community's perspective on cultural tourism?
   The local community's perspective on cultural tourism is generally shaped within an environment where agricultural activities are predominant. Concerns about the impact and future of tourism on the regional economy lead to a negative attitude towards tourism. However, it is anticipated that there could be changes in this attitude with increased awareness of cultural heritage and tourism potential.
3. Despite having rich cultural potential, why are tours to Anamur so infrequent?
   Anamur faces transportation challenges both by air, sea, and land routes. Additionally, inadequate promotion of Anamur, limited accommodation and service infrastructure, as well as insufficient lodging, dining, and entertainment venues, limit its capacity to accommodate tourist groups.
4. What are your recommendations for promoting the cultural values of Anamur?
   The prioritized promotion of Anamur should involve the effective utilization of the internet and various social media platforms, emphasizing its agricultural products, cultural, and economic significance, which should prominently feature in documentaries. Improving transportation facilities could be achieved by supporting promotional activities through participation in tourism fairs and providing incentives for large accommodation establishments. Local governments and sector representatives need to establish a shared vision. Various events should be organized, and promotional materials such as printed publications and brochures should be prepared. In short, advertising and marketing efforts need to be more comprehensive.
5. What are the prominent aspects of cultural tourism in Anamur?
   Anamur boasts a diverse range of historical and natural elements. It is enriched with various historical structures such as the ancient city of Anemurium and Mamure Castle. Additionally, the Yörük culture prevalent in the region forms a significant part of cultural tourism with its traditional lifestyle, handicrafts, and cuisine. Furthermore, historical sites like caves,

the Ak Mosque, and various natural areas including the sea, forests, plateaus, and caves contribute to the cultural tourism landscape.

6. In your field, how is cooperation and coordination ensured between the public sector and stakeholders regarding cultural tourism?

Cooperation and coordination between the public sector and stakeholders in the realm of cultural tourism are typically facilitated through official approvals, supportive organizations, protocols, and events, inter-agency meetings, communication and correspondence, as well as projects and initiatives carried out at the provincial and district levels. Particularly crucial for the development of cultural tourism is the collaboration between government agencies and civil society organizations in promoting and preserving historical and cultural assets. Furthermore, the importance of cooperation and coordination in cultural tourism is emphasized through inter-agency meetings and communication channels, enabling knowledge sharing and joint efforts.

7. What efforts should be made for the development and sustainability of cultural tourism in Anamur?

Primarily, the efforts should be directed towards facilitating transportation, improving roads, and ensuring more efficient utilization of facilities such as Gazipaşa Airport. Utilizing media channels such as the internet, press, television, as well as events like tourism fairs, is recommended to promote the region's tourism potential to a wider audience. Emphasizing that the region is a safe and peaceful tourism destination and highlighting the availability of facilities catering to tourist needs such as shopping and dining are also important. Additionally, recommendations have been made for enhancing the quality of facilities and services.

8. According to you, what are the strengths of Anamur in terms of cultural tourism?

Historical structures such as the ancient city of Anemurium and Mamure Castle constitute the rich cultural heritage of the region. The Yörük culture prevalent in Anamur, along with its traditional festivals and lifestyle, is intriguing for tourists. The region's beautiful sea and climate, especially its prominence in banana and strawberry production and its provision of opportunities for extreme sports, complement cultural tourism.

9. According to you, what are the weaknesses of Anamur in terms of cultural tourism?

Foremost are transportation issues. The distance of Anamur from Antalya and Mersin, inadequate transportation facilities, and the weakness of the hotel infrastructure negatively affect tourism. The lack of full awareness of tourism potential, the prioritization of agricultural production, and insufficient importance given to tourism are other weaknesses. The scarcity of various organizations such as promotion and fairs, inadequate facilities, restoration deficiencies, lack of qualified personnel, and the local population's lack of education and enthusiasm towards tourism can be listed as well as problems like environmental cleanliness, infrastructure deficiencies, and neglect.

10. What are the opportunities for cultural tourism in Anamur?
Infrastructure projects such as shortening the Mersin-Antalya highway with tunnels, opening Gazipaşa Airport, and constructing a new port pier can increase accessibility to tourism. Scientific excavations and restoration of historical artifacts, along with agricultural products, can support cultural tourism. Additionally, the continued presence of the Yörük community is an opportunity.

11. What are the threats to cultural tourism in Anamur?
Especially prominent threats include environmental pollution and marine pollution, air pollution from greenhouses, and indiscriminate disposal of agricultural chemicals. Inadequate knowledge of cultural tourism among the local population and indiscriminate agricultural land use leading to haphazard urbanization are also threats.

Results

The findings obtained from the discussions conducted to evaluate the cultural tourism potential of Anamur reveal the richness of the district's potential. However, in order to fully capitalize on this potential, certain steps need to be taken. Anamur's overall tourism potential relies on its natural and historical assets. Many structures, particularly Anemurium ancient city and Mamure Castle, possess qualities that would make the region appealing. However, the tourism potential is not fully realized. Among the reasons for this are transportation issues (land, sea, and air), as well as deficiencies in all forms of promotion and inadequate facilities. Resolving these issues and strengthening tourism infrastructure will enhance Anamur's tourism potential.

The local community's perspective on cultural tourism is also an important matter. Discussions on how their agricultural livelihoods and economic concerns influence their attitudes towards tourism are essential for understanding the impact of tourism on the region. Conversations regarding this aspect shed light on the region's economic impact. Increasing awareness among the local population about cultural heritage and tourism potential will significantly contribute to the region's economy. Events such as the Anamur Banana Festival (It planned to be held for the 16th time in 2024) seem to have achieved desired goals agriculturally (Mersin Gazetesi, 2024). However, it appears challenging to say the same for cultural tourism dimensions.

In conclusion, Anamur holds significant potential for cultural tourism, which gives hope for the future. When looking at the SWOT analysis, its strengths and opportunities outweigh its weaknesses and threats. With the dominance of a long-lasting summer season in this district, increasing awareness among the local population about cultural riches will naturally boost cultural tourism. The operation of Gazipaşa Airport, the commencement of sea voyages, and the conversion of the intercity highway into a dual carriageway will accelerate this momentum. Furthermore, with the ongoing construction of the Çukurova Airport (between Mersin and Adana), interest in the area will likely increase further. Efforts in this regard are progressing rapidly, and it is expected that within 2-3 years, the issues will be minimized. Consequently, all types of tourism in Anamur (Karakeçili, 2018), especially cultural tourism, will become even more attractive. However, it is crucial for public and tourism stakeholders to come together and develop a balanced development plan.

References

Aksu, M. (2004). *Turistler ve destinasyonlarda sunulan kültürel miras arasındaki ilişkiyi belirlemeye yönelik bir araştırma: Troia örneği* (Yüksek Lisans Tezi). Çanakkale On Sekiz Mart Üniversitesi, Sosyal Bilimler Enstitüsü Turizm İşletmeciliği Anabilim Dalı, Çanakkale.

Artal-Tur, A. (2018). Culture and cultures in tourism. *Anatolia: An International Journal of Tourism and Hospitality Research*, 29(2), 179–182. https://www.tandfonline.com/doi/full/10.1080/13032917.2017.1414433

Besculides, A., Lee, M. E., & McCormick, P. J. (2002). Residents' perceptions of the cultural benefits of tourism. *Annals of Tourism Research*, 29(2), 303–319. https://doi.org/10.1016/S0160-7383(01)00066-4

Çetinsöz, B. C., & Altınbıçak, A. (2008). Anamur İlçesinin turizm potansiyeli ve halkın turizme yaklaşımı üzerine bir araştırma. In Y. Özdemir (Ed.), *Mersin Sempozyumu, 18-22 Kasım 2008, Bildiriler* (vol. 1, pp. 2586-2602). Mersin: Mersin Valiliği.

Çetinsöz, B. C., & Subaşı, B. (2014). Anamur ilçesinde turizm sektörünün gelişim sorunları ve çözüm önerileri. In B. Zengin & K. Ö. Özer (Eds.), *1. Uluslararası Turizm ve Yönetim Araştırmaları Kongresi Bildiri Kitabı* (pp. 108–117). Sakarya: Sakarya Üniversitesi & Nişantaşı Üniversitesi.

Çetinsöz, B. C., & Atsan, M. (2018). Anamur ilçesinin marka kimliği ve kişiliği üzerine bir araştırma. In K. Birdir (Ed.), *2. Uluslararası Turizmin Geleceği Kongresi: İnovasyon, Girişimcilik ve Sürdürebilirlik Kongresi (Futourism 2018) Bildiriler Kitabı* (pp. 630–639). Mersin: Mersin Üniversitesi Yayınları.

Çetinsöz, B. C., & Temiz, G. (2018). Sürdürülebilir kültür turizmi bağlamında Anemurium antik kenti. In K. Birdir (Ed.), *2. Uluslararası Turizmin Geleceği Kongresi: İnovasyon, Girişimcilik ve Sürdürebilirlik Kongresi (Futourism 2018) Bildiriler Kitabı* (pp. 119–126). Mersin: Mersin Üniversitesi Yayınları.

Çoban, B., & Karakaya, Y. E. (2010). Geleceği planlamada stratejik yönetim ve SWOT analizi: kavramsal yaklaşımlar. *e-Journal of New World Sciences Academy*, 5(4), 342–352. https://doi.org/10.12739/10.12739

Dragouni, M., Fouseki, K., & Georgantzis, N. (2017). Community participation in heritage tourism planning: Is it too much to ask?. *Journal of Sustainable Tourism*, 26(5), 1–23. https://doi.org/10.1080/09669582.2017.1404606

Gölgeli, Ü. K. (2016). *Yerel yiyeceklerin gastronomi turizmindeki yeri ve önemi: Anamur örneği* (Yüksek Lisans Tezi). Mersin Üniversitesi, Sosyal Bilimler Enstitüsü Turizm İşletmeciliği Anabilim Dalı, Mersin.

Göçmen, İ. (2021). *Cilicia Bölgesi Köprüleri* (Doktora Tezi). Mersin Üniversitesi, Sosyal Bilimler Enstitüsü Arkeoloji Anabilim Dalı, Mersin.

Karakeçili, G. (2018). Anamur'un biyoçeşitliliği ve eko turizm potansiyeli. In B. C. Çetinsöz & G. Temiz (Eds.), *Marka Kent Stratejileri ve Anamur* (pp. 213–241). Mersin: Mersin Büyükşehir Belediyesi Kültür Yayınları.

Kayabaşı, O. (2008). *Anamur folkloru* (Yüksek Lisans Tezi). Selçuk Üniversitesi, Sosyal Bilimler Enstitüsü Halk Bilim Anabilim Dalı, Konya.

Kozak, M. (2021). *Bilimsel araştırma: tasarım, yazım ve yayım teknikleri*. Ankara: Detay Yayıncılık.

Kuzu, F. P. (2010). *Anamur folkloru* (Yüksek Lisans Tezi). Atatürk Üniversitesi, Sosyal Bilimler Enstitüsü Halk Bilimi Anabilim Dalı, Erzurum.

Nas, F. (2018). Anamur Kent Kimliği. In B. C. Çetinsöz & G. Temiz (Eds.), *Marka Kent Stratejileri ve Anamur* (pp. 80–101). Mersin: Mersin Büyükşehir Belediyesi Kültür Yayınları.

Onan, G. (2018). Anamur Kent Vizyonu, Stratejiler ve Hedefleri. In B. C. Çetinsöz & G. Temiz (Eds.), *Marka Kent Stratejileri ve Anamur* (pp. 36–60). Mersin: Mersin Büyükşehir Belediyesi Kültür Yayınları.

Saygın, M. (2018). Kentlerin markalaşması, marka kentler ve Anamur. In B. C. Çetinsöz & G. Temiz (Eds.), *Marka Kent Stratejileri ve Anamur* (pp. 22–35). Mersin: Mersin Büyükşehir Belediyesi Kültür Yayınları.

Subaşı, B. (2018). Anemurion Antik Kenti. In B. C. Çetinsöz & G. Temiz (Eds.), *Marka Kent Stratejileri ve Anamur* (pp. 164–177). Mersin: Mersin Büyükşehir Belediyesi Kültür Yayınları.

Şahin, S., Özcan, S., & Sol, S. A. (2021). Son Dönem Çalışmaları Işığında Anamur Mamure Kalesi. *Amisos*, 6(11), 267–284. https://doi.org/10.48122/amisos.987156.

Temiz, G. (2018). Sürdürülebilir kültür turizmi: Mamure Kalesi örneği. In B. C. Çetinsöz & G. Temiz (Eds.), *Marka Kent Stratejileri ve Anamur* (pp. 195–212). Mersin: Mersin Büyükşehir Belediyesi Kültür Yayınları.

Yıldırım, A. & Şimşek, H. (2021). *Sosyal Bilimlerde Nitel Araştırma Yöntemleri*. Ankara: Seçkin yayıncılık.

Internet References

Kültür ve Turizm Bakanlığı (2024, April 4). *UNESCO Dünya Mirası Geçici Listesi'ne Giren Mersin'deki Kültür Varlıkları*. https://mersin.ktb.gov.tr/TR-73385/unesco-dunya-mirasi-gecici-listesine-giren-mersinde-ki-.html

Kültür Varlıkları ve Müzeler Genel Müdürlüğü (2024, April 4). *Anamur Müzesi*. https://kvmgm.ktb.gov.tr/TR-44123/mersin-anamur-muze-mudurlugu.html

Mersin Gazetesi (2024, April 4). *Anamur Belediyesi Kültür ve Muz Festivali*. https://mersingazetesi.com/index.php/2023/07/04/

Havva-Gözgeç Mutlu[1] and Volkan Akgül[2]

Chapter 5 Wellness Hotel Practices: Flexibility and Mind

Introduction

Wellness *"as the essence of a healthy lifestyle, encompassing a harmonious combination of physical and mental health, proper nutrition, appropriate physical activity, and the elimination of bad habits"* (Sylchuk, Kyrpichenkova & Druz, 2023, p. 231). Wellness has become a thriving industry. It consists of sub-sectors such as "spas, thermal/mineral springs, wellness tourism, workplace wellness, wellness real estate, physical activity, mental wellness, personal care & beauty; healthy eating, nutrition, & weight loss; public health, prevention, & personalized medicine; and traditional & complementary medicine" (Global Wellness Economy Monitor 2023).

In 2022, the wellness industry had $5.6 trillion in revenue. This substantial income was sourced from different areas, such as physical activity ($976 billion), wellness tourism ($651 billion), spas ($105 billion), and mental wellness ($181 billion) services. International wellness tourists spent an average of $1,764 per trip, which is 41 % more than typical international tourists (Global Wellness Economy Monitor, 2023).

The increasing revenue from the wellness sector and the higher average spending of wellness tourists, along with consumers' growing health awareness, pursuit of healthy lifestyles, increased wealth, and aging population (Li & Gao, 2023), are factors contributing to the rise in the number of wellness hotels in recent years among various stakeholders (Chi, Chi & Quyang, 2020). These hotels aim for guests to sustain healing strategies that leave a lasting impact even after their stay (Kim & Yang, 2021). Additionally, these hotel customers have been segmented by researchers based on their motivations into "socially aspirational, holistic, budget-minded, and discretionary wellness seekers" (Chi, Chi, Deng & Price, 2024).

---

1 PhD., Bandırma Onyedi Eylül University, Erdek Vocational School, Health Tourism Management Program, hgozgec@bandirma.edu.tr
2 Lecturer, Bandırma Onyedi Eylül University, Erdek Vocational School, Health Tourism Management Program, vakgul@bandirma.edu.tr

When studies on wellness hotels are evaluated, it is observed that their number is limited (Chi, Chi & Ouyang, 2020; Kim & Yang, 2021; Li & Gao, 2023; Kessler & Kim, 2023; Chi, Chi, Deng & Price, 2024). These studies have not comprehensively examined the practices offered to guests for mental health and physical flexibility within the scope of wellness hotels. Therefore, this study aims to fill that gap. In this context, it delves into the conceptual explanations and health benefits of practices such as *yoga, meditation, breathing therapy, Emotional Freedom Technique, inversion table, private fitness sessions, trampoline, and RedCord,* implemented in wellness hotels. A general framework has been drawn regarding flexibility and mind-related wellness hotel practices.

Yoga and Its Benefits

"Yoga is an ancient Indian system of philosophy, culture, tradition, and way of maintaining better life, established in India thousands of years ago" (Saha et al., 2014, p. 169). Yoga is considered an ancient science dating back to 5000 years ago (Choudhury & Bordhan, 2015) and consists of eight steps. These steps include Yama (moral guidelines), Niyama (personal etiquette), Asana (physical postures), Pranayama (breath control), Pratyahara (sensory control), Dharana (concentration), Dhyana (meditation), and Samadhi (unity consciousness) (Andelkar, Kanzode & Shamkuwar, 2018).

Regular yoga practice has been shown to enhance both physical and mental well-being (Saha et al., 2014; Dubey & Tiwari, 2023). It is beneficial in managing specific symptoms, especially type 2 diabetes mellitus and pain (Alla & Santhosha Mrudula, 2022) for instance, it was noted to reduce joint pain (Juneja & Kaur, 2022). Additionally, it has the potential to lower blood pressure, promote weight loss, increase muscle mass, and potentially be effective in cardiopulmonary functions (Kaur, 2022; Rathore & Yadav, 2022). In addition to physiological benefits, it had positive effects on mental health (Zadrozna et al., 2022).

Regular yoga practice reduced cortisol levels, stress, anxiety, and depressive symptoms (Kaur, 2022; Zadrozna et al., 2022). It particularly promoted relaxation for both the mother and the baby during pregnancy (Patni & Sinha, 2023) and reduced maternal anxiety during childbirth (Hutasori et al., 2021). Yoga practices had positively influenced the personality traits of women over 40, such as self-esteem, body image, and overall well-being, anxiety control (Coco et al., 2020). Liu, Gao, Huang, & Zhou, (2023) highlighted the benefits of self-focus through breathwork, self-awareness from sensory developments, self-confidence from improved appearance, self-efficacy from advanced postures, and self-regulation from body alignment.

Mediation Therapy and Its Benefits

Meditation is a spiritual practice that originated in India and China (Wang, Rawat, & Panda, 2022). This practice is based on the traditions of Buddhism and Hinduism (Rao, 2017). It is defined as *"being in the present moment or being alert in the present moment, instead of constantly struggling to change or to become"* (Bista et al., 2023, p. 1).

Jamil et al. (2023) emphasized the importance of meditation practices for mental and physical health. Santaella (2021) highlighted that meditation enhances brain functions. Jamil et al. (2023) stated that meditation had positive effects on diseases such as diabetes and hypertension. Bista et al. (2023) noted that meditation positively affects brain systems associated with attention, awareness, memory, sensory integration, emotion regulation, and higher cognitive functions.

Some researchers (Arora & Gupta, 2021; Thero, Kataria & Suman, 2022; Jamil et al., 2023) have pointed out the therapeutic effects of meditation on social anxiety disorder, post-traumatic stress disorder, and appearance.

Breath Therapy and Its Benefits

Breath Therapy is Breathwork (Aideyan, Martin & Beeson, 2020). It encompasses various techniques to enhance individuals' physiological and psychological well-being (Aldridge, 2001). Breath therapy is a treatment method developed in Germany over the last 90 years, just like in Eastern spiritual practices (Mehling, 2001).

In research, it has been stated that breath therapy has beneficial effects in the treatment of conditions such as anxiety and depression, asthma, chronic back pain, cardiac dysfunction, and chronic low back pain (Lalande, Bambling, King ve Lowe, 2012; Mehling et al., 2005; Stutz & Schreiber, 2017). Additionally, it was used during childbirth to direct the mother's attention to different points during pain, increase the pain threshold, promote relaxation, ease uteroplacental circulation, and reduce the perception of pain (Yıldırım & Şahin, 2003). Furthermore, it was observed that various types of breath therapy such as Integral Breath Therapy, Slow Deep Breathing, and Dhikr Breathing Relaxation Therapy were effective for various diseases.

Integral breath therapy has been found effective in reducing symptoms of grief following a traumatic loss (Turner, Wooten, & Chou, 2019). Slow deep breathing was effective in reducing pain levels in head-injured patients (Abdullah, Thalib, & Nurhalisa, 2023). Lastly, the dhikr breathing relaxation therapy had a positive effect on improving sleep quality in insomnia disorders (Purwanto, Anganti, & Yahman, 2022).

Emotion Freedom Technique (EFT) and Its Benefits

EFT (Emotional Freedom Techniques) is a psychological acupressure method that uses energy meridians to tackle physical symptoms, pain, anxiety disorders, and depression, fostering mental clarity and resilience (Alwan, Muhammad, & Makhfudli, 2018).

An increasing amount of research highlights that EFT can deliver significant benefits to a wide range of people in different situations. Studies have consistently demonstrated its efficacy in reducing students' anxiety and improving psychological well-being (Ghasemzadeh, Ghamari & Hosseinian, 2019), as well as enhancing emotional stability in adolescent prisoners (Harbottle, 2019) and mood and quality of life in older adults (Khoeriyah, Utami, & Istichomah, 2018). Furthermore, EFT has shown effectiveness in addressing specific issues such as phobias and nurse burnout (Salas, Brooks, & Rowe, 2011; Wati, Mirayanti & Juanamasta, 2019). From beginner radio broadcasters to postmenopausal women, from veterans with Post-Traumatic Stress Disorder to pregnant women with hypertension, EFT has been found to have significant positive effects on a range of psychological and physiological symptoms (Adhriani, & Yusra, 2014; Mehdipour, Abedi, Ansari, & Dastoorpoor, 2022; Church, Sparks, & Clond, 2016; Pujiastuti, & Mulyantoro, 2020).

Inversion Table and Its Benefits

Inversion therapy has been used in the treatment of spinal disorders since approximately 2,400 years ago by Hippocrates (Vasiliadis, 2009), and later defined concerning Hippocrates by Vidus Vidius in 1544 (Mendelow et al., 2021). This therapy is a type of traction that utilizes gravity; the individual lies on a table while bent, thereby alleviating pressure on the spine.

Modern inversion therapy is conducted using an inversion table, a medical device specially designed to invert individuals. Typically, it consists of components such as support elements, weight elements, and control wires to facilitate controlled inversion maneuvers. Additionally, the device may include features such as adjustable ankle enclosures, posture adjustments, and features for decompression purposes (Zbinden, 2016).

The inversion table is a tool that relaxes the body for the relief of intense back pain, relaxation, and stretching while also removing toxins. During this therapy, the individual hangs safely upside down while the top of the table supports their back. Inversion therapy is necessary for the body to stretch and relax. It has been indicated that inversion therapy can reduce spinal nerve root pain caused by spinal separation and normalize neurological deficiencies (Krause et al., 2000).

Private Fitness Session and Its Benefits

According to a study conducted by the World Health Organization (WHO), physical activity can reduce the likelihood of developing chronic diseases such as heart disease, stroke, diabetes, and certain types of cancer by approximately 30–40 % (Marcos-Pardo et al., 2024). Furthermore, by mitigating changes in body composition, such as increased body fat and decreased lean muscle mass, physical activity can contribute to preventing a range of health issues, including metabolic disorders and diminished functional capacity (Oh et al., 2021).

Regular exercise is the fundamental means of preventing physical and bodily disorders caused by a sedentary lifestyle and enhancing the physiological capacity to maintain health (Zorba, 2001).

Fitness denotes being physically fit and healthy (Oxford, 2018). It encompasses various sports disciplines and primarily aims at toning and strengthening muscles through both single and instrument-assisted exercises. Fitness exercises are utilized to improve the conditioning of muscle groups used in sports programs (Eurogymstar, 2018). Key elements for achieving success in physical activities include motivation, discipline, and consistency. However, ensuring these elements individually can often be challenging. Additionally, due to the monotonous structure of many sports programs, desired outcomes may not be attained or sustained. Therefore, specialized fitness trainers can facilitate individuals in reaching their goals more efficiently and minimizing the risk of injury by tailoring personalized programs to their specific needs and objectives.

One advantage of exercising in fitness centers is the opportunity to work with professional trainers, which reduces the risk of incorrect movements and minimizes the risk of injury (Curup, 2023).

Trampoline and Its Benefits

The trampoline is an entertainment and sports apparatus consisting of a frame connected with springs. It was first used in the early 20th century and has become particularly popular in the United States. Trampolines, which have maintained their popularity to the present day, vary in their usage purposes. Especially full-size trampolines are commonly preferred in entertainment activities (Bortoleto,

Carrara & Roveri, 2018). This sports apparatus not only contributes significantly to flexibility, bone, and joint development but also to muscle development. Therefore, it enables individuals to enjoy sports and engage in sports for longer periods.

The benefits of trampoline exercises have been proven through studies. The flexible springs of the trampoline increase the tone levels of sensory systems by providing the opportunity for vigorous bouncing, while also creating a rocking effect on muscles, which can reduce tone (Dufek and Bates, 1991). Additionally, trampoline use has the potential to reduce the mechanical load on the muscle system and enhance the effect of the muscle-tendon complex (Arabatzi et al., 2018).

However, along with the benefits of trampoline use, there are potential risks that need to be considered. Trampoline sessions performed without warming up and without consciousness can lead to accidents that may prevent individuals from engaging in sports for a long time. Therefore, warming up before sports is important, and movements should be performed consciously. If possible, movements should be learned and applied under the guidance of an expert.

RedCord and Its Benefits

RedCord therapy, also known as Neurac (neuromuscular activation), is a rehabilitation technique that combines suspending and vibrating a segment of the body to treat various conditions. It has been shown to positively impact patients with arterial hypertension, chronic pain, acute subacromial impingement syndrome, and chronic neck pain (Kirkesola, 2009; Calvillo, Racz, & Noe, 2016; Soo et al., 2015).

Developed by Norwegian physiotherapists in the early 1990s, RedCord is a suspension-based therapy concept. The use of body weight and straps forms the basis of this therapy. RedCord therapies are used worldwide by individuals and athletes for rehabilitation, injury prevention, and performance enhancement. By using elastic cords, RedCord aims to identify symptoms caused by pain or immobility in the body, relax overloaded tense muscles, restore muscles to normal and pain-free functions, and increase control power in the individual's movements. During RedCord Therapy, the focus is on addressing problems that cause negative interactions between muscles (Redcord, 2024).

Kim et al. (2014) stated that Neurac exercises are an effective method to reduce pain, adjust postural balance, and normalize muscle response patterns in patients with chronic low back pain.

Conclusion

The study has considered some wellness practices on physical and mental health in the hospitality industry. Studies and statistical data have shown the size and importance of the wellness industry. The significance of wellness hotels is also growing in this context. Various wellness services are provided in these hotels, and different practices are applied for physical and mental health. Mind and flexibility are some of these services. In wellness hotels, physical exercise, yoga, breath therapy, meditation, and EFT are widespread and practiced by different categories of people. Some of the practices studied have multiple positive physical and mental effects.

Regular yoga has a beneficial impact on physical wellness and reduces stress, contributing to mental wellness. Breath therapy is effective in addressing conditions such as anxiety and depression. Meditation shows significant benefits in reducing stress and enhancing mental health. EFT has significant positive effects on various psychological symptoms such as anxiety disorders, depression, and physiological symptoms such as pain.

Trampoline, personal fitness sessions, and RedCord therapy also have positive effects on physical health. The use of inversion tables stands out as an effective method for reducing back pain and improving spinal health. Private fitness sessions are important for preventing chronic diseases and increasing physical capacity. Trampoline exercises have positive effects on flexibility, bone, and muscle development. RedCord therapy is used in rehabilitation, injury prevention, and performance enhancement.

This study emphasizes the overall impact of wellness practices on mental and physical health within the hospitality industry and encourages further research in this area.

References

Abdullah, R., Thalib, A. H. S., & Nurhalisa, S. (2023). Slow deep breathing therapy for reducing pain in patients with head injury. *Jurnal Ilmiah Kesehatan Sandi Husada*, 12(1), pp. 104–110.

Adhriani, V., & Yusra, Z. (2014). Efektifitas emotional freedom techniques (eft) untuk menurunkan kecemasan pada penyiar radio pemula. *Jurnal Student Psikologi Universitas Negeri Padang*, 1, pp. 1–11.

Aideyan, B., Martin, G. C., & Beeson, E. T. (2020). A practitioner's guide to breathwork in clinical mental health counseling. *Journal of Mental Health Counseling*, 42(1), pp. 78–94.

Aldridge, D. (2001). Philosophical speculations on two therapeutic applications of breath. *Subtle Energies & Energy Medicine Journal Archives*, 12(2).

Alla, D., & Santhosha Mrudula, A. S. (2022). A comprehensive review on the role of yoga in the management of type- 2 diabetes mellitus and its benefits over physical exercise in type 2 DM. *International Journal of Research in Medical Sciences*, 10(10), pp. 2334–2338.

Alwan, R., Muhammad, A., & Makhfudli, M. (2018). Emotional Freedom Technique (EFT) for Physiological Symptoms, Pain, Anxiety Disorders and Depression: a Systematic Review.

Andelkar, A., Kanzode, S. P., & Shamkuwar, J. (2018). Basic concept of yoga and its health benefits-a short review. *International Journal of Research in AYUSH and Pharmaceutical Sciences*, 2(2), 217–221.

Arabatzi, F., Tziagkalou, E., Kannas, T., Giagkazoglou, P., Kofotolis, N., & Kellis, E. (2018). Effects of two plyometric protocols at different surfaces on mechanical properties of Achilles tendon in children. *Asian Journal of Sports Medicine*, 9(1), 67–97

Arora, R., & Gupta, R. (2021). Effectiveness of meditation programs in empirically reducing stress and amplifying cognitive function, thus boosting individual health status: A narrative overview. *Indian Journal of Health Sciences and Biomedical Research Kleu*, 14(2), pp. 181–187.

Bista, S., Ghimire, B., Sapkota, V., Poudel, L., & Khadka, R. (2023). Meditation for human mind and brain: Findings from functional neuroimaging investigations. *Annals of Advanced Biomedical Sciences*, 6(1), pp. 1–7.

Bortoleto, M., Carrara, P., & Roveri, M. G. (2018). Trampoline gymnastics: the Brazilian participation at international championships-the Olympic games still a dream. *Science of Gymnastics Journal*, 10(3).

Calvillo, O., Racz, G.B., & Noe, C. (2016). Theory and Mechanisms of Action of Neuroaugmentation. In: Racz, G., Noe, C. (eds) Techniques of Neurolysis. Springer, Cham.

Chi, C. G. Q., Chi, O. H., & Ouyang, Z. (2020). Wellness hotel: conceptualization, scale development, and validation. *International Journal Of Hospitality Management*, 89, 102404.

Chi, O. H., Chi, C. G., Deng, D. S., & Price, M. M. (2024). Wellness on the go: Motivation-based segmentation of wellness hotel customers in North America. *International Journal of Hospitality Management*, 119, 103725.

Choudhury, M. K. & Bordhan, S. (2015). Yoga: an art of living. *International Journal of Applied Research*, 1(7), pp. 740–743.

Church, D., Sparks, T., & Clond, M. (2016). EFT (Emotional Freedom Techniques) and resiliency in veterans at risk for PTSD: a randomized controlled trial. *Explore*, 12(5), pp. 355–365.

Coco, M., Buscemi, A., Sagone, E., Pellerone, M., Ramaci, T., Marchese, M., & Musumeci, G. (2020). Effects of yoga practice on personality, body image and lactate. pilot study on a group of women from 40 years. *Sustainability*, 12(17), pp. 6719.

Curup H, (2023). Rekreatif amaçlı fitness egzersizi yapan bireylerin sosyal medya kullanımları ile mutluluk düzeyleri arasındaki ilişki, Yüksek Lisans Tezi, Eğitim Bilimleri Enstitüsü, Mersin Üniversitesi.

Dubey, J. P., & Tiwari, A. (2023). The role of practicing yoga and its physiological benefits. *International Research Journal of Ayurveda & Yoga*, 6(3), pp. 114–116.

Dufek, J. S., & Bates, B. T. (1991). Biomechanical factors associated with injury during landing in jump sports. *Sports Medicine*, 12, 326–337.

Eurogymstar. (2018). Fitness Faydaları [online] http://www.eurostargym.com

Ghasemzadeh, A., Ghamari, M., & Hosseinian, S. (2019). The impact of emotional freedom techniques on Students' anxiety reduction and psychological well-being increase. *Education Strategies in Medical Sciences*, 12(4), pp. 135–145.

Global Wellness Economy Monitor, (2023). 2023-gws/gws2023-the-globalwellness-economy https://www.globalwellnesssumm

Global Wellness Industry Report, (2023). 2023-gws/gws2023-the-global-wellness-economy https://www.globalwellnesssummit.com

Harbottle, L. (2019). Potential of emotional freedom techniques to improve mood and quality of life in older adults. *British Journal of Community Nursing*, 24(9), 432–435.

Hutasori, E. S., Yanti, N., Hayati, S., Azwar, Y., Noviyanti, N., & Utami, K. (2021). Pemantauan Kecemasan dan Lama Persalinan Kala I pada Ibu Bersalin dengan Pelaksanaan Yoga Kehamilan. *Jurnal ABDIMAS-HIP Pengabdian Kepada Masyarakat*, 2(2), 96–101.

Jamil, A., Gutlapalli, S. D., Ali, M., Oble, M. J., Sonia, S. N., George, S. & Ali Sr, Z. (2023). Meditation and its mental and physical health benefits in 2023. *Cureus*, 15(6).

Juneja, R., & Kaur, M. (2022). The health and disease effects of yogic practices or transcendental meditation. *International Journal of Innovative Research in Engineering & Management*, 9(1), 457–460.

Kaur, M. (2022). A comprehensive study of yoga for major depressive illness. *International Journal of Innovative Research in Engineering & Management*, 9(1), 461–464.

Kessler, D., & Kim, S. H. (2023). Segmenting tourists by wellness hotel attributes and demographics: a study of North American wellness tourists. The Academy of Korea Hospitalty & Tourism, 25(12), 109–125.

Khoeriyah, S. M., Utami, D. P., & Istichomah, I. (2018). Effect of emotional freedom technique for emotional stability in adolescent prisoners. *Indonesian Nursing Journal of Education and Clinic (INJEC)*, 3(1), 15–21.

Kim, B., & Yang, X. (2021). "I'm here for recovery": the eudaimonic wellness experiences at the Le Monastère des Augustines Wellness hotel. *Journal of Travel & Tourism Marketing*, 38(8), 802–818.

Kim, E.R, Oh, J.S, & Yoo, W.G. (2014). Effect of vibration frequency on serratus anterior muscle activity during performance of the push-up plus with a redcord sling. *The Journal of Physical Therapy Science*, 26, 1275–1276.

Kirkesola, G. (2009). Neurac a new treatment method for long-term musculoskeletal pain. *J Fysioterapeuten*, 76, 16–25.

Krause, M, Refshauge, K.M, Dessen, M., & Boland, R. (2000). Lumbar spine traction: evaluation of effects and recommended application for treatment. *Manual Therapy*, 5(2), 72–81.

Lalande, L., Bambling, M., King, R., & Lowe, R. (2012). Breathwork: an additional treatment option for depression and anxiety? *Journal of Contemporary Psychotherapy*, 42(2), 113–119.

Li, Z., & Gao, Y. (2023). Better wealth, better health: wellness hotel attributes and consumer preferences in China. *Journal of China Tourism Research*, 20(2), 333–355.

Liu, H., Gao, M., Huang, Y., & Zhou, Y. (2023). An exploration of the associations between perceived physical and psychological benefits of Chinese yoga leisure participants a qualitative approach. *Journal of Leisure Research*, 54(4), 472–492.

Marcos-Pardo, P., Espeso-García, A., Vaquero-Cristóbal, R., Abelleira-Lamela, T., & González-Gálvez, N. (2024). The effect of resistance training with outdoor fitness equipment on the body composition, physical fitness, and physical health of middle-aged and older adults: a randomized controlled trial. *Healthcare*, 12(7), 726.

Mehdipour, A., Abedi, P., Ansari, S., & Dastoorpoor, M. (2022). The effectiveness of emotional freedom techniques (EFT) on depression of postmenopausal women: a randomized controlled trial. *Journal of Complementary and Integrative Medicine*, 19(3), 737–742.

Mehling, W. E. (2001). The experience of breath as a therapeutic intervention– psychosomatic forms of breath therapy. A descriptive study about the actual situation of breath therapy in Germany, its relation to medicine, and its application in patients with back pain. *Forschende Komplementärmedizin und Klassische Naturheilkunde/Research in Complementary and Classical Natural Medicine*, 8(6), 359–367.

Mehling, W. E., Hamel, K. A., Acree, M., Byl, N., & Hecht, F. M. (2005). Randomized controlled trial of breath therapy for patients with chronic low-back pain. *Alternative Therapies İn Health And Medicine*, 11(4), 44–53.

Mendelow AD, Gregson BA, Mitchell P, Schofield I, Prasad M, Wynne-Jones G, Kamat A, Patterson M, Rowell L, & Hargreaves G. (2021). Lumbar disc disease: the effect of inversion on clinical symptoms and a comparison of the rate of surgery after inversion therapy with the rate of surgery in neurosurgery controls. *The Journal of Physical Therapy Science*, Nov; 33(11) pp. 801–808.

Oh, Y.H.; Choi, S.; Lee, G.; Son, J.S.; Kim, K.H. & Park, S.M. (2021). Changes in body composition are associated with metabolic changes and the risk of metabolic syndrome. *Journal of Clinical Medicine*, 10(4), 745.

Oxford, (2018). "Fitness" https://en.oxforddictionaries.com/definition/ fitness

Patni, K., & Sinha, G. (2023). An analytical review on health benefits of prenatal yoga for mother and fetus. *International Journal of Ayurveda and Pharma Research*, 11, 46–52.

Pujiastuti, R. S. E., & Mulyantoro, D. K. (2020). Spiritual emotional freedom technique (seft) intervention on blood pressure among pregnancy with hypertension. *International Journal of Nursing and Health Services (IJNHS)*, 3(3), 402–410.

Purwanto, S., Anganti, N. R. N., & Yahman, S. A. (2022). Validity and effectiveness of dhikr breathing relaxation model therapy on insomnia disorders. *Indigenous: Jurnal Ilmiah Psikologi*, 7(2), 119–129.

Rao, K. (2017). What is meditation? in: foundations of yoga psychology. *Springer*, Singapore.

Rathore, V., & Yadav, N. (2022). Yoga as a complementary and alternative therapy for cardiopulmonary functions. *Yoga Mimamsa*, 54(2), 133–139.

RedCord 2024, https://www.redcord.com

Saha, M., Halder, K., Tomar, O.S., Pathak, A., Pal, R. (2014). Yoga for preventive, curative, and promotive health and performance. ın: singh, s., prabhakar, n., pentyala, s. (eds) translational research in environmental and occupational stress. Springer, New Delhi.

Salas, M. M., Brooks, A. J., & Rowe, J. E. (2011). The immediate effect of a brief energy psychology intervention (Emotional Freedom Techniques) on specific phobias: A pilot study. *Explore*, 7(3), 155–161.

Santaella, D. F. (2021). Neurobiology of meditation. ın s. telles & r. gupta (eds.), handbook of research on evidence-based perspectives on the psychophysiology of yoga and ıts applications (pp. 61–71). IGI Global.

Soo, Y., You, L, K., & Suk, M, L. (2015). The effect of neurac training in patients with chronic neck pain. *Journal of Physical Therapy Science*, 27(5),1303–1307.

Stutz, R., & Schreiber, D. (2017). Die therapeutische wirksamkeit westlicher atemtherapiemethoden: ein systematischer review. *Complementary Medicine Research*, 24(6).

Sylchuk, T., Kyrpichenkova, O., & Druz, T. (2023). Wellness tourism as service innovation of the hospitality industry. *Black Sea Economic Studies*, 79, 231–235.

Thero, V. S., Kataria, H. B., & Suman, A. (2022). Meditation for skin aging, reduces wrinkles and change your appearance. *International Journal of Clinical & Experimental Dermatology*, 7(1), 12, 16.

Turner, R., Wooten, H. R., & Chou, W. M. (2019). Changing suicide bereavement narrative through integral breath therapy. *Journal of Creativity in Mental Health*, 14(4), 424–435.

Vasiliadis, E.S., Grivas, T.B., & Kaspiris, A. (2009). Historical overview of spinal deformities in ancient Greece. *Scoliosis*, 4(6).

Vidius V: Chirurgia e Graeco in Latinum conversa, Vido Vidio Florentino interprete, cum nonnullis eiufdem Vidij comentarijs. Excudebat Petrus Gallerius, 1544, Comment III: 179.

Wang, Z., Rawat, V., Yu, X., & Panda, R. C. (2022). Meditation and its practice in Vedic scriptures and early Taoism scriptures. *Yoga Mimamsa*, 54(1), 41–46.

Wati, N. M. N., Mirayanti, N. W., & Juanamasta, I. G. (2019). The effect of emotional freedom technique therapy on nurse burnout. *Jurnal Medicoeticolegal dan manajemen rumah sakit*, 8(3), 173–178.

World Health Organization. (2015). Physical activity: global recommendations on physical activity for health consequences of physical ınactivity; WHO regional office for Europe: Copenhagen, Denmark,

Yıldırım, G., & Hotun, Ş. N. (2003). Doğum ağrısının kontrolünde hemşirelik yaklaşımı, *C. Ü. Hemşirelik Yüksek Okulu Dergisi*, 7(1).

Zadrozna, K., Wysokińska, O., Żyga, J., Małek, A., Fabiś, M., Wójcik, B., & Iwaniszyn-Zapołoch, K. (2022). Effects of regular yoga practice on neurological conditions and mental health. *Journal of Education, Health and Sport*, 12(12), 35–41.

Zbinden, A. (2016) https://patents.justia.com/patent/20160361582

Zorba, E. (2001). Fiziksel Uygunluk. Ankara: Gazi Kitabevi.

Didem Kutlu[1]

Chapter 6 A Conceptual Overview of Regenerative Tourism

Introduction

Tourism, with its unique attributes, serves as a significant instrument for development. It bolsters productive capabilities by catalyzing commercial activities and creating employment opportunities linked to the tourism value chain. Developing nations stand to gain from tourism, especially because of their resources such as natural surroundings, weather, cultural heritage, and human capital. Nonetheless, tourism is also a sector that harms the environment, generates pollution, depletes scarce resources, and can lead to adverse societal changes (UNWTO, 2013). Environmental repercussions, such as biodiversity loss, landscape alteration, waste production, and dwindling water resources, along with social consequences like overtourism, gentrification, and social discord, are indicative of the negative transformations brought about by tourism. The COVID-19 pandemic has highlighted the adverse impacts of the tourism sector, revealing the unsustainability of traditional tourism models for the future. This has necessitated a reassessment of tourism with the concept of 'building a better future' (CBI, 2022) and the demand for a more robust tourism model that relies on alternative tourism systems, such as the regenerative economy, transcending capitalist practices (Becken & Kaur, 2022; Dredge, 2022; Fusté-Forné & Hussain, 2022; Mathiesen et al., 2022; McEnhill et al., 2020; Pung et al., 2024; Sheldon, 2021). The necessity to transcend sustainability towards a more qualitative growth that enhances human health and well-being via ecosystem health has given rise to the concept of regenerative tourism (CBI, 2022).

Regenerative tourism is an approach where stakeholders in the tourism sector collectively demonstrate care and protection through their decisions and actions, aiming to enhance and enrich the natural, social, and man-made elements when visiting, residing, or dwelling in a destination (Earthcheck Research Institute, 2023). With the objective of enhancing the sustainability and vitality of destinations, regenerative tourism integrates indigenous and Western scientific

---

1 Assoc. Prof., Akdeniz University, Vocational School of Social Sciences, Department of Tourism and Travel, didemkutlu@akdeniz.edu.tr

perspectives and knowledge. It seeks to transform tourism through life systems in tourism that establish connections between people and places and foster mutually beneficial relationships (Bellato et al., 2023). Moreover, regenerative tourism aims to improve and maintain positive environmental conditions while assisting in the restoration and preservation of the social and cultural fabric of the communities where it is implemented (McEnhill et al., 2020). Seen as an advanced progression of sustainable tourism, regenerative tourism focuses on promoting long-term social, cultural, environmental, and economic growth through the transformation of travelers, locals, and other stakeholders in the sector.

The Concept of Regenerative Tourism

The concept of regenerative tourism has garnered considerable attention in recent years, alongside related concepts such as regenerative agriculture, regenerative design, and regenerative development. It is contended that these regenerative concepts have evolved in response to concerns about the efficacy of implementing the sustainability paradigm (McEnhill et al., 2020). In the realm of tourism, the regenerative concept was initially probed by C. Owen in her study aimed at identifying the potential of ecotourism facilities for a more regenerative transformation. Owen (2007) asserts that regenerative design diverges from sustainability in three fundamental ways. Firstly, it transitions the frame of reference from minimal impact to positive impact. Secondly, it contests the Cartesian distinction between subject and object, which forms the basis of traditional human/environment relations. Cartesian dualism posits that humans are separate from nature and in competition with it. Moreover, it seeks to reconnect environmentalism with a socio-political dimension that is absent in the sustainability discourse.

To delve deeper into the concept of regenerative tourism, it's useful to draw a comparison with sustainability. Sustainable tourism primarily views tourism as an industry and often prioritizes top-down, standardized, and compartmentalized interventions. On the other hand, regenerative tourism bases its interventions on enhancing the regenerative capacity of entire systems, focusing on infinite economic growth rather than managing socio-ecological impacts (Bellato et al., 2023). Regenerative tourism is not seen as a result, a plan, or a single output, but as a journey of capacity building. It aims to effect change by altering individual understanding and mindset (Dredge, 2022). Both sustainable and regenerative tourism recognize that the standard business practices within the scope of tourist activities will lead to the degradation of the natural, cultural, and social resources that form the basis for the realization and promotion of tourism.

However, to address this issue, while sustainable tourism aims to preserve the current state of our world, regenerative tourism strives to enhance the current state of the world (Coll-Barneto & Fusté-Forné, 2023). Regenerative tourism is characterized by three main features: living systems and nature, local communities, and holistic knowledge (Pung et al., 2024). As delineated by Mathisen et al. (2022), the regenerative paradigm is rooted in an ontology that asserts the inseparability and interconnectedness of society and nature, conceptualised as a dynamic web of life. All living entities are part of nature, and humans possess the ability to take responsibility for restoring and maintaining harmony within this web of life. The stakeholders that constitute the living system undertake transformative roles that contribute to the regeneration of themselves, destinations, and communities (Bellato et al., 2023). In the context of local communities, regeneration holds the potential to fortify local communities, establish stronger connections with agriculture and food supply systems, implement circular economy principles, and contribute to human health and well-being (Dredge, 2022). Regenerative tourism is a concept that seeks to amalgamate various forms of knowledge, including indigenous, experiential, and spiritual knowledge, with the aim of enhancing tourism management in specific locations. The integration of local and Western scientific knowledge presents a unique opportunity to customise tourism management strategies to the specific characteristics of a given destination. Regenerative tourism requires the application of comprehensive decision-making and understanding strategies based on shared cultural values and principles, such as mutual aid, downsizing, and localisation (Becken & Kaur, 2022). Sustainable tourism involves community participation and sharing of benefits, allowing local communities to reap economic and social benefits from tourism. However, the goal is to maintain balance rather than actively contribute to community empowerment and regeneration. Regenerative tourism places a significant emphasis on community empowerment and participation. The goal is to involve local communities in decision-making processes, promote entrepreneurship, and leave a lasting positive impact on the social fabric of destinations (Goa Tourism Development Corporation, n.d.). Dredge (2022) suggested that regenerative tourism is underpinned by seven principles: holistic nature, care and respect, agency, dynamic and evolutionary, and collaborative and continuous learning. To achieve the desired transformation, it is necessary to implement system change, a change in mindset, and the necessary measures to effect this change. The principles of sustainability relate to the environmental, economic, and socio-cultural facets of tourism development. They mandate a suitable balance between these three dimensions to ensure long-term sustainability. The principle of economic sustainability involves the generation of prosperity at various societal levels and the analysis of the cost-effectiveness of all economic transactions. The concept of social sustainability includes the principles of respect for human rights and equal opportunities for all society segments. It underscores the protection and enhancement of local communities' life support systems, the recognition and respect for diverse cultures, and the avoidance of all forms of exploitation. The concept of environmental sustainability involves the protection and management of resources, particularly focusing on those that are non-renewable or valuable for life support (UNEP & WTO, 2005). The most significant critique of the principles of sustainability is that they prioritize the conservation of resources to meet social and economic needs, while neglecting the ecological needs of other species. The Bruntland report's assertion that species and ecosystems of economic value for development and human welfare should be protected indicates a policy that tends to adopt anthropocentric and instrumental value systems. A sustainable development approach should be considered in light of ethical codes that impose moral values on both humans and non-humans. An ethically responsible attitude will not overlook the potential consequences for other living beings in a sustainable development process (Imran et al., 2014). The regenerative tourism strategy challenges the existing growth model and seeks to create net positive impacts on social-ecological systems. In contrast to this approach, sustainability-oriented strategies are confined to actions aimed at minimizing social-ecological damage (Inversini et al., 2024).

Indeed, based on the discussions above, regenerative tourism can be characterized as a process where stakeholders in the tourism sector collectively demonstrate care and stewardship through their decision-making and practices. This is done with the objective to augment and enrich the natural, human, and man-made (constructed) elements when they are moving, visiting, living, or traveling (ERI, 2023). This approach emphasizes the importance of a holistic and sustainable approach to tourism that benefits both the environment and the communities involved.

Regenerative Tourism: Impacts and Challenges

The most significant positive consequences of regenerative tourism can be grouped into four categories: economic resilience, environmental protection, cultural enrichment, and social inclusion. Regenerative tourism enhances economic resilience by fostering active participation of local communities in tourism activities. A more equitable distribution of tourism income improves the economic well-being and quality of life of communities by creating a more balanced and long-term sustainable economic structure for communities (GTDC, n.d.). The Earthcheck Research Institute has ensured an equitable distribution of economic benefits, with more than 75 % of the benefits remaining in the local area. Regenerative tourism is a concept that aims to achieve a net increase in ecosystem rehabilitation with native species. This is achieved by promoting ecosystem restoration and development for biodiversity conservation. In contrast to the conventional approach to tourism development, which is primarily focused on profit and volume, regenerative tourism is value-driven and prioritises the environmental and social needs of a destination. Regenerative tourism has the potential to transform the industry by enabling tourists to experience a thriving host community and better reflect the relevant local environmental context and community aspirations (Qi et al., 2024). In terms of social inclusion, it promotes inclusive and responsible tourism through the implementation of equitable economic models that empower vulnerable groups, including youth, women, indigenous peoples, and people with disabilities (GTDC, n.d.).

Regenerative tourism presents a promising perspective on sustainable travel solutions. However, this area still faces barriers such as resistance to change, lack of awareness, and the need for coordinated efforts among stakeholders. Active involvement from governments, businesses, and tourists is necessary in the process of adopting a regenerative approach to the tourism sector. Future travel experiences will be shaped by a harmonious balance between economic growth and environmental and cultural conservation, where regenerative practices are prioritised (GTDC, n.d.). Dredge (2022) stated that the challenge of transitioning to a regenerative mindset in tourism stems from tourism's deep commitment to scientific thinking and strategic management. Additionally, economic profit is prioritised over social, cultural and ecological value creation and the value generated is often not distributed equally. This makes it difficult to develop a regenerative value proposition for tourism. In Cave et al.'s (2022) study on regenerative tourism, participants stated that part of the challenge of regenerative tourism is that some of its foundations are not well known and change is long-term. The authors emphasised the importance of transformational leadership to overcome this resistance and meet the need to transform the next generation of professionals, stakeholders, and communities by creating tools and actions based on regenerative tourism principles. From a practical perspective, 'Tourism Colab', a social enterprise community, develops and supports regenerative leadership through collaborative learning, coaching, and mentoring (The Tourism CoLab, n.d.). Another challenge for regenerative tourism is that there is no single method or way to make tourism communities regenerative. For each community, different methods may need to be used to achieve the goal. Acceptance of regenerative tourism requires a change in mindset, co-operation, community, viewing tourism as a system rather than an industry, a different marketing approach, suitability and conscious travel habits (Coll-Borneto & Fusté-Forné, 2023).

Regenerative Hospitality

In the hospitality sector, the concept of regenerative hospitality goes beyond merely preserving or restoring existing ecological, social, or climate systems. It implies that a company's societal engagement leads to improvements in these systems. Regenerative hospitality promotes a cooperative relationship between humans and nature, incorporates an inclusive approach to economic systems, and aims for long-term enhancement of the ecosystem in which a business operates (Legrand et al., 2024). The nature and scope of regenerative hospitality were evaluated by Inversini et al. (2024) from the perspectives of academics, consultants, and hoteliers. Regenerative hospitality requires a shift in understanding, moving from a revenue-oriented to a purpose-oriented approach, with a focus on generating net positive impacts beyond mere conservation. According to the 2023 sustainability report by Booking.com, due to concerns about climate change, travellers are increasingly adopting a regenerative approach to travel. They are redefining their understanding of 'value' by creating regenerative experiences and seeking more meaningful ways to spend their resources. These regenerative experiences positively impact destinations, benefiting wildlife, conservation efforts, and local communities, and involve participation in local philanthropy and ethical experiences.

In terms of its reach, regenerative hospitality is classified as both local and systemic, embodying an inside-out approach to growth, location, and human intellect. As per Inversini et al. (2024), the internal viewpoint represents human intelligence, which is associated with employees (who are frequently community members and hence have a connection with the external environment), suppliers who furnish the necessary goods and services for hotel operations, and guests who reap the benefits of regeneration and, ultimately, transformation. The external viewpoint pertains to nature and the community, embodying the intelligence of the location through practices such as integrating architecture with nature, optimising local supply chains, and regenerating nature. Dredge (2022) suggests that the transformation of tourism will be feasible with the shift from scientific thinking to integrated intelligence. Integrated intelligence acknowledges the existence of multiple methods of comprehension and knowledge, originating from the head, heart, and brain, and recognises that knowledge can also be socialised and passed down through generations.

The Regenerative Tourism Community is a network of independent hotels committed to the practice of regenerative tourism. One member of this community, Hamanasi Resort, has been designed with sustainability in mind. This includes the use of natural lighting, reduction of energy consumption, provision of eco-friendly items to guests, and bulk purchasing of these items to reduce the resort's transport footprint. The resort has also undertaken the reforestation of a portion of the threatened coastal forest. Additionally, the organization conducts extensive outreach activities within the local community, focusing on health, welfare, and children's education. Guests at Hamanasi Resort have the opportunity to learn about the local flora and fauna from guides certified by the Belize Tourism Board (Northrop et al., 2022).

The ability to influence and ideally enchant, enliven, and enrich the lives of visitors depends on the capacity of hosts who are deeply connected to their place to feel an intimate, co-creative connection with the place that fosters a deep love for it and a desire to care for all life within it. A life-affirming regenerative tourism that works with wholes rather than parts, with permeable rather than rigid boundaries, allows tourism to better integrate with different life support systems - whether hydrological, food-based, socio-cultural, political, or economic. These systems are expressed in place and enable tourism to function as a broader life support system (Pollock, 2019).

Regenerative Tourism Practices

The literature has yet to establish a standard regarding the principles of regenerative tourism practices. Bellato et al. (2022) have classified practices within the context of seven theoretical principles. These principles encompass the following: utilising an ecological worldview; employing living systems thinking; exploring the unique potential of a regenerative tourism destination; harnessing the capacity of tourism life systems to catalyse transformation; embracing recovery strategies that foster cultural rejuvenation and land restitution, and giving precedence to the viewpoints, knowledge, and practices of indigenous and marginalised populations; creating regenerative places and communities; collaborating to develop and implement regenerative tourism approaches.

One example of remediation that encourages the privileging of indigenous practices is the Hawaiian Islands. The islands are implementing a more advanced, post-pandemic tourism strategy rooted in local Hawaiian values, such as malama (caring) and kuleana (responsibility). They have adopted regenerative strategies that include comprehensive processes to gather community input in determining the future path of tourism (Sheldon, 2021).

In Aotearoa (New Zealand), a regenerative tourism model is being implemented to ensure that visitors leave as storytellers who can share the destination's culture and heritage with the world, creating a positive impact on communities. This tourism model aims to transform the foundations of the tourism system, build destination branding and improve the visitor experience, support communities to manage tourism and work in partnership with others (Tourism Projects, ty). Indeed, sharing stories and culture builds resilience to crises in tourism and is inherently regenerative for local communities and cultural heritage. It also provides an important source of income (Northrop et al., 2022). Becken and Kaur (2022) have attempted to develop a values-based tourism framework for regenerative tourism under the New Zealand Ministry of Conservation's 'Papatuanuku Thriving' objective. In doing so, they have drawn on critical studies, existing New Zealand tourism strategies, stakeholder input and discussions with government staff. Using the analogy of a tree, the root system symbolises values as priorities and the agreements and tools in place to protect them. Knowledge forms part of the root system, drawing wisdom from both indigenous (Matauranga Maori) and Western Science and providing the opportunity to manage tourism in a way that is unique to New Zealand. At the top of the tree are societal values - environmental, social, cultural and economic. This values-based system creates feedback loops for tourism to 'give back' and create relational value with people and place. It is thought that the most important reason for finding more examples of regenerative tourism, particularly in New Zealand, is that there are regenerative initiatives such as the New Zealand Sustainability and Resilience Institute's (SRI) 'Regenerative Tourism Project', "New Zealand Awaits" and "Seventh Generation Tours" (Fusté-Forné & Hussain, 2022).

Mathiesen et al. (2022) conducted an analysis of regenerative tourism using the metaphor of 'earth, spirit, and society', based on the interconnection of society and nature in line with the regenerative paradigm. Their study centred on a tourism company located in Alta, in northern Norway. The spirit of the business owners, symbolising internal sustainability, is characterised by a commitment to values, interests, and passions, a dedication to their work, and a pursuit of regenerative development to enhance the well-being of future generations. In the context of the earth and society, the aim is to impart knowledge and a passion for nature to guests through nature-centric activities. The objective is to learn from nature in all business operations and contribute to the guests' learning.

In the realm of coastal and marine tourism, a regenerative approach is adopted, focusing on four key areas: restoration of coastal and marine ecosystems; protection and repopulation of marine life; enhancement of the skills capacity of local communities; and development and revitalisation of cultural heritage, traditional knowledge, and local identity. These efforts aim to lay the foundation for a sustainable ocean economy by shaping a more prosperous and holistic destination that prioritises the well-being of the entire ecosystem (Northrop et al., 2022). For instance, the Shangri-La Yanuca Island Resort organises a marine education campaign that produces ecological, social, and financial benefits. The campaign aims to utilise stakeholders' knowledge and understanding of the current and future impacts on ecosystems to develop management actions that will sustain ecosystems and human well-being. Guests invest their time and energy to construct fish houses in the reef areas surrounding the hotel and can monitor these fish houses via Google Earth using GPS coordinates (Northrop et al., 2022).

Goa has taken the initiative in India by introducing regenerative tourism practices. Under the banner of regenerative tourism, Goa has focused its efforts on promoting green tourism, digitalisation, skill development, support for micro, small and medium enterprises, and destination management in a regenerative manner. Their goal is to develop a sustainable, resilient, and responsible tourism sector, harness the power of digitalisation to enhance competitiveness, inclusiveness, and sustainability in the tourism sector, equip the youth with business and entrepreneurial skills in the tourism sector, and re-envision destination management from a comprehensive perspective (GTDC, n.d.).

In addition to public institutions, accommodation providers, residents, and other industry representatives, travellers are increasingly adopting more sustainable behaviours. As a result, travellers are creating regenerative experiences that have a positive impact on destinations, benefiting wildlife, conservation efforts, and the local community, and participating in local philanthropy and ethical experiences. More mindful travellers are translating this into action by turning off air conditioning in their accommodations, reusing towels, opting for more sustainable modes of road transport, and shopping locally (booking.com). This shift in traveller behaviour is a promising sign for the future of regenerative tourism.

Results

Tourism is a sector that acts as a catalyst for national economic development, generating significant employment, economic growth, and cultural interaction. However, in addition to its benefits, it has several negatives, including environmental damage, cultural erosion, and potential exploitation. The negative effects of tourism have prompted the development of sustainable tourism, which is described as "tourism that fully considers current and future economic, social, and environmental impacts, caters to the needs of visitors, and ensures the sustainability of tourism" (UNWTO, 2013). Nevertheless, given the insufficiency of sustainable tourism's objective of curbing the damage inflicted by human activities on economic, environmental, and socio-cultural systems, a strategy surpassing sustainable tourism was necessitated. This strategy is embodied in the concept of regenerative tourism, which denotes a holistic approach aiming to enhance the entire life and economic model by fostering conditions conducive to life-enhancing life. Regenerative tourism is a transformational approach that seeks to unlock the potential of tourism destinations for development and net positive impacts by augmenting the regenerative capacity of communities and ecosystems.

It is important to recognise that regenerative tourism is not a universal solution, given that each destination has its own characteristics and local communities have different needs. Nevertheless, it is of paramount importance that all stakeholders, including the local community, local administrations, tourism businesses and tourists, collaborate in locations where regenerative tourism is to be implemented. Given that regenerative tourism can only be achieved through a shift in mindset (Dredge, 2022), it is imperative that community leaders, businesses, policy makers and other stakeholders undertake training on the adoption of a regenerative approach. There is a body of research in this area, with New Zealand serving as a case study in the adoption and implementation of regenerative tourism at both the institutional and academic levels. 'The Tourism CoLab' social enterprise community offers courses on topics such as regenerative development, regenerative leadership, storytelling, creative thinking through collaborative learning, coaching, and mentoring (The Tourism CoLab, n.d.). The implementation of analogous approaches in disparate destinations is of paramount importance for the sustainability of tourism.

References

Becken, S., & Kaur, J. (2022). Anchoring "tourism value" within a regenerative tourism paradigm – a government perspective. *Journal of Sustainable Tourism*, 30 (1), 52–68. https://doi.org/10.1080/09669582.2021.1990305.

Bellato, L., Frantzeskaki, N., Briceño Fiebig, C., Pollock, A., Dens, E., & Reed, B. (2022). Transformative roles in tourism: adopting living systems' thinking for regenerative futures. *Journal of Tourism Futures*, 8 (3), 312–329. https://doi.org/10.1108/JTF-11-2021-0256.

Bellato, L., Frantzeskaki, N., & Nygaard, C. A. (2023). Regenerative tourism: a conceptual framework leveraging theory and practice. *Tourism Geographies*, 25 (4), 1026–1046. https://doi.org/10.1080/14616688.2022.2044376.

Booking.com (2023). Sustainable travel report. Available at: https://news.booking.com/download/31767dc7-3d6a-4108-9900-ab5d11e0a808/booking.com-sustainable-travel-report2023.pdf.

Cave, J., Dredge, D., van't Hullenaar, C., Koens Waddilove, A., Lebski, S., Mathieu, O., Mills, M., Parajuli, P., Pecot, M., Peeters, N., Ricaurte-Quijano, C., Rohl, C., Steele, J., Trauer, B., & Zanet, B. (2022). Regenerative tourism: the challenge of transformational leadership. *Journal of Tourism Futures,* 8 (3), 298–311. https://doi.org/10.1108/JTF-02-2022-0036.

CBI (2022). Regenerative tourism: moving beyond sustainable and responsible tourism. Available at: https://www.cbi.eu/market-information/tourism/regenerative-tourism#:~:text=%E2%80%9CRegenerative%20tourism%E2%80%9D%20is%20the%20idea%20that%20tourism%20should,in%20other%20words%2C%20not%20causing%20any%20extra%20damage%29.

Coll-Barneto, I., & Fusté-Forné, F. (2023). Understanding environmental actions in tourism systems: ecological accommodations for a regenerative tourism development. *Journal of Tourism, Sustainability and Well-being,* 11(4), 239–253. https://doi.org/https://doi.org/10.34623/0v3w-7192.

Earthcheck Research Instute-ERI (2023). Regenerative tourism. Discussion paper. Available at: https://earthcheck.org/wp-content/uploads/2023/07/0723_EarthCheck_RegenerativePaper_FINAL.pdf.

Dredge, D. (2022). Regenerative tourism: Transforming mindsets, systems and practices. *Journal of Tourism Futures,* 8(3), 269–281. https://doi.org/10.1108/JTF-01-2022-0015.

Imran, S., Alam, K., & Beaumont, N. (2014). Reinterpreting the definition of sustainable development for a more ecocentric reorientation. *Sustainable development,* 22(2), 134–144. https://doi.org/10.1002/sd.537

Inversini, A., Saul, L., Balet, S. and Schegg, R. (2024). The rise of regenerative hospitality. *Journal of Tourism Futures,* 10 (1), 6–20. https://doi.org/10.1108/JTF-04-2023-0107

Fusté-Forné, F., & Hussain, A. (2022). Regenerative tourism futures: a case study of Aotearoa New Zealand. *Journal of Tourism Futures,* 8 (3), 346–351. https://doi.org/ 10.1108/JTF-01-2022-0027.

Goa Tourism Developmet Corporation (n.d.). India's first state to launch Regenerative Tourism. Available at: https://goa-tourism.com/regenerative-tourism/ (access date: 06.05.24).

Legrand, W., Kuokkanen, H., Marucco, F., Hazenberg, S., & Fischer, F. (2024). Survival of the fittest? a call for hospitality to incorporate ecology into business practice and education. *Cornell Hospitality Quarterly,* 65(1), 68–87. https://doi.org/10.1177/19389655231182083.

Mathisen, L., Søreng, S.U., & Lyrek, T. (2022). The reciprocity of soil, soul and society: the heart of developing regenerative tourism activities. *Journal of Tourism Futures*, 8 (3), 330–341. https://doi.org/10.1108/JTF-11-2021-0249.

McEnhill, L., Jorgensen, E.S., & Urlich, S. C. (2020). Paying it forward and back: regenerative tourism as part of place. Centre of Excellence for Sustainable Tourism Internal Report, 2020/101, Lincoln University.

Northrop, E., Schumann, P., Burke, L., Fyall, A., Alvarez, S., Spenceley, A., Becken, S., Kato, K., Roy, J., Some, S., Veitayaki, J., Markandya, A., Galarraga, I., Greño, P., RuizGauna, I., Curnock, M., Epler Wood, M., Yue Yin, M., Riedmiller, S., . . .Godovykh, M. (2022). Opportunities for transforming coastal and marine tourism: Towards sustainability, regeneration and resilience. https://oceanpanel.org/ publication/opportunities-for-transforming-coastal-andmarine-tourism-towards-sustainability-regeneration-andresilience/

Owen, C. (2007). Regenerative tourism: A case study of the resort town Yulara. *Open House International*, *32*(4), 42–53. https://doi.org/10.1108/OHI-04-2007-B0005.

Pollock, A. (2019). Regenerative tourism: the natural maturation of sustainability. Available at: https://medium.com/activate-the-future/regenerative-tourism-the-natural-maturation-of-sustainability-26e6507d0fcb.

Pung, J. M., Mackenzie, S. H., & Lovelock, B. (2024). Regenerative tourism: Perceptions and insights from tourism destination planners in Aotearoa New Zealand. *Journal of Destination Marketing & Management*, 32, 100874, https://doi.org/10.1016/j.jdmm.2024.100874.

Qi, F., Pforr, C., & Justin Dit, T. (2024). Exploring the regenerative potential for community-based ecotourism in the Niah National Park in Sarawak, Malaysia. *Journal of Ecotourism*, 1–9. https://doi.org/10.1080/14724049.2024.2317315.

Sheldon, P. J. (2021). The coming-of-age of tourism: embracing new economic models. *Journal of Tourism Futures*, 8(2), 200–207. https://doi.org/10.1108/JTF-03-2021-0057.

The Tourism CoLab (n.d.). Courses. Available at: https://www.thetourismcolab.com.au/online-courses.

Tourism Projects (n.d.). Transitioning to a regenerative tourism model. Available at: https://www.mbie.govt.nz/immigration-and-tourism/tourism/tourism-projects/governments-tourism-snapshot/transitioning-to-a-regenerative-tourism-model/.

UNEP & WTO (2005). Making tourism more sustainable. A guide for policy makers. Available at: https://wedocs.unep.org/bitstream/handle/20.500.11822/8741/-Making%20Tourism%20More%20Sustainable_%20A%20Guide%20for%20Policy%20Makers-2005445.pdf?sequence=3&amp%3BisAllowed=.

UNWTO (2013). Sustainable tourism for development guidebook. Enhancing capacities for Sustainable Tourism for development in developing countries.

Münevver Çiçekdaği[1], Ayşe Cabi Bilge[2] and Seda Özdemir Akgül[3]

Chapter 7 What Is Türkiye's Tourism Industry's Place in the Digital Divide?

Introduction

In order to bring an inclusive, transparent and governance-based understanding to the institutional and social structure, it is necessary to ensure equal and fair distribution of Information and Communication Technologies (ICT) to all segments of society. In other words, there should not be a digital divide between the society and the system (Yıldırım & Öner, 2004: 50). Even in societies known as the information society, the digital divide brings with it the danger of becoming an information-rich and poor-separated society (Öztürk, 2005).

The concept of the digital divide, which refers to inequality and differences in the use of technology and access to digital opportunities, is used to emphasize the digital inequality between developed and developing countries. This concept refers not only differences between countries, but also between individuals and groups with various social and economic structures within a country. Ratio of the number of internet users to the population, ratio of the number of social media users to the population, internet access (internet penetration rate of the population, ratio of the number of individuals living in rural and urban areas to the population), use of internet-enabled devices (ratio of the number of cellular mobile connections to the population), ability to have digital skills (ratio of the number of individuals over and under the age of 65, the female and male population ratio), gross national product, internet infrastructure (median and fixed internet connection speed) are among the digital divide indicators. The digital divide is important as a concept that reveals inequality between regions. In this study, first of all, what the concept of digital divide is and its importance for the tourism sector will be discussed. In addition, Turkey's digital divide level is evaluated in terms of the tourism sector by comparing the digital divide

---

1 Asst. Prof., Selçuk University, Tourism Faculty, Department of Tourism Guidance, mcicekdagi@selcuk.edu.tr
2 Asst. Prof., Selçuk University, Vocational of Social Science, Department of Hotel, Restaurants and Catering Sevice, cabi@selcuk.edu.tr
3 Assoc. Prof., Selçuk University, Tourism Faculty, Department of Tourism Management, sedaozdemir@selcuk.edu.tr

indicators among the top ten countries that attract the most tourists according to UNWTO's 2023 data.

The Concept of Digital Divide

Developing technology is profoundly affecting every area of our lives and is gaining importance every day. However, the advantages of technological developments are not equally accessible to everyone. In this context, the concept of the digital divide, which represents inequalities in accessibility and use of digital Technologies (Xue & Yang, 2023), has entered the literature. The concept is defined as a social problem that indicates inequalities in access to the internet and ICT among different social and demographic groups (Ramalingam & Kar, 2014). The digital divide is a phenomenon that refers to the difference in the level of digitalization, especially between rural and urban regions. The digital divide, which is linked to differences in access to digital technologies, society's level of digital literacy and ICT use, negatively affects balanced regional development (Arion et al., 2024). On the one hand, the digital divide refers to the inability of individuals to access the Internet, which is called "technological exclusion", especially due to economic reasons, and on the other hand, the lack of digital knowledge of individuals (Val & Bueno, 2024). In other words, the digital divide includes not only inequalities in physical access to technology, but also imbalances in having skills as a digital user (Thorvaldsen & Richards, 2011). Based on the given definitions, the digital divide in the context of ICT can be described as an imbalance caused by differences in access to technology and digital skills due to inadequate infrastructure, social, economic, and cultural factors.

Causes of the Digital Divide

A more comprehensive look at the causes of the digital divide, which defines the difference between access to and use of digital technology, although it is often understood as the difference between those who have access and those who do not, or those who use digital technology and those who do not (Flynn, 2024), is important for the steps to be taken. In this section, the causes of the digital divide are discussed in headings.

>*Infrastructure Inequalities*: One of the main causes of the digital divide is the unequal distribution of technological infrastructure. The lack of infrastructure, such as high-speed internet access, restricts individuals' ability to fully utilize digital resources and services, thereby exacerbating the digital divide (Flynn, 2024).

Economic Inequalities: Economic factors emerge as an important element in the formation of the digital divide. Individuals' or groups' access to resources such as technological devices and internet connection varies according to their socio-economic status (Sezgin & Fırat, 2024).

Digital Skills: Digital skills are necessary to be able to take full advantage of technology. Lack of digital skills can create barriers to the effective use of online services (Flynn, 2024). Digital skills are often associated with demographic variables.

Education Inequalities: Inequalities in educational opportunities contribute to the digital divide. Individuals with low-quality education lack access to current technology and resources, while individuals with high-qualification education are more likely to have digital skills and access to the Internet (Xue & Yang, 2023).

Geographical Isolation: Individuals in certain geographical areas may have easier internet access and technology, while others may have limited or no access (Sezgin & Fırat, 2024). Geographical isolation, especially in rural and remote areas, can cause digital divide by limiting access to internet infrastructure and digital services.

Measuring the Digital Divide: Key Indicators

While the digital divide was previously evaluated as the opportunity to access ICT, it has started to be evaluated with dimensions such as the quality of this access, the state of benefiting from technologies and the level of technology management. This concept, which was initially expressed as the access opportunities of developed and developing countries to these technologies, has later turned into a concept that covers the access opportunities of individuals, businesses and communities with different socio-economic structures to ICT as well as their purposes of use (Taşkıran, 2017).

The integration of ICT into economic and social life requires personal computer ownership and internet access, as well as an environment of democratic rights and civil liberties, respect for the rule of law and property rights, investment in human capital and government stability (Rodriguez & Wilson, 2000).

When measuring the digital divide, the focus is on the widespread of digital phenomena on the one hand and the socio-economic effects of these phenomena on the other (Corrocher and Ordanini, 2002). While it is not an accepted index for the digital divide, measurements have been carried out using various indicators (Öztürk, 2002). In order to measure the ICT development of countries, the International Telecommunication Union (ITU) has developed the ICT

Development Index (IDI). Indicators of the IDI, classified as access, use and skill sub-indexes; number of fixed telephone lines, mobile phone ownership, international internet bandwidth per user, household owning a computer, household with internet access, individual internet user, fixed broadband internet subscriber, active mobile broadband internet subscriber, number of adult literates, secondary education schooling rate and higher education schooling rate. The World Economic Forum uses the Networked Readiness Index (NRI). Indicators of NRI, which have sub-indexes of environment, use, readiness and impact; political and regulatory environment, business and innovative environment, infrastructure and digital content, affordability, skill, individual use, business use, public use, economic impacts and social impacts (Kalaycı, 2013). Öztürk (2005) lists the commonly used international digital divide indicators as the number of personal computers, teledensity, number of internet hosts, number of internet users, number of websites, languages of websites, languages of users, bandwidth, labor force employed in ICT, share of ICT in total exports. Wilson (2006) evaluates ICT in eight dimensions: physical access, material access, skills, usability, content access, production access, institutional access and access to governmental institutions. Aytun (2006) also lists the digital divide indicators as the number of computers and the amount of internet access as well as infrastructure. Norris (2001) discusses the digital divide in terms of social, global and democratic dimensions. Ay & Kılıç (2023) list the main indicators of the digital divide as computer ownership, broadband internet subscription, internet speed and the number of internet cafes. According to Pınar (2024), the determining dynamics of the digital divide are the education gap, gender gap, economic gap, language gap and access gap.

The distribution of the indicators identified in measuring the digital divide according to demographic data is also examined. Aytun (2006) categorizes demographic data comparison as income, education, household size, age, gender, language, location and race.

Since measuring the digital divide between countries is a complex issue, researchers have to create their own methodologies. The lack of a theoretical framework, the lack of appropriate data, the lack of standard indicator acceptance, and the fact that the subject falls within the scope of several different disciplines are among the problems experienced in measuring the international digital divide (Öztürk, 2002).

Digital Divide in Tourism Industry

Considering that the Internet has become the main source of information for tourists, it is very important for tourists to have access to information about the places to travel, accommodation, local cuisine, transportation facilities. With the widespread use of the Internet, countries experience ease in reaching potential tourists, creating their brand image, promoting their touristic attractions, and increasing the visitor experience. ICT is used in all stages of the touristic activity process and provides advantages in providing information to tourists. If consumers have sufficient knowledge and skills in technology, they can benefit from the convenience of technology. However, lack of access to ICT, inadequate usage skills, and adaptation difficulties cause the digital divide.

The diversity of technological tools utilized in the creation and consumption of touristic products and the differences arising from the ability to use these tools create a digital divide (Minghetti & Buhalis. 2010). Capurro et al. (2004) in their study; refer to the unequal distribution of knowledge and skills between men and women (gender gap), urban and rural areas (geographical gap), young and old (age gap), ethnological groups (ethnographic gap), different levels of education (educational gap) and different income groups (income gap). In terms of the age gap, inequalities in the consumption of touristic products among the elderly are noteworthy. The elderly face more technical difficulties in using digital tools and this affects their motivation to travel. When the elderly receive support from their spouses, children or friends in these matters, the impact of the digital divide on tourism consumption decreases. When digital tasks are performed on behalf of the elderly in line with the support received, their anxiety decreases (Wu and Yang, 2023). Compared to young people who can use technological tools more easily, the elderly find it difficult to understand the complexity of these technologies and this leads to high transaction costs. Elderly may have increased concerns about digital technology in terms of information search, online communication, security concerns (Omotayo, 2020), and digital payment transactions (Ihm & Hsieh, 2015). Compared to younger generations, older individuals are disadvantaged in terms of access to digital devices and digital skills (Augner, 2022). In other words, the gap between those with high and low digital access is also reflected in travel consumption (Minghetti & Buhalis, 2010; Reverte & Luque, 2021).

When evaluated in terms of geographical gap, it is thought that the awareness of those living in rural areas regarding internet access and ICT use will be more limited. According to Akça et al. (2007), the establishment and development of village websites is necessary and very useful to ensure the interaction of people in rural areas and to reduce the digital divide between rural and urban areas. According to Khoo et al. (2024), women do not have digital infrastructure, digital devices, internet access and quality education compared to men. Additionally, economic inability to receive appropriate education, gendered barriers to technology adoption, and responsibilities at home are seen as obstacles women face in gaining digital skills. These barriers can be exemplified by gender, education and income gaps.

The digital divide in tourism has many dimensions for tourism destinations, tourists and tourism businesses (Wu & Yang, 2023). Minghetti & Buhalis (2010) categorized tourism destinations and tourists in terms of digital access levels. In both groups, there are four different levels: high, upper, medium and low. When digital access levels are evaluated in terms of tourism destination; destinations with high digital access have high-quality websites and are easy to access online through multiple channels. In addition, the number of online reservations and e-marketing efforts of these destinations is at a good level. Upper digital access destinations have a good online presence but are more complex and less efficient. In addition, e-commerce is only implemented for specialty products. Destinations with medium-sized digital access level have weak infrastructure, the website is used only as an information platform but not for online booking. Low digital access destinations have minimal access to ICT use and are highly dependent on intermediaries to market products to customers in high or upper digital access markets. When digital access levels are evaluated in terms of tourists, high digital access tourists have high socioeconomic status and the ability to use and accept technological tools. Tourists with upper digital access have a good level of technological knowledge and have sufficient usage skills. Tourists with medium digital access have medium or low level of intention to use technological tools and have low participation due to access difficulties. Tourists with low digital access have no technological access at all. These individuals are part of marginalized social groups. According to Maurer & Lutz (2012), if there are differences between the business and consumer sides in terms of technology adoption and implementation, this will lead to a digital divide. Different levels of digital access can lead to a digital divide for tourists and tourism destinations and thus an information gap between the demand and supply side of tourism.

Methodology

In this section, the process and findings of the implementation study are presented. In this context, according to UNWTO 2023 data, France, Spain, USA, China, Italy, Turkey, Mexico, Thailand, Germany and the UK, which are the top ten countries attracting the most tourists, are discussed. In the analysis, 14 indicators were determined as a result of the literature review. These indicators are; *Internet Users (IU%), Social Media Users (SMU%), Cellular Mobile connections percent (CMC%), population by gender male (Male%), population by age under*

65 (Under65 %), internet penetration rate (IP%), Internet Users Increase (IUI%), Median mobile internet connection speed via cellular networks (MCS Mbps), Median fixed internet connection speed (FCS Mbps), median mobile internet connection speed increase (MCSI%), median fixed internet connection speed increase (FCSI%), population lived in urban areas (Urban%), GDP (USD Billion), Education Index (EI). Education index data is taken from the Human Development Index prepared by UNDP (hdr.undp.org), GDP data is taken from the IMF Report (imf.org), and other indicators are taken from Digital 2024 Global Overview Report – published in partnership between We Are Social and Meltwater (Rodriguez, F. & Wilson, 2024). The data obtained are shown in Table 7.1.

Table 7.1. Digital Divide Indicator Data by Country

	France	Spain	USA	China	Italy	Türkiye	Mexico	Thailand	Germany	UK
IU%	0,937	0,959	0,971	0,762	0,877	0,864	0,832	0,879	0,932	0,977
SMU%	0,782	0,835	0,701	0,741	0,728	0,668	0,699	0,683	0,814	0,828
CMC%	1,157	1,272	1,161	1,230	1,387	0,937	0,972	1,361	1,452	1,309
Male%	48,3	49	49,5	51	48,8	50,1	48,8	48,5	49,4	49,4
Under65 %	77,8	79	82,2	14,4	75,3	91,9	91,3	16,3	77,1	80,4
IP%	93,8	96	97,1	76,4	87,7	86,5	83,2	88	93,3	97,8
IUI%	0,2	-0,1	0,5	1	-0,3	0,5	6,6	0,1	0,2	0,3
MCS (Mbps)	79,6	41,5	106,2	161,5	46,0	33,9	25,1	40,7	57,4	48,4
FCS (Mbps)	207,4	201,0	219,7	248,9	71,4	41,49	60,28	216,2	89,93	92,1
MCSI%	33,4	15,2	42	47,7	17,3	2,03	4,4	7,5	2,2	5,2
FCSI%	38	20,5	15,9	16	18,5	9,61	20,8	5,1	11,7	25,8
Urban%	81,9	81,7	83,4	64,9	72,1	77,6	81,7	53,9	77,9	84,8
GDP	3	1,6	26,9	17,7	2,2	1,2	1,8	0,5	4,4	3,3
EI	0,762	0,717	0,883	0,573	0,727	0,68	0,623	0,608	0,917	0,901

Accordingly, the relevant destinations were compared in terms of digital divide with multidimensional scaling analysis and a scatter plot was created. Multidimensional scaling analysis is a technique widely used in exploratory data analysis with multiple variables (Gürsoy et al., 2009). With this technique, the positions of destinations in terms of digital divide issues can be interpreted by reducing from multidimensional variables to two dimensions. The distance matrix data of the ten countries included in the study and their positions in space according to their similarities are presented visually in the scatter plot.

	France	Spain	USA	China	Italy	Türkiye	Mexico	Thailand	Germany	UK
France		2,097	3,268	5,190	2,848	4,186	4,138	4,457	3,399	2,614
Spain	2,097		3,399	5,286	2,365	3,594	4,080	3,798	2,131	1,721
USA	3,268	3,399		4,758	3,732	4,267	4,973	4,865	3,603	3,526
China	5,190	5,286	4,758		4,699	5,104	5,592	4,575	5,432	5,919
Italy	2,848	2,365	3,732	4,699		2,735	3,443	2,896	2,127	2,641
Türkiye	4,186	3,594	4,267	5,104	2,735		2,823	3,989	3,499	3,725
Mexico	4,138	4,080	4,973	5,592	3,443	2,823		4,621	4,330	4,294
Thailand	4,457	3,798	4,865	4,575	2,896	3,989	4,621		3,884	4,644
Germany	3,399	2,131	3,603	5,432	2,127	3,499	4,330	3,884		1,595
UK	2,614	1,721	3,526	5,919	2,641	3,725	4,294	4,644	1,595	

Figure 7.1. Distance Matrix

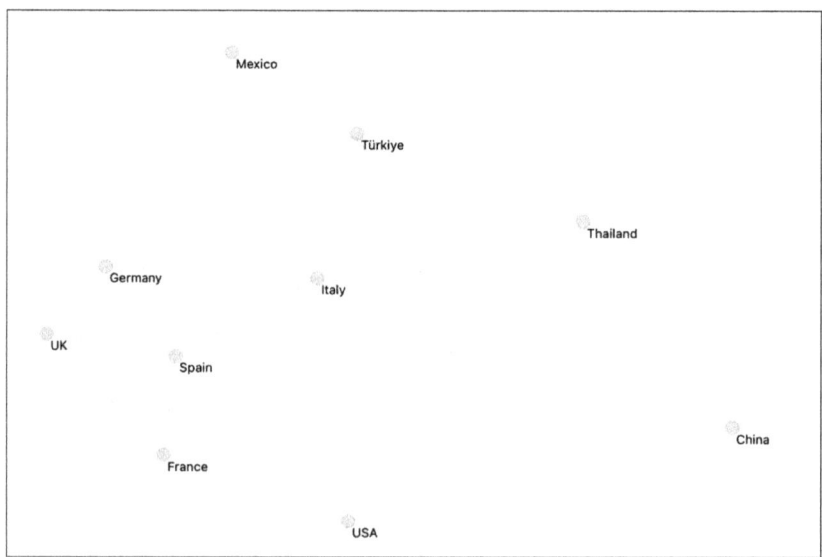

Figure 7.2. Scatter Plot

According to the figure, it is observed that the UK, Germany, Spain, France and Italy are located close to each other. Mexico and Turkey are positioned relatively closer to these five countries.

America is positioned differently compared to all other countries in the figure. China is positioned relatively close to the United States on the vertical axis. On the other hand, Thailand is located closer to the group of five countries than Mexico and Turkey. Looking at the data, it can be said that America, which is in one of the extreme positions, is the country furthest away from the digital divide. In the figure, it is possible to say that France, one of the two countries close to America, is behind in terms of technological infrastructure and economic development, and China is behind in terms of its elderly population and education variable. Other countries are located closer to the digital divide than America due to the lower values of different variables included in the analysis.

Results

It can be said that America is far from the digital divide in the scatter plot due to the high values it has in 14 indicators in general. It is seen that Mexico is close to the digital divide compared to other countries, with low social media usage (0.69) and low mobile connection speed (25.14).

It is seen that Mexico's and Turkey's education index, GDP, urban population, internet penetration, population under 65 and social media usage data are close. This places these two countries close in the scatter plot.

Since the mobile connection speeds of the USA and China are higher than other countries (106.28 and 161.56), they are far from the digital divide. However, China's education indicator (0.573) and population rate under 65 years of age (14.4) have a low value compared to other countries, which has moved China away from America in the scatter plot. While Thailand's fixed connection speed rate (216.26) is close to the USA and China; It can be said that it has moved away from China due to the data on population under the age of 65 (16.3), mobile connection speed increase (7.5), fixed connection speed increase (15.1), GDP (0.5) and social media users (0.68).

UK, Germany, Italy, Spain and France appear to be clustered because they have similar values. This can be interpreted as their digital divide level being close.

According to 2023 UNWTO data, Turkey, which ranks 6th in the number of incoming tourists, is rich in natural and cultural attractions and provides an advantage in terms of purchasing power of tourists due to the exchange rate, making Turkey stand out compared to the other four countries behind it. While the digital divide levels of the UK, Thailand and Germany are lower than Turkey, it can be said that Turkey ranks higher than them, thanks to these advantages. It is seen that the five countries that come before Turkey in the ranking are in a good situation in terms of digital divide, and it is thought that it would be beneficial to make improvements in the indicators that bring Turkey closer to the digital divide in order for Turkey to rank higher in terms of the number of incoming tourists.

In order to eliminate the digital divide in tourism, it would be appropriate to strengthen the digital infrastructure to be used in all tourism activities, create interfaces for the elderly to use digital technologies more easily, educate the society on digital literacy, prevent gender inequality in digital use and ensure that women use ICT more, to create applications that will increase technology access in rural areas.

References

Akca, H., Sayili, M., & Esengun, K. (2007). Challenge of rural people to reduce digital divide in the globalized world: Theory and practice. *Government Information Quarterly*, 24(2), 404–413.

Arion, F. H., Harutyunyan, G., Aleksanyan, V., Muradyan, M., Asatryan, H., & Manucharyan, M. (2024). Determining Digitalization Issues (ICT Adoption, Digital Literacy, and the Digital Divide) in Rural Areas by Using Sample Surveys: The Case of Armenia. Agriculture, 14(2), 249.

Augner, C. (2022). Digital divide in elderly: Self-rated computer skills are associated with higher education, better cognitive abilities and increased mental health. *The European Journal of Psychiatry*, 36(3), 176–181. https://doi.org/10.1016/j.ejpsy.2022.03.003

Ay, S., & Kılıc, T. (2023). Coğrafi dijital uçurum: Türkiye'de dijital dönüşümün kentsel-kırsal, bölgesel ve cinsiyet eşitsizlikleri. *Coğrafya Dergisi*, 46, 111–122. https://doi.org/10.26650/JGEOG2023-1169477

Aytun, C. (2006). Enformasyon toplumu sürecinde dijital bölünme kavramının anlamı ve önemi. XI. Türkiye'de İnternet Konferansı, TOBB Ekonomi ve Teknoloji Üniversitesi, 21–23. Ankara

Capurro, R., Hausmanninger, T., & Scheule, R. (2004) 'Vernetzt gespalten. Ein Trialog' in Capurro, R. et al. (Eds): Vernetzt gespalten. Der Digital Divide in ethischer Perspektive, München, Wilhelm Fink Verlag, pp. 15–24.

Corrocher, N., & Ordanini, A. (2002). Measuring the digital divide: a framework for the analysis of cross-country differences. Journal of Information technology, 17, 9–19. DOI: 10.1080/02683960210132061

Flynn, S. (2024). Keeping up with the times in Ireland: Older adults bridging the age-based digital divide together?. *Studies in the Education of Adults*, 1–19.

Gursoy, D., Baloglu, S. ve Chi, C. G. (2009), destination competitiveness of middle eastern countries: An examination of relative positioning, *Anatolia: An International Journal of Tourism and Hospitality Research*, 20 (1), ss. 151–163.

Human Development Report. (March 13, 2024). Human Development Report 2023–24. https://hdr.undp.org/content/human-development-report-2023-24, Access Date: March 18, 2024.

Ihm, J., & Hsieh, Y. P. (2015). The implications of information and communication technology use for the social well-being of older adults. *Information, Communication & Society*, 18(10), 1123–1138. https://doi.org/10.1080/1369118X.2015.1019912

imf.org. (n.d.). Real GDP growth. imf.org/external/datamapper/ngdp_rpch@weo/oemdc/advec/weoworld, Access Date: March 18, 2024.

Kalaycı, C. (2013). Dijital bölünme, dijital yoksulluk ve uluslararası ticaret. *Atatürk Üniversitesi İktisadi ve İdari Bilimler Dergisi*, 27(3), 145–162.

Khoo, C., Yang, E. C. L., Tan, R. Y. Y., Alonso-Vazquez, M., Ricaurte-Quijano, C., Pécot, M., & Barahona-Canales, D. (2024). Opportunities and challenges of digital competencies for women tourism entrepreneurs in Latin America: a gendered perspective. *Journal of Sustainable Tourism*, 32(3), 519–539.

Maurer, C., & Lutz, V. (2012). Strategic implications for overcoming communication gaps in tourism caused by digital divide. *Journal of Information Technology & Tourism*, 13(3), 205–214.

Minghetti, V., & Buhalis, D. (2010). Digital divide in tourism. *Journal of Travel Research*, 49(3), 267–281.

Norris, P. (2001). Digital divide: Civic engagement information poverty, and the Internet Worldwide. Cambridge, New York: Cambridge University Press.

Omotayo, F. O. (2020). Use and non-use of Internet banking among elderly people in Nigeria. International *Journal of Social Sciences and Management*, 7(2), 42–54. https://doi. org/10.3126/ijssm.v7i2.28597

Öztürk, L. (2002). Dijital uçurumun küresel boyutları. *Ege Academic Review*, 2(1), 1–10.

Öztürk, L. (2005). Türkiye'de dijital eşitsizlik-tübitak-bilten anketleri üzerine bir değerlendirme. *Erciyes Üniversitesi İktisadi ve İdari Bilimler Fakültesi Dergisi*, 24(1), 111–131.

Pınar, A. (2024). Determinants and Impacts of Technological Inequalities: A Review on the Digital Divide. *Uluslararası Sosyal Siyasal ve Mali Araştırmalar Dergisi*, 4(1), 28–43.

Ramalingam, A., & Kar, S. S. (2014). Is there a digital divide among school students? an exploratory study from Puducherry. *Journal of Education and Health Promotion*, 3.

Reverte, F. G., & Luque, P. D. (2021). Digital divide in E-tourism. In X. Zheng, M. Fuchs, U. Gretzel, & W. Hpken (Eds.), Handbook of e-tourism (pp. 1–14). Springer press.

Sezgin, S., & Fırat, M. (2024). Exploring the Digital Divide in Open Education: A Comparative Analysis of Undergraduate Students. *The International Review of Research in Open and Distributed Learning*, 25(1), 109–126.

Taşkıran, A. (2017). Açık ve uzaktan öğrenmede dijital bölünme. AUAd, 3(4), 108–124.

Thorvaldsen, S., & Richards, G. (2011). Some frontiers in open and distance learning in the North. Athabasca University Press, 12(4). https://doi.org/10.19173/irrodl.v12i4.986

Val, S., & Bueno, H L. (2024). Analysis of Digital Teacher Education: Key Aspects for Bridging the Digital Divide and Improving the Teaching–Learning Process. *Multidisciplinary Digital Publishing Institute*, 14(3), 321–321. https://doi.org/10.3390/educsci14030321

Wilson, E.J., 2006. The Information Revolution and Developing Countries. MIT Press, Cambridge, MA.

Wu, X., & Yang, Y. (2023). Digital divide and senior travel consumption: An empirical study from China. *Asia Pacific Journal of Tourism Research*, 28(4), 306–322.

Xue, W., & Yang, Y. (2023). Digital divide and senior travel consumption: an empirical study from China. *Routledge*, 28(4), 306–322. https://doi.org/10.1080/10941665.2023.2228938

Yıldırım, U., & Öner, S. (2004). Bilgi toplumu sürecinde yerel yönetimlerde eğitim-bilisim teknolojisinden yararlanma: Türkiye'de e-belediye uygulamaları. *TOJET: The Turkish Online Journal of Educational Technology*, 3(1).

Savaş Yildiz[1] and Zafer Yildiz[2]

Chapter 8 Possible Trend in Tourism of Türkiye: Treasure Hunting Tourism

Introduction

As Heraclitus of Ephesus (500 BC) stated, "The only thing that is constant is change." From past to present, many scientific, technological, cultural, social and political developments have influenced each other and caused various changes in both social and individual life. When the issue of change is evaluated in terms of tourism, it is seen that many different types of tourists and therefore types of tourism have emerged until today. Because the emergence of new types of tourism is caused by people's personal differences, wishes and expectations. This differentiation is the main source of change in the field of tourism. Considering these explanations, it is seen that traditional mass tourism is insufficient to meet people's holiday expectations. For this reason, people tend to choose different types of tourism. Treasure hunting tourism, which is closely related to culture and adventure tourism, is actually a type of tourism that emerged as a result of treasure hunting, which is carried out in an amateur way or professionally in many different countries of the world, within the scope of tourism. Considering the impact of those who are interested in seeing and examining different plant species on the emergence of botanical tourism; the impact of those interested in bird species on the emergence of bird watching tourism; the impact of those interested in space on the emergence of space tourism; the impact of those who are interested in different cultural and local cuisines on the emergence of gastronomy tourism and the impact of those who are interested in golf on the emergence of golf tourism, it is clear that people who are interested in treasure and treasure hunting will have an impact on the emergence of treasure hunting as a type of tourism.

There is no research in the literature that deals with treasure hunting within the scope of tourism. From this perspective, the fact that it is the first study in the

---

1 PhD., National Defence University, Military Academy, Department of Defence Studies, savasyildiz77@gmail.com
2 Ass.Prof., Karamanoğlu Mehmetbey University, Faculty of Economics and Administrative Sciences, Department of Economics, zyildiz@kmu.edu.tr

literature in the field of treasure hunting tourism increases the originality and importance of this study. In the study, which aims to shed light on the studies to be carried out in this field, the basic concepts related to treasure and treasure hunting, the legal framework of treasure hunting in Türkiye, treasure hunting in terms of tourism and its relationship with cultural tourism and adventure tourism are evaluated

Basic Concepts Related to Treasure Hunting

Treasure: It is possible to come across different definitions of "treasure" in the literature. According the Turkish Civil Code Article-772; "Valuable things that have been buried or stored long before they were discovered, and that no longer have the owner, are deemed treasure." (www.tusev.org.tr). Özdoğan (2014: 189) describes the concept of "treasure" in his study as "non-cultural but materially valuable findings that are estimated/imagined to be found anywhere other than an archaeological site." Akkaya (1993: 259) defines treasure in his study as; "Items consisting of jewellery, money and precious stones that were buried or hidden before the time they were found and whose ownership cannot be proven." According to the provision of Article 4/b of the Regulation on Treasure Hunting amended on 21.07.2020 treasure is defined as "Movable properties that are buried or hidden and that it is clearly understood that there is no owner anymore depending on the situation, that are outside the Article-6 and Article-23 of Law No. 2863 and that do not have scientific value." (Kuşçu, 2022: 101).

Treasure Hunter: In his study, Akkaya (1993: 259) defines a treasure hunter as a person who digs or has someone else do it, with or without permission, in order to find treasures. Briefly, a treasure hunter can be defined as a person who digs or has it done by legal or illegal means in order to find treasure.

Treasure Hunting: Although treasure hunting is seen as an illegal activity in daily use, it is an activity that can be done legally. Özdoğan (2014: 189) defines "treasure hunting" as "Finding and uncovering the properties of the person(s) who left or abandoned their location for any reason." In his study, Karaduman (2005: 29) defines treasure hunting as "the dream of people who lack cultural awareness to become rich by finding treasures illegally for a better economic life and with the influence of religious elements, the search for imaginary treasures that mature over time and become legendary by being passed on from generation to generation."

Cultural Property: According to Article 3/a-1 of the Law on the Protection of Cultural and Natural Property; "Cultural property" shall refer to movable and immovable property on the ground, under the ground or under the water pertaining to science, culture, religion and fine arts of before and after recorded history or that is of unique scientific and cultural value for social life before and after recorded history. (https://kvmgm.ktb.gov.tr/).

Natural Property: According to Article 3/a-2 of the Law on the Protection of Cultural and Natural a Property; "Natural property" shall refer to all properties on the ground, under the ground or under the water pertaining to geological periods, prehistoric periods until present time, that are of unique kind or require protection due to their characteristics and beauty (https://kvmgm.ktb.gov.tr/).

Conservation Zone: According to Article 3/a-5 of the Law on the Protection of Cultural and Natural Property; "Conservation zone" shall mean anarea to be protected mandatory with activities to conserve its cultural and natural property or its historical environment. (https://kvmgm.ktb.gov.tr/).

Characteristics of the Treasure

Treasure(s) can be found as a result of treasure hunting activities, agricultural activities or works such as digging foundations for any construction, highway, bridge or dam. Sometimes natural disasters such as floods and landslides can also cause treasure to emerge. Any object that emerges must have certain characteristics in order to be considered a treasure. Erdoğan (2013: 517–521) itemized the characteristics of treasure as follows, based on the definition of the concept of treasure in the first paragraph of Article 772 of the Turkish Civil Code:

(a) *Its being buried or hidden*: This characteristic brings to mind two possibilities. The first of these indicates that the treasure was buried by someone. The other one means that the treasure remains under the surface due to natural reasons such as landslides and floods, being buried under snow and ice, being covered with tree branches and leaves over time and the area where the treasure is located being flooded. Anything that requires a clearing or excavation effort to uncover for whatever reason is considered as buried. For treasure, "hidden" and "buried" have different meanings. As a matter of fact, since it is not possible for the treasure to hide itself, hiding the treasure is an act that requires deliberate action by someone.

(b) *Its beign hidden or buried a long time ago*: When evaluated within the scope of property law, for an item to be considered as treasure, it must have been buried or hidden long before it was found. However, here the phrase "a long time ago" brings up the question "How long ago?" A century ago? Five years ago? Erdoğan (2013: 518) here referred to Article-166 of Mecelle: "The one/thing whose previous is not known is ancient." In fact, the structure of the treasure found, the way it is stored and how it was found can give an idea about whether it is a treasure from ancient times.

(c) *Its being portable*: This feature means that the thing(s) can be separated from where it is stored or buried. Whether it is hidden/buried in an immovable property or a movable property, the treasure must be separable (Eren, 2012: 509). The "separation" in question here refers to physical separation.

(d) *Its being something valuable*: An object that can be described as a treasure may have only material or moral value, or it may have both material and moral value. The material value of the object is actually related to the fact that it has an economic value that can be measured in money and can be bought and sold. Although some objects do not have economic value, being intangable makes them worth preserving. For example, objects such as a canteen, bayonet or a button on the uniform of a Turkish soldier from the Gallipoli War may not have a real economic value, but for a Turkish citizen, the intangible value of the object in question can be more meaningful than its economic value, as he/she will describe it as "priceless". In addition to the material and intangible value of an object that is considered as a treasure, its scientific and cultural value is also an issue that must be taken into consideration.

(e) *Its being unowned*: In order for an object buried or hidden a long time ago to be considered as a treasure, it must have no owner. However, the ownership here refers to the ownership status of the object at the time it was found, not the ownership at the time the object was buried or hidden. Therefore, after the object is discovered, if someone claims ownership over this object and proves this claim, the object in question will not be considered within the scope of treasure. In such a case, the "found goods" provisions will apply to the object that has been found, and different provisions will be applied depending on whether the statute of limitations for acquiring ownership of the object has passed or not (Feyzioğlu, 1955: 186–187).

In summary, these five characteristics, have an important role in determining whether any asset that emerges as a result of human effort or naturally is considered a treasure.

Legal Treasure Hunting Process in Türkiye

Two laws stand out in the legal treasure hunting process. The first of these is the Law on the Protection of Cultural and Natural Resources. The other is the Regulation on Treasure Hunting, which was published in the Official Gazette

No. 18294 dated January 27, 1984, and various articles of which were amended on 21.07.2020. The legal treasure hunting process consists of the stages explained below.

(a) *Application*: At this stage, it is important that those who will apply to search for treasure have the capacity to act, that is, the capacity to exercise their civil rights. Because according to the Turkish Civil Code Article-9 and Article-10, a person's ability to acquire rights and incur obligations depends on the person in question having the capacity to act. Each adult who has the ability to distinguish and has no mental disability has the capacity to act (Edis, 1979: 30).

(b) *Treasure hunting areas*: Criteria for places where treasure hunting will take place are determined by law. The treasure may be in the "protected area" defined in Article-3/a-5 of the Cultural and Natural Property Protection Law, or in areas that are not defined as a "protected area" by the law. Law on the Conservation of Cultural and Natural Property states in detail what real estate and cultural properties need to be protected, with examples. Therefore, permission to search for treasure can be given in places other than those that are considered places that need to be protected (https://kvmgm.ktb.gov.tr/).

(c) *Issuance of license*: The Governorship can issue a treasure hunting license to the applicant if it sees no harm in granting the license. The validity period of the license is limited to the year in which the application is made. Therefore, the license will be automatically canceled on 31 December of the relevant year. Since the treasure hunting license is given to an individual, it cannot be transferred to anyone else, that is, it can only be used by the person who received the license. A license that is not used properly may be cancelled. For example, if the officials appointed by the governorship determine that the 50m^2 area allowed for excavation is exceeded, the license can be cancelled.

(d) *Officials who should be present during treasure search and expenses*: Article-11 of the Regulation on Treasure Hunting clarifies this stage of the legal treasure hunting process. Accordingly, during treasure hunting, two specialized personnel assigned by the nearest museum directorate, as well as a local representative of the Ministry of Environment and Urbanization and the Ministry of Internal Affairs, who are responsible for law enforcement services, must accompany the work. The travel expenses and daily wages of the officials who will supervise the treasure search, the expenses incurred to restore the excavation site and the expenses of possible damages must be paid by the person who will search for treasure (https://www.mevzuat.gov.tr/).

(e) *Ending the treasure hunt:* If the treasure hunting activities are legally stopped by the relevant authorities, the treasure seeker cannot claim compensation for his/her losses and damages. If cultural and natural properties are found during the treasure search, the search must be stopped immediately and the Ministry must be informed of the situation. If cultural and natural properties are found, the person searching for treasure cannot claim any rights over them. It is possible that no treasure will be found as a result of treasure hunting. However, if a treasure is found, the objects must be examined by an expert team consisting of at least 3 people appointed by the Ministry, according to Article-17 of the Regulation on Treasure Search.

Treasure Hunting Tourism: The Intersection Point of Treasure Hunting and Tourism

The rapid change and development in science and technology causes rapid changes in people's needs and desires both in their business and private lives. A product that was in great demand yesterday may lose its appeal today. This situation also causes the definition of the concept of "holiday/vacation", which is one of the basic elements of tourism, to change. For example, while reading a book on the beach means a holiday for one person, ice climbing, participating in a safari or camping may mean a "holiday/vacation" for another person. Depending on the meaning people attach to the concept of "holiday/vacation", their expectations are also affected. This year, a tourist who prefers a completely relaxation-oriented holiday that includes the sea-sand-sun trio in an all-inclusive accommodation facility may demand a holiday that includes activities with very high risk and danger levels that require going out of the comfort zone for the next holiday.

Developments in science and technology directly or indirectly lead to the emergence of new tourism types. For example, thanks to the point reached today in aviation and space technologies, concepts such as "space tourism" and "space tourist" are mentioned in the literature today (Yıldız, 2021). Technological developments make people's work and private lives easier and allow them to be less physically tired. This situation suggests that "resting"-one of the main purposes of tourism- is not as effective as it used to be in people's participation in tourism activities. In other words, today people have begun to prefer types of tourism that include activities where they can be more physically active and have different experiences. Although "treasure hunting tourism" has not emerged due to scientific and technological developments like "space tourism", it is similar to space tourism in terms of including features such as "being the first", "excitement" and "adrenaline".

Throughout history, treasures and treasure hunting have attracted people's attention. Today, as in the past, treasure hunters search for treasures not only for financial reasons but also for adventure. Türkiye, which has hosted many civilizations due to its geographical location, is among the few countries in the world in terms of historical and cultural richness. Due to this prominent feature of Türkiye, treasure hunters who want to find treasure, cause irreparable damage to historical and cultural riches through illegal excavations.

What tourism has in common with treasure and treasure hunting is the existence of ancient civilizations that previously lived in a geography. Discovered and undiscovered historical and cultural properties belonging to the civilizations that lived in a region, are the common element that attracts tourists and treasure hunters. However, the point that needs to be emphasized here is that the goals of treasure hunters and tourists are different.

Various legal regulations have been made in Türkiye in order to prevent natural and cultural properties, which are the raw materials of the tourism industry, from being damaged due to illegal treasure hunting and to legalize treasure hunting. Carrying out treasure hunting within the framework of the law will make a significant contribution to the realization of treasure hunting within the scope of tourism, which is an important source of income for many countries. Since there is no study focusing directly on treasure hunting tourism in the literature, it is not possible to come across a clear definition of treasure hunting tourism. However, in the light of the explanations made so far, "treasure hunting tourism" can be defined as "a type of tourism that is associated with the active or passive participation of adventure enthusiasts in every stage of a legally permitted treasure hunting activity and/or visiting places where there are natural or man-made signs used in detecting a treasure."

Treasure hunting has two dimensions: active and passive. These two dimensions are related to the active or passive participation of tourist(s) in treasure hunting. The physical labor of the tourist(s) in the team of the person who has a treasure hunting license to search for and unearth the treasure, expresses the active dimension of treasure hunting tourism. Although the tourist does not actively take part in the search or extraction of the treasure, witnessing the stages of the treasure search and discovery process, expresses the passive dimension of treasure hunting tourism. Another aspect of treasure hunting tourism that can be considered within the scope of its passive dimension is visiting and seeing places where there are various signs used to locate the treasure. "Reading the land" is as important as the advanced detectors used in treasure hunting. The expression "reading the land" actually refers to various signs in the field that many people cannot easily notice and are used to determine where the treasure is. These signs, which are used to locate the treasure, are actually like today's safe passwords or keys. In the past, due to the lack of banks or safes, people had to hide or bury their valuable object(s) to protect them from thieves. They used natural or man-made signs in the area to easily find the valuable objects(s) they had hidden when they needed them. Signs used in determining the location of treasures can be of different types such as piles of soil that do not fit the natural structure of the land, stones placed in a symmetrical manner, various animal figures engraved on a natural rock, arrow signs, small cavities and religious motifs. Tourists visiting the land accompanied by a tourguide and someone specialized in treasure hunting is an issue that can be considered as passive dimention of treasure hunting tourism.

Examining the Relationship of Treasure Hunting Tourism with Culture Tourism and Adventure Tourism

New tourism types that emerge as a result of people's changing preferences over time are directly or indirectly related to each other. This situation is no different for treasure hunting tourism. Because, considering the explanations made about treasure hunting tourism, it is understood that treasure hunting tourism is related to cultural tourism and adventure tourism in various aspects (see Figure 8.1).

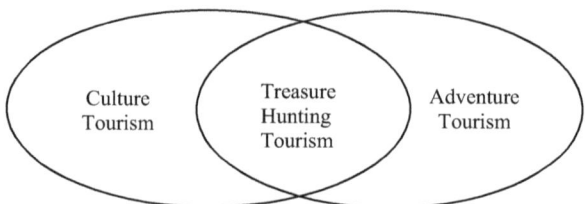

Figure 8.1. The Relationship between Treasure Hunting Tourism, Cultural Tourism and Adventure Tourism
Source: Created by the author.

Treasure Hunting Tourism and Culture Tourism: Culture, which is the basic component of cultural tourism is defined as the style of living and thinking that makes a society different from other societies, that continues to change from past to present, and that constitutes its identity with its unique art, beliefs, customs and traditions, understanding and behavior (Mercan, 2016: 34). The fact that societies have different cultural characteristics has caused people to be interested in the cultures of other contemporary societies and ancient civilizations. This interest lies at the basis of culture tourism. Therefore, the travel that tourists make to get to know different cultures is called cultur tourism (Uygur and Baykan, 2007: 33). Especially its close relationship with culture and natural assets allows treasure hunting tourism to be associated with culture tourism.

Within the scope of treasure hunting tourism, tourists will have the chance to learn more about the culture of the community that lived in that region through objects such as coins, ornaments, small statues, containers of different sizes and qualities with decorations on them, found as a result of a treasure hunting in any region. On the other hand, even if there is no treasure hunting activity, man-made signs used in locating a treasure will also contribute to tourists' knowledge about the civilization lived in that region. Therefore, its cultural aspect expresses the common point of treasure hunting tourism and culture tourism. From this perspective, it is obvious that tourists who are interested in culture tourism will also be a potential source for treasure hunting tourism.

Treasure Hunting Tourism and Avdenture Tourism: Since tourists participating in treasure hunting within the scope of treasure hunting tourism cannot have economic expectations, it can be said that the search for adventure is the most important motivation behind treasure hunting tourism. The word "adventure", which came from Arabic to Turkish, is defined in the Turkish Dictionary as "an interesting event or chain of events" (https://sozluk.gov.tr). Based on the dimension of "uncertain outcome", Paul Zweig (1974) defined the concept of "adventure" as "ad venio- whatever comes" in his book "The Adventurer: The fate of adventure in the Western World" (as cited in Dickson and Dolnicar, 2004: 5). According to Swarbrooke et al. (2003: 7–8), the concept of "adventure" evokes concepts such as excitement, discovery, struggle, excursion, adrenaline, enthusiasm, risk, success, effort, extreme, pride, courage, victory, majestic, horror. Considering the explanations about the concept of "adventure", it is understood that "adventure" has characteristics such as uncertain outcome, danger, risk, struggle, challenge, expected reward, arousal and excitement, escape and distance, research and discovery, dedication and focus, emotional contrast (Swarbrooke et al., 2003: 9–14). These characteristics of "adventure" contribute to the emergence of a new type of tourist and adventure tourism.

United Nations World Tourism Organization-UNWTO (2015: 36) defines adventure tourism as a type of tourism that generally takes place in destinations with certain geographical features and landscapes and tends to be associated with physical activity, cultural interaction and exchange, and interaction with nature. In the study of Fabrizio (2014: 4), defines adventure tourism as; "The least talked about but fastest growing type of tourism that focuses on the natural environment, requires special equipment, experience and knowledge, and involves risks for the tourists involved." Adventure Tourism Development Index published by Adventure Travel Trade Association-ATTA states that "adventure experience may involve some form of real or perceived risk and may require significant physical and/or mental effort." According to ATTA, a travel is considered within the scope of adventure tourism if it includes at least two of the following elements: a) adventure is its essence, b) interaction with nature, c) interaction with culture, d) a physical activity (ATTA, 2011: 6). Characteristics of the concept of "adventure" that Swarbrooke et al. (2003: 9–14) mentioned in their study, it also reflects the characteristics of adventure tourism. Considering the relationship between adventure tourism and treasure hunting tourism, the common aspects of both types of tourism are examined below.

The uncertain outcome: This one is one of the main characteristics of adventure tourism, is also acceptable for treasure hunting tourism. For example, there is a possibility that an adventure-enthusiast tourist who will do ice climbing within the scope of adventure tourism may not be able to do this climb, or a tourist participating in a safari may not be able to see an animal he/she hopes to see. Similarly, there is a possibility that the treasure may not be found as a result of the treasure hunting activity, or that the treasure may not be found as a result of the authorities stopping the treasure hunting activity for a valid reason. Therefore, "the uncertain outcome" can be considered as one of the factors that constitute the common aspect of treasure hunting tourism and adventure tourism.

Excitement: People become excited when they are exposed to environments and/or situations that stimulate their senses, emotions, mind and body physiology, whether or not they are personally involved in any activity. For example, in a stadium full of spectators, it is possible for the spectators watching the football match to be as excited as the player playing football. When this situation is evaluated in terms of treasure hunting tourism, the expectation of finding a treasure hidden years ago can cause excitement in people. Therefore, the excitement it causes in tourists is another prominent common aspect of adventure tourism and treasure hunting tourism.

Research and discovery: Increased knowledge and self-awareness that contribute to the discovery of new places, cultures and skills are one of the rewards awaited by adventurers. Research and discovery are also the basic elements of treasure hunting tourism. As a matter of fact, locating and finding the treasure requires a research and discovery process in itself.

Dedication: Among the activities within the scope of adventure tourism, especially hard adventure activities force the adventure-enthusiast(s) to go out of their comfort zone. Therefore, experiencing a hard adventure activity may require a person to endure difficult conditions in terms of accommodation, physical endurance and food and beverage. A person's ability to step out of her/his comfort zone is closely related to her dedication. When dedication -which is one of the main features of adventure tourism- is evaluated in terms of treasure hunting tourism, it is similar especially for tourists who personally participate in treasure hunting activities. Because the process of searching for treasure, especially when it is carried out in remote places far from settlements, requires the person to go out of his comfort zone, get dirty and spend the night in environments (such as tents) that are not particularly comfortable in terms of accommodation.

Travel motivation, expectations from travel, experiences, preferences, education level, socio-economic status, age, gender and marital status of individuals traveling for tourism purposes are different from each other. In addition to these differences between people, changes in their expectations and wishes have led to the emergence of different types of tourism and different types of tourists (Yıldız, 2019: 23). Its relationship with culture tourism and adventure tourism also gives clues about the type of potential tourists who may be interested in treasure hunting tourism. When the studies on tourist types in the literature are examined; "explorer tourists" determined by Cohen in 1972 (Kozak and Bahçe, 2009: 25); "extrovert tourist" type determined by Plog in 1977 (Kozak et all, 2015: 11); "adventurous tourists" defined by Perreault et al. in 1979; "explorer tourists" defined by Smith in 1989 (as cited in Avcıkurt, 2015: 27); The "modern idealist tourist" and "traditional idealist tourist" types determined in the study conducted by Dalen in 1989; "adventurous tourist" type, determined as a result of the research conducted in collaboration with Gallup and American Express (1989), are among the potential tourist types that may be interested in treasure hunting tourism.

Results

Every potential that is not utilized in both private and business life means a "loss". This is the same for the tourism sector. Today, as in the past, many countries are working, determining policies and developing strategies to increase their income from tourism in order to benefit more from all kinds of tourism potential they have. Therefore, having different types of tourism that can appeal to every type of tourist is an effective factor in increasing the share of countries in the global tourism market. Türkiye is among the countries that strive to increase its share of the world tourism market, which grows every year. In addition to its historical and cultural riches and natural beauties, Türkiye is one of the rare countries where different climatic conditions can be experienced at the same time. This situation allows different types of tourism to be implemented in Türkiye. Its history dating back to ancient times and the fact that it has hosted many civilizations make Türkiye a country with significant potential in terms of treasure hunting tourism. However, this potential, which has not been utilized in terms of tourism, means a "loss" for Türkiye.

Although it is possible for those interested to carry out treasure hunting activities legally in Türkiye, they are generally carried out illegally. The main reason for this is the legal procedures that must be followed to get a treasure hunting license are quite difficult and complicated. Unfortunately, illegal treasure hunting by treasure hunters can also have negative consequences, such as damage to cultural and natural properties, illegal trade of historical objects or smuggling them abroad. In order to prevent the negative consequences of illegal treasure hunting and to bring Türkiye's hidden cultural and natural properties to tourism of Türkiye through treasure hunting tourism, it is important to take the following steps:

- Carrying out promotional activities regarding legal treasure hunting, especially on social media,
- Increasing academic studies dealing with treasure hunting from a legal perspective,
  - Tourism actors and government officials come together to create a legal infrastructure that will enable local and foreign tourists to participate actively or passively in treasure hunting activities,
  - Appointing a single ministry to get the necessary legal permission for treasure hunting activities or ensuring that those who want to get a treasure license have to deal with a single authority,
  - Since there are no trained human resources for treasure hunting tourism, tourist guides should be coordinated with people experienced in treasure hunting and land reading, etc.
- Live broadcast of some excavations under the supervision of experts.

References

Adventure Travel Trade Association (ATTA). (2011). Adventure Tourism Development Index 2011 Report. http://www.adventureindex.travel/downloads.html (Accesed: 18.03.2024)

Akkaya, M. (1993). Eski Eser Tahribatı ve Defineciler. *Ankara Üniversitesi Dil ve Tarih - Coğrafya Fakültesi Dergisi*, 36(1-2), 259-262. https://dergipark.org.tr/tr/download/article-file/2153102 (Accesed: 11.03.2024)

Gallup & American Express (1989). *Unique four national travel study reveals traveller types.* New release. American Express. https://www.courseh ero.com/file/82020548/GALLUP-ADN-AMERICAN-EXPRESSdocx/ (Accesed: 24.03.2024)

Avcıkurt, C. (2015). *Turizm Sosyolojisi; Genel ve Yapısal Yaklaşım.* (4. Basım), Ankara: Detay Yayıncılık.

Dalen, E. (1989). Reseacrh into values and consumer in Norway. *Tourism Management*, 10(3), 183-186.

Dickson, T. & Dolnicar, S. (2004). No risk, no fun: The role of perceived risk in adventure tourism. *13th International Research Conference of the Council of Australian University Tourism and Hospitality Education.*https:// ro.uow.edu.au/cgi/viewcontent.cgi?article=1256&context=commpapers (Accesed: 02.02.2024)

Edis, S. (1979). *Medeni Hukuka Giriş ve Başlangıç Hükümleri.* Anakara: Sevinç Matbaası.

Erdoğan, İ. (2013). Hukuki Açıdan Define. *Gazi Üniversitesi Hukuk Fakültesi Dergisi*, 17(1-2), 513-533

Eren, F. (2012). *Mülkiyet Hukuku.* (2. Baskı). Yetkin Yayınları, Ankara: Yetkin Yayınları.

Fabrizio, N., (2014). *Adventure Tourism Management.* (Honor College Master Theses). Page University. https://digitalcommons.pace.edu/cgi/viewcontent. cgi?article=1146&context=honorscollege_theses (Accesed: 03.02.2024)

Feyzioğlu, N. F. (1955). Lükata ve Define. *İstanbul Üniversitesi Hukuk Fakültesi Mecmuası.* 20(1-4), 167-196. https://dergipark.org.tr/tr/download/article-file/96440 (Accesed: 04.05.2024)

Karaduman, H. (2005) Metal Arayıcı Detektörlerin Kullanımı Kısıtlanmalıdır. *İDOL, Arkeologlar Derneği Dergisi*, (7-25), 29-33.

Kozak, M. A. & Bahçe, S. (2009). *Özel İlgi Turizmi.* Ankara: Detay Yayıncılık.

Kozak, N., Kozak A. M. & Kozak, M. (2015). *Genel Turizm İlkeler Kavramlar.* (18. Baskı). Ankara: Detay Yayıncılık.

Kuşçu, D. (2022). Define Arama Yönetmeliğinde Yapılan Değişikliklere Dair Değerlendirmeler. *Türkiye Adalet Akademisi Dergisi*, 13(50), 93-114. DOI: 10.54049/taad.1093110 (Accesed: 14.01.2024)

Mercan, N. (2016). Çok kültürlü ortamlarda kültürlerarası farklılıkları yönetme sanatı: kültürel zekâ. *Açıköğretim Uygulamaları ve Araştırmaları Dergisi*, 2(2), 32-49.

Özdoğan, M., (2014). *50 Soruda Arkeoloji, Bilim ve Gelecek Kitaplığı.* İstanbul: Kayhan Matbaacılık.

Swarbrooke, J,. Beard, C., Leckie, S. & Pomfret, G. (2003). *Adventure Tourism: The New Frontier.* USA: Elsiver Science Ltd.

T.C. Kültür ve Turizm Bakanlığı. (n/a). *Law on the Conservation of Cultural and Natural Property (2863).* https://kvmgm.ktb.gov.tr/TR-43249/law-on-the-conservation-of-cultural-and-natural-property-2863.html (Accesed: 07.04.2024)

Turkish Language Association. (n/a). *Adventure.* https://sozluk.gov.tr/, Erişim Tarihi: 06.02.2024.

TÜSEV. (n/a). *Turkish Civil Code.* https://www.tusev.org.tr/usrfiles/files/Turkish_Civil_Code.pdf (Accesed: 24.04.2024)

United Nations World Tourism Organisation (UNWTO). (2015). Understanding Tourism: Basic Glossary, http://statistics.unwto.org/sites/all/files/docpdf/glossaryterms.pdf (Accesed: 24.02.2024)

Uygur, S. M. & Baykan, E. (2007). Kültür Turizmi ve Turizmin Kültürel Kültür Turizmi Ve Turizmin Kültürel Varlıklar Üzerindeki Etkileri. *Ticaret ve Turizm Eğitim Fakültesi Dergisi*, 2, 30–49.

www.mevzuat.gov.tr. (n/a). *Define Arama Yönetmeliği.* https://www.mevzuat.gov.tr/mevzuat?MevzuatNo=17238&MevzuatTur=7&MevzuatTertip=5 (Accesed: 21.04.2024)

Yıldız, S. (2019). *Macera Turizmi ve Macera Rehberliği.* Ankara: Gece Kitaplığı.

Yıldız, S. (2021). Turizmin Yeni Şafağı: Uzay Turizmi. *EJSER 8th International Symposium on Social Sciences.* (pp. 61–83). 20–22. November, Hamburg-Germany.

Cevat Ercik[1]

Chapter 9 Touristic Motivation in the World of Flavors

Introduction

The tourism sector is constantly evolving with dynamics such as globalization, technological advancements, and changing consumer preferences. These changes lead to a profound transformation in tourists' travel habits, preferences, and expectations (Cracolici & Nijkamp, 2008). Particularly, with the development of alternative and sustainable tourism alongside traditional tourism destinations, tourists' demands and expectations have diversified. This diversification has increased the significance of gastronomic tourism, which is a prominent trend in tourism (Hjalager, 2010).

Gastronomic tourism refers to tourists' inclination to enjoy local cuisine, experience local culinary cultures, and participate in gastronomic events during their travels (Tikkanen, 2007). In this context, the importance tourists place on gastronomic elements in destination selection is not only limited to individual preferences but also influenced by tourists' socio-economic and psychological profiles (Tse & Crotts, 2005).

Gastronomy is a significant part of a destination's cultural heritage and identity. Local cuisines provide tourists with an in-depth understanding of the geographical region's culture, history, and traditions (Ignatov & Smith, 2006; Mitchell & Hall, 2006). Therefore, gastronomy serves not only as a dietary practice but also as a cultural heritage and a determinant of a destination's identity (Harrington & Ottenbacher, 2010).

The evolving trends in the tourism sector and the evolution of tourists' demands increasingly emphasize the role of gastronomic elements in destination selection. Hence, destination managers need to evaluate their gastronomic potentials, introduce local culinary cultures, and enhance gastronomic experiences to strengthen their destinations.

The motivating role of gastronomic experiences on tourists in touristic destinations has been extensively studied by tourism scholars, and various studies

---

1 Instructor Dr. Mersin University, Tourism Faculty, Department of Gastronomy and Culinary Arts cevatercik@mersin.edu.tr Mersin/Türkiye.

have made significant contributions to this field (Hjalager & Richards, 2004). These studies demonstrate that local culinary culture influences tourists' destination preferences and plays a decisive role in their travel motivations. Additionally, they emphasize the importance of promoting gastronomic experiences in tourism literature in enhancing the competitiveness of destinations (Henderson, 2009).

Understanding tourists' travel motivations is crucial because most tourists, while traveling, seek to fulfill their basic physiological needs by visiting restaurants and dining places within Maslow's hierarchy of needs framework. However, the ways in which tourists fulfill their dining needs and their interest in gastronomy can vary (Hjalager, 2004).

This study employs a literature review and a conceptual approach for a detailed examination and evaluation of the concept of gastronomic tourism. The research model is constructed as shown in Figure 9.1. By employing a conceptual approach, the concept of gastronomic tourism is examined in various dimensions, and different perspectives and definitions from the literature are synthesized to form the research model.

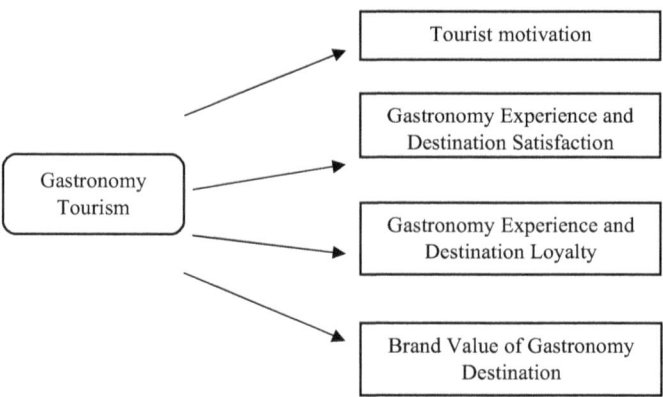

Figure 9.1. Research Model

The research was conducted through document analysis within the framework of a qualitative research approach. The theoretical universe of the research consisted of articles and theses related to gastronomic tourism conducted worldwide. These studies addressed topics such as tourist motivation, gastronomic experience, destination satisfaction, gastronomic experience and destination loyalty, and the brand value of gastronomy destinations.

Sampling was not utilized in the research; instead, the aim was to reach the entire accessible universe. As a result of the literature review, 23 articles and 6 master's theses related to gastronomic tourism were identified and included in the study. Research meeting the criteria at the doctoral thesis level was not found. The studies included in the research were obtained from search systems such as Google search engine, Google Scholar, Taylor & Francis, Wiley-Blackwell, Springer Science & Business Media, Elsevier, Oxford University Press, Cambridge University Press, Mersin University Library, and the National Thesis Center of YÖK (Council of Higher Education).

Various academic sources and research articles were examined to determine the definition of gastronomic tourism, its importance, its role in tourism destinations, its impact on tourist motivation, the effect of gastronomic experiences on tourist satisfaction and loyalty, and the importance of gastronomic tourism in the tourism market. Findings from research on the importance and impact of gastronomic tourism were synthesized and analyzed during the literature review. These analyses were organized to provide a comprehensive understanding of the place and role of gastronomic tourism in the tourism industry.

Gastronomic Tourism

Terms such as gastronomy tourism, culinary tourism, gastro-tourism, food tourism, wine tourism, and gourmet tourism are commonly used to describe tourism activities centered around food and beverage culture (Pavlidis & Markantonatou, 2020). These terms represent an experiential journey to a gastronomic region and encompass visits to primary and secondary food and beverage producers, gastronomy festivals, food fairs, events, farmers' markets, cooking demonstrations and shows, tasting of quality food products, or any tourism activity related to food.

According to UNWTO (2012), gastronomy tourism includes tourists and visitors who partially or entirely plan their trips to taste local cuisine or engage in gastronomy-related activities. Thus, gastronomy tourism includes various holiday products and services such as restaurants, bars, culinary schools, specialized culinary travel packages, food guides, cookbooks, vineyards, wineries, breweries, distilleries, farms, farmers' markets, and more.

Cuisine and gastronomy are considered elements of a broader cultural heritage system by Timothy and Ron (2013). In this context, Minihan (2014) and Richard (2015) note that local cuisines constitute one of the most prominent and defining signs of a destination's culture and heritage. The methods of preparing and preserving foods and the ingredients used are specific to a community's natural resources, climate, and way of life, leading to cuisines being seen as an expression of a society's lifestyle.

Ellis et al. (2018) highlight three main factors explaining the impact of food on tourism: the diversity that different cuisines bring, tourists' tendency to learn about a region's culture through food, and the attractiveness of destinations offering culinary heritage. Mason and O'Mahony (2007) note the importance of cuisine and food offerings for tourist destinations, while Hjalager and Richards (2004) emphasize that local cuisines are a significant factor in tourists' destination choices.

In conclusion, gastronomic tourism is an important strategy for marketing and branding tourist destinations, utilizing the destination's culinary heritage for the development of tourism (Ellis et al., 2018).

The Impact of Gastronomic Experiences on Tourist Motivation

Gastronomic experiences have a significant impact on tourists' preferences and travel decisions in today's tourism landscape. These experiences not only involve eating but also the desire to explore different cultural practices and local flavors (Gheorghe et al., 2014). Tourists consider the attractiveness of local cuisines when choosing a destination, thus increasing the influence of gastronomic experiences on tourists.

When examining the impact of gastronomic experiences on tourists, it is observed that these experiences are not limited to dining activities alone but also shaped from various perspectives such as experiencing regional cuisines, participating in gastronomy events, and joining gastronomy tours (Berbel-Pineda et al., 2019). Factors that drive tourists to experience these activities include the diversity of local cuisines, the quality of food offered, and the atmosphere of the destination.

There is a noticeable increase in tourists' travel motivation due to gastronomic experiences (Mengual-Recuerda et al., 2020). Dining activities have become a fundamental source of motivation for tourists, leading to the rise of gastronomy tourism (Saeed et al., 2022). The role of gastronomic experiences in tourists' destination choices is becoming increasingly important (López-Guzmán et al., 2018).

In the tourism industry, it is acknowledged that gastronomic experiences are a key factor in enhancing destinations' attractiveness and attracting tourists (Galvez et al., 2017: 254). When tourists enjoy local food during their holidays, they develop a more positive attitude towards these destinations and increase their desire to return. Additionally, gastronomic experiences contribute to increasing tourism revenues and benefiting local economies.

Gastronomic tourism has attracted the attention of researchers examining destinations' images and tourist experiences (Lin et al., 2022). Interest in local foods increases tourists' interest in gastronomic experiences and becomes a determining factor in destination choices. Gastronomic experiences are considered to enhance tourists' holiday satisfaction and are recognized as a primary motivation factor in destination choices (Folgado-Fernández et al., 2017; Agapito et al., 2017). Therefore, there is a strong relationship between tourists' satisfaction with destinations and gastronomic experiences.

The Relationship between Gastronomic Experience and Destination Satisfaction

Travel experiences are part of a colorful and rich palette, each telling its unique stories. Every journey triggers the sparkle in the eyes of travelers, the tastes left on their palates, and the emotions felt in their souls (Kotler et al., 2010). However, discrepancies between pre-travel expectations and the reality of the destination can sometimes lead to feelings of dissatisfaction among travelers. In this context, there are many different views in the literature regarding the concept of satisfaction (Barroso & Martín, 1999). Satisfaction typically involves evaluating specific features against a standard, but each travel experience is personal and unique.

The connection between gastronomy and tourism has garnered significant attention in the travel industry. Food and beverages not only nourish the stomach but also feed the soul, making them an integral part of the tourist experience (Pertile & Gastal, 2012). Gastronomic experiences can have a profound impact on travelers and shape their perceptions and satisfaction with the destination (Diaconescu et al., 2016). In this context, gastronomic experience plays a critical role in evaluating a destination and encompasses many factors such as food preparation, culinary culture, cuisine style, and gastronomic traditions (Babolian et al., 2016). However, gastronomic experiences do not always provide a positive experience; sometimes an unexpected taste or a lack of service quality can lead to dissatisfaction among travelers (Nield et al., 2000). Nevertheless, gastronomy is generally considered a significant component of a destination's attractiveness.

Gastronomy tourism is a significant factor that influences travelers' motivations, experiences, and consequently, their satisfaction (López-Guzmán et al., 2017; Pérez Gálvez et al., 2015). Emphasizing local flavors, culinary cultures, and gastronomic events can enhance the competitiveness of destinations and make travelers' experiences richer and more memorable.

In conclusion, gastronomic experiences are considered a strategic tool to enhance the attractiveness and competitiveness of destinations, along with their critical role in travelers' holiday experiences. Therefore, destination managers should focus on developing the gastronomic experience by highlighting local flavors and culinary cultures. This approach is of critical importance for the sustainable tourism industry.

Gastronomic Experience and Destination Loyalty

The critical role of gastronomic experiences in strengthening a destination's image and increasing tourist loyalty is emphasized as an important topic in contemporary tourism research. Studies by Tussyadiah et al. (2018) demonstrate that destination image directly influences tourists' behaviors and shapes their perceptions of destinations.

Research by Khan et al. (2022) indicates that tourists' loyalty to a particular destination increases after experiencing a high-quality gastronomic experience. This can strengthen tourists' intentions to revisit the same destination and may even extend their stays.

According to Chen and Rahman (2018), tourists experiencing positive gastronomic experiences in a particular destination are more likely to recommend these experiences to others. This enhances the destination's image and attracts more tourists.

In this context, it is highlighted that gastronomic experiences are of vital importance for creating and strengthening a destination's image, and destination managers can gain a competitive advantage in the tourism sector by focusing on improving these experiences.

Gastronomic experience is emphasized as an important element in creating a strong brand in the tourism sector (Kivela & Johns, 2003). Academic research has shown that tourists' enjoyment of gastronomic experiences at holiday destinations can affect their satisfaction and influence their intentions to return (Sparks et al., 2003).

Loyalty in tourism is characterized by repeat visits to a destination and recommendations (Lee et al., 2006). In this context, loyalty is an important factor for destination management because loyal tourists tend to extend their stays (Oppermann, 2000). Additionally, it is noted that the recommendation of destinations by loyal tourists can lead to greater and more stable revenues for destinations (Huang et al., 2010).

There is ample research indicating that gastronomic experience influences destination loyalty. Kim et al. (2010) suggested that gastronomic experience affects destination loyalty. Similarly, Tse and Crotts (2005) confirmed in their study that repeat visits are influenced by gastronomic experience. Other researchers also note the positive impact of gastronomic experience on destination loyalty (Folgado-Fernández et al., 2017).

In particular circumstances, Molina (2021) claimed that destination loyalty is directly influenced in specialized tourism areas such as wine tourism. Investments are needed to motivate tourists to recommend their destinations to develop consumer loyalty. For example, a study by Correa et al. (2015) showed that tourists satisfied with the local gastronomy in Portugal tend to revisit the destination.

In conclusion, the literature indicates a strong interaction between gastronomic experience and destination loyalty. Tourists satisfied with their choice of local food may return to destinations through positive word-of-mouth messages, which can play a significant role in destination marketing strategies (Kim et al., 2010; Alderighi et al., 2016).

Gastronomy Destination Brand Value

The brand value of gastronomy destinations is composed of consumers' perceptions of the brand name, symbols, and values associated with that destination (Aaker, 1991). These perceptions are shaped by elements such as the quality and variety of food offered, the uniqueness of local flavors, and the authenticity of gastronomic experiences.

For gastronomy destinations, brand value enhances their attractiveness to tourists. Particularly, as the brand value of a destination increases, tourists' intentions to visit that destination may also increase. This contributes to the economic growth of the destination and the development of the local gastronomy industry.

Brand value is directly related to the perceived value of the destination by tourists, its promotional activities, and its image (Kapferer, 2008). Elements such as local culinary culture, the quality of restaurants, and the preservation of traditional flavors are among the significant factors determining the brand value of a gastronomy destination.

Moreover, dimensions contributing to brand value include brand image, perceived quality, brand loyalty, and brand awareness (Oh and Hsu, 2014). These dimensions influence tourists' perceptions of the destination and their attitudes toward it. Therefore, enhancing brand value for gastronomy destinations is an important strategy to shape tourists' perceptions and increase interest in the destination.

In conclusion, the brand value of gastronomy destinations is the sum of the values associated with the destination by tourists, contributing to the tourism potential and the local economy (Liu, 2016). Therefore, effective marketing strategies and efforts to preserve and promote local flavors are crucial for increasing the brand value of gastronomy destinations.Formun ÜstüFormun ÜstüFormun Üstü

Results

Gastronomy tourism has become increasingly important within the tourism industry. Research indicates that gastronomic experiences influence tourists' destination preferences, enhance travel motivations, and strengthen destination loyalty. Additionally, it has been found that local cuisines and gastronomic events serve as significant attraction factors for tourists, and the economic impact of gastronomy tourism on destinations is noteworthy. However, sustainability concerns should also be taken into account. The preservation of local food resources and the sustainability of natural-cultural heritage are crucial for the long-term success of gastronomy tourism.

Gastronomy tourism constitutes a significant part of a destination's cultural heritage and identity. Local cuisines provide tourists with an in-depth understanding of the geography, history, and traditions of a region. The importance tourists attribute to gastronomic elements in destination selection is not limited to individual preferences but is also influenced by tourists' socio-economic and psychological profiles.

Gastronomy tourism reflects tourists' inclination to savor local cuisines, experience local culinary cultures, and participate in gastronomic events during their travels. In this context, the importance tourists place on gastronomic elements in destination selection is not confined to individual preferences alone. It is also influenced by tourists' socio-economic and psychological profiles.

Gastronomic experiences are now recognized as a significant component of tourist motivation. These experiences are shaped not only by tasting local foods and cuisines but also by factors such as meal prices, the atmosphere of the destination, and high-quality customer service. Gastronomic experiences attract an increasing number of tourists to tourist destinations and motivate them to visit.

In addition to gastronomic experiences, various other factors contribute to tourists' desire to experience a destination's gastronomic appeal. Dann (1981) categorized visitor motivations into two main categories: escape desire and exploration desire. With the rise in gastronomic motivation, the role of gastronomic experiences in tourists' travel and destination choices becomes even more significant.

Numerous studies suggest that gastronomic experiences influence destination loyalty. Kim et al. (2010) argue that gastronomic experiences affect destination loyalty, and similarly, Tse and Crotts confirm in their study that repeat visits are influenced by gastronomic experiences. In this context, it is suggested that gastronomic experiences can enhance a destination's image and increase the likelihood of tourists returning to these destinations.

In light of these findings, the following recommendations are proposed for more effective management and evaluation of gastronomy tourism:

i. Strategic plans for promoting local food culture should be developed and implemented in destinations. These plans should include identifying, branding, and marketing local cuisines.
ii. Collaborations and joint projects with local businesses should be developed for gastronomy tourism. Collaborations with local businesses such as restaurants, farms, markets, and handicraft workshops can stimulate the local economy and contribute to local culture.
iii. Gastronomic experiences for tourists should be diversified and enhanced. Activities such as gastronomy-themed events, food tours, cooking classes, and interactions with local chefs can increase tourists' attachment to a destination.
iv. Collaboration and communication between local governments, civil society organizations, and tourism stakeholders should be strengthened. Multi-stakeholder approaches should be adopted for the sustainability of gastronomy tourism and local development.

In conclusion, gastronomy tourism serves as a strategic tool not only for a culinary journey but also for the sustainability of the tourism sector, local economic development, and preservation of cultural heritage. However, coordinated and sustainable efforts are required to fully harness this potential.

References

Aaker, D. A. (1991). *Managing brand equity*. New York, NY: Free Press.

Agapito, D., Pinto, P., & Mendes, J. (2017). Tourists' memories, sensory impressions, and loyalty: In locomotion and post-visit study in Southwest Portugal. *Tourism Management*, 58, 108–118. https://doi.org/10.1016/j.tourman.2016.10.013

Alderighi, M., Bianchi, C., & Lorenzini, E. (2016). The impact of local food specialties on the decision to (re)visit a tourist destination: Market expansion or business theft? *Tourism Management, 57*, 323–333. https://doi.org/10.1016/j.tourman.2016.06.004

Babolian, H. R. (2016). The effect of dining experience on tourist satisfaction: The case of Indonesia. INTERNATIONAL JOURNAL OF CULTURAL TOURISM AND HOSPITALITY RESEARCH, 10(3), 272–282. https://doi.org/10.1108/IJCTHR-04-2015-0030

Barroso, C., & Martin, E. (1999). *Relational marketing*. Madrid: ESIC.

Berbel-Pineda, J. M., Palacios-Florencio, B., Ramírez-Hurtado, J. M., & Santos-Roldán, L. (2019). *Gastronomic experience as a motivation factor in tourist movements. International Journal of Gastronomy and Food Science, 18*, 100171. https://doi.org/10.1016/j.ijgfs.2019.100171

Chen, H., & Rahman, I. (2018). Cultural tourism: Analysis of participation, cultural contact, memorable tourism experiences, and destination loyalty. *Tourism Management Perspectives, 26*, 153–163.

Correa, A., Zius, A. H., & Silva, F. (2015). Why do tourists insist on visiting the same destination? *Tourism Economics, 21*(1), 205–221. https://doi.org/10.5367/te.2014.0443

Cracolici, M. F., & Nijkamp, P. (2008). Attractiveness and competitiveness of tourist destinations: A study on Southern Italian regions. *Tourism Management, 30*, 336–344. https://doi.org/10.1016/j.tourman.2008.07.006

Dann, G. M. (1981). Tourist motivation and evaluation. *Annals of Tourism Research, 8*(2), 187–219.

Diaconescu, D. M., Moraru, R., & Stanciulescu, G. (2016). Thoughts on gastronomy tourism as a component of sustainable local development. *Amfiteatru Economic, 18*(10), 999–1014. https://www.econstor.eu/handle/10419/169051

Ellis, A., Park, E., Kim, S., & Yeoman, I. (2018). What is food tourism? *Tourism Management, 68*, 250–263. https://doi.org/10.1016/j.tourman.2018.03.025

Folgado-Fernández, J. A., Hernández-Mogollón, J. M., & Duarte, P. (2017). Destination image and loyalty development: The impact of tourists' food experience in gastronomic events. *Tourism and Hospitality Research, 17*(1), 92–110. https://doi.org/10.1080/15022250.2016.1221181

Galvez, J. C. P., López-Guzmán, T., Buiza, F. C., & Medina-Viruel, M. J. (2017). Gastronomy as an attraction factor in a tourism destination: The case of Lima, Peru. *Journal of Ethnic Foods, 4*(4), 254–261. https://doi.org/10.1016/j.jef.2017.08.002

Gheorghe, G., Tudorache, P., & Nistoreanu, P. (2014). Gastronomy tourism, a new trend for contemporary tourism. *Cactus Tourism Journal, 9*, 12–21. Retrieved April 26, 2024, from https://www.cactus-journal-of tourism.ase.ro/Pdf/vol9/nistorescu.pdf

Harrington, R. J., & Ottenbacher, M. C. (2010). Culinary tourism: A case study on the gastronomic capital. *Journal of Culinary Science & Technology, 8*(1), 14–32. http://doi.org/10.1080/15428052.2010.490765

Henderson, J. C. (2009). Food tourism evaluated. *Food Journal, 111*(4), 317–326. https://doi.org/10.1108/00070700910951470

Hjalager, A. M. (2004). *Tourism and gastronomy* (pp. 21–35). London: Routledge.

Hjalager, A. M. (2010). A review of innovation research in tourism. *Tourism Management, 31*, 1–12. https://doi.org/10.1016/j.tourman.2009.11.008

Hjalager, A. M., & Richards, G. (2004). Tourism and gastronomy. London: Routledge.

Huang, S., Hsu, C. H., & Chan, A. (2010). Tour guide performance and tourist satisfaction: A study of the package tours in Shanghai. *Journal of Hospitality & Tourism Research, 34*(1), 3–33.

Ignatov, E., & Smith, S. (2006). Dividing Canadian culinary tourists into segments. *Current Issues in Tourism, 9*(3), 235–255. https://doi.org/10.2167/cit229

Kapferer, J. N. (2008). *The new strategic brand management: Creating and sustaining brand equity long term*. London, UK: Kogan Page.

Khan, J., Osman, M., Saeed, I., Ali, A., & Nisar, H. (2022). Does workplace spirituality affect knowledge-sharing behavior and organizational commitment? The mediating role of trust. *Journal of Managerial Sciences, 12*, 51–66.

Kim, Y. G., Suh, B. W., & Eves, A. (2010). The relationships between food-related personality traits, satisfaction, and loyalty among visitors attending food events and festivals. *International Journal of Hospitality Management, 29*, 216–226. https://doi.org/10.1016/j.ijhm.2009.10.009

Kivela, J., & Crotts, J. (2003). Gastronomy tourism: A meaningful travel market segment. *Journal of Culinary Science & Technology, 4*(2/3), 39–55. https://doi.org/10.1300/J385v04n02_03

Kotler, P., Bowen, J. T., & Makens, J. C. (2010). *Marketing for hospitality and tourism*. Upper Saddle River, NJ: Pearson Education Inc.

Lee, S., Kim, W. G., & Kim, H. J. (2006). The impact of co-branding on post-purchase behaviours in family restaurants. *International Journal of Hospitality Management, 25*(2), 245–261. https://doi.org/10.1016/j.ijhm.2005.04.004

Lin, M.-P., Marine-Roig, E., & Llonch-Molina, N. (2022). Creating a gastronomic experience (together): Evidence from Taiwan and Catalonia. *Tourism Recreation Research, 47*(2), 277–292. https://doi.org/10.1080/02508281.2020.1857895

Liu, C. H. S. (2016). The relationships among brand equity, culinary attraction, and foreign tourist satisfaction. *Journal of Travel & Tourism Marketing*, 33, 1143–1161.

López-Guzmán, T., Lotero, C. P. U., Galvez, J. C. P., & Rivera, I. R. (2017). Gastronomy festivals: Tourist attitude, motivation, and satisfaction. *British Food Journal*, 119(2), 267–283. https://doi.org/10.1108/BFJ-06-2016-0246

López-Guzmán, T., Torres Naranjo, M., Perez-Galvez, J. C., & Carvache Franco, W. (2018). Gastronomic perception and motivation in a touristic destination: The case of Quito City, Ecuador. *GeoJournal of Tourism and Geosites*, 21(1), 61–73. Retrieved from April 08, 2024, https://d1wqtxts1xzle7.cloudfront.net/80714358/271_Guzmanlibre.pdf?1644751316

Mason, R., & O'Mahony, B. (2007). In search of food and wine: Tracking meaningful experiences for the tourist. *Leisure Studies Annual Review*, 10(3–4), 498–517. https://doi.org/10.1080/11745398.2007.9686778

Mengual-Recuerda, A., Tur-Viñes, V., & Juárez-Varón, D. (2020). Neuromarketing in haute cuisine gastronomic experiences. *Frontiers in Psychology*, 11, Article 1772. https://www.frontiersin.org/journals/psychology/articles/10.3389/fpsyg.2020.01772/full

Minihan, C. (2014). Investigating the culinary tourism experience: An examination of the supply sector for brewery and restaurant owners. *Colorado State University Press*. Retrieved April 04, 2024, from https://www.proquest.com/docview/1552970249?

Mitchell, R., & Hall, C. M. (2006). Wine tourism research: The state of play. *Tourism Review International*, 9(4), 307–332. https://doi.org/10.3727/154427206776330535

Molina-Gómez, J. (2021). New perspectives on satisfaction and loyalty in festival tourism: The function of tangible and intangible attributes. Plos One, 16(2), Article e0246562. https://journals.plos.org/plosone/article?id=10.1371/journal.pone.0246562

Nield, K., Kozak, M., & Le Grys, G. (2000). The role of food service in tourist satisfaction. *Hospitality Management*, 19, 375–384. https://www.sciencedirect.com/science/article/pii/S0278431900000372

Oh, H., & Hsu, C. H. (2014). Assessing equivalence of hotel brand equity measures in cross-cultural contexts. *International Journal of Hospitality Management*, 36, 156–166. https://www.sciencedirect.com/science/article/pii/S0278431913001229

Oppermann, M. (2000). Tourism Destination Loyalty, *Journal of Travel Research*, 39: 78–84.

Pavlidis, G., & Markantonatou, S. (2020). Gastronomy tourism in Greece and beyond: A comprehensive review. *International Journal of Gastronomy and Food Science*, 21, 100229. https://doi.org/10.1016/j.ijgfs.2020.100229

Pérez-Gálvez, J. C., Muñoz-Fernández, G. A., & López-Guzmán, T. (2015). Motivation and tourist satisfaction in wine festivals: XXXI ed. wine tasting Montilla-Moriles, Spain. *Tourism & Management Studies*, 11, 7–13. Retrieved from https://dialnet.unirioja.es/servlet/articulo?codigo=5181246

Pertile, K., & Gastal, S. (2012). Gastronomy and tourism: The presence of ingredients in traditional dishes. Presented at the 5th Congress of Tourism Research - CLAIT, São Paulo.

Richards, G. (2015). Creative tourism: New opportunities for destinations worldwide? Paper presented at the World Travel Market Conference on Creative Tourism, November 3, London, United Kingdom.

Saeed, I., Khan, J., Zada, M., Ullah, R., Vega-Muñoz, A., & Contreras-Barraza, N. (2022). Exploring the connection between workplace spirituality and workforce agility: Unveiling higher education institutions. *Psychology Research and Behavior Management*, 15, 31. https://www.tandfonline.com/doi/full/10.2147/PRBM.S344651

Sparks, B., Bowen, J., & Klag, S. (2003). Restaurants and the tourist market. *International Journal of Contemporary Hospitality Management*, 15(1),6–13. https://www.emerald.com/insight/content/doi/10.1108/09596110310458936/full/html?fullSc=1

Tikkanen, I. (2007). Maslow's hierarchy and culinary tourism in Finland: Five cases. *British Food Journal*, 109(9), 721–734. https://doi.org/10.1108/00070700710780698

Timothy, D. J., & Ron, A. S. (2013). Understanding heritage cuisines and tourism: Identity, image, authenticity, and change. *Journal of Heritage Tourism*, 8(2–3), 99–104. https://doi.org/10.1080/1743873X.2013.767818

Tse, P., & Crotts, J. C. (2005). Precursors to innovation seeking: International visitors' predisposition to try Hong Kong's culinary traditions. *Tourism Management*, 26, 965–968. https://doi.org/10.1016/j.tourman.2004.07.002

Tussyadiah, I. P., Wang, D., Jung, T. H., & Tom Dieck, M. C. (2018). Virtual reality, presence, and attitude change: Evidence from tourism. *Tourism Management*, 66, 140–154. https://doi.org/10.1016/j.tourman.2018.03.025

UNWTO. (2012). Global report on food tourism. Retrieved April 20, 2024, from http://dtxtq4w60xqpw.cloudfront.net/sites/all/files/pdf/global.report_on_food_tourism.pdf

Begüm Ilbay Vatan[1]

Chapter 10 Volunteer Tourism Proposal for the Post-Kahramanmaraş Earthquakes Recovery

Introduction

The first examples of volunteer tourism are the journeys of the missionaries and healers to provide voluntary services, and the sending off the people to the different regions by religious and health organizations to provide education or health services and also for spiritual reasons (Tourism Research & Marketing, 2008). Volunteer tourism, which has become an important sub-branch of the tourism industry with increasing interest in recent years, means that individuals voluntarily participate in social, environmental or humanitarian projects during holiday periods. Volunteer tourism, which provides versatile benefits to both communities and volunteers, goes beyond being a holiday activity and becomes a tool that increases social solidarity and awareness.

In this section of the book, the volunteer tourism model is proposed as an alternative for the regions affected by the two major earthquakes that happened in the center of Kahramanmaraş province in Turkey on February 6, 2023. The purpose is to draw attention to its applicability as a model that will contribute to the reconstruction process of these provinces after the earthquakes that deeply affected eleven provinces both physically and socially. In this context, firstly the conceptual framework of volunteer tourism was drawn and then the relevant literature was examined and information was presented on how volunteer tourism was implemented in different regions and with which strategies the successful results were achieved. The planning, coordination and sustainability elements required for the successful implementation of volunteer tourism have been analyzed based on the literature.

---

1 Asst. Prof., Kahramanmaraş İstiklal University, Faculty of Tourism, Department of Gastronomy and Culinary Arts, begumilbay@hotmail.com

The Concept of Volunteer Tourism

Volunteer tourism, which is considered an alternative type of tourism that promotes sustainable and responsible travel experiences and can be called voluntourism (Wearing, Young & Everingham, 2017), is a rapidly growing niche tourism type around the world and has attracted great attention from academics and tourism industry practitioners in the last two decades (Lee, 2011; Easton & Wise, 2015; Yea, Sin & Griffiths, 2018). It is known that volunteer tourism projects increased especially before the pandemic period and this increase continues after 2022. For example, it is noteworthy that more than 1600 volunteer tourism projects have been published in just one of the most important websites that publish volunteer tourism projects (Volunteer Abroad, n.d.).

Volunteer tourism is a type of tourism in which people travel to a region outside their residence for activities such as helping people in need, contributing to communities, participating in environmental renewal studies or conducting research, by allocating certain time during holiday periods and using their own financial means if desired (Wearing, 2001; McGehee & Santos, 2005). Some of the individuals participating in volunteer tourism are motivated by the idea of traveling and participating in the tourism movement. For those who participate in volunteer tourism on a travel basis, volunteering is a small part of their trip. On the other hand, some individuals are motivated by volunteer activities. These individuals travel on a voluntary basis and the basis of their travels is volunteer work. Nowadays, tour operators and travel agencies, whether based on volunteering or traveling, plan volunteer tourism trips where volunteer tourists can interact culturally with the local people and engage in voluntary work (Brown, 2005).

The basic dynamic of volunteer tourism is the concept of "volunteer". Volunteering is not limited to volunteer tourism, but is a concept that is considered from a broader perspective such as social services. In volunteer tourism, it is seen that both volunteer and tourist definitions come together (İlbay, 2014). The concept of volunteer is used for people who undertake tasks such as helping societies in financial poverty during their holidays, restoration of the environment, or research on society and the environment (Wearing, 2002). Volunteers, of their own free will, take part in activities such as cultural studies, anti-racism projects, education and environmental protection, which generally last between 6–18 months. These activities provide significant contributions not only to the region but also to the volunteers. Through these activities, volunteers enter the social and intercultural learning process and experience social change and development (Association of Volunteer Service Organizations, n.d.).

According to some researchers who use the concept of volunteer tourists instead of volunteers, volunteer tourists are people who seek tourist experiences not only for individual development but also for mutual benefit in the social, natural and economic context in which they are directly and positively involved. These people are people who aim to gain new skills or improve their skills, while restoring their environment or participating in research assistance groups (Wearing, 2004; Rowe & Hall, 2003 as cited in Tourism Research & Marketing, 2008).

Volunteer tourism, which contributes to societies, takes tourists beyond visiting the region they visit by providing them with an alternative experience (Lyons, Hanley, Wearing & Neil, 2012). Interaction, one of the most important elements of volunteer tourism, plays a major role in this. The interaction between volunteer tourists and the local people living in the region that hosts them initiates social change and enables people to establish relationship with each other (Wearing, 2002). In addition, volunteer tourism activities affect the lives of local people and volunteer tourists because they are aimed at common goals such as protecting local values (McGehee & Santos, 2005). On the other hand, volunteer tourism also includes efforts to heal the wounds of regions affected by natural and human disasters. For this reason, it is seen that tourism models such as volunteer tourism are recommended in the relevant literature for implementation in disaster-affected regions (Lin, Kelemen, & Tresidder, 2018; Wright & Sharpley, 2018; Wearing, Beirman & Grabowski, 2020).

According to the Ministry of Interior Disaster and Emergency Management Presidency Report (2023), the earthquake that happend on February 6, 2023 at 04.17 had a magnitude of 7.7, the epicenter of which was the Pazarcık district of the Kahramanmaraş province in Turkey, and a magnitude 7.6 earthquake, whose epicenter was the Elbistan district of the Kahramanmaraş province, at 13.24 on the same day. Two magnitude earthquakes caused great destruction in 11 provinces. These two earthquakes were recorded as the most destructive terrestrial "double" earthquakes in the history of the Republic of Turkey (Şen, 2023; Kılıç Ekici, 2023). While the devastating effects of the earthquakes still continue, it is thought that volunteer tourists, who are seen as the unsung heroes of development with their volunteer tourism activities (Eddins, 2013), will play important roles in healing the wounds of the provinces affected by the earthquakes and contributing to their development.

Volunteer Tourism Model

Volunteer tourism, which is considered as an indicator of socio-cultural change, is an alternative type of tourism that is frequently emphasized academically and has an increasing global trend. While volunteer tourism was a phenomenon that first emerged in Britain and Europe, it has now spread to wide areas including Australia, the United States, Asia and Africa (Lo & Lee, 2011; Alexander, 2012).

It can be said that volunteer tourism, which focuses on communities in need, develops programs for them, and aims to find solutions to social and environmental problems (Wearing & McGehee, 2013), is one of the most suitable types of tourism for the regions affected by the Kahramanmaraş earthquakes. For this reason, a model proposal has been presented by overview the literature in order to realize volunteer tourism in the relevant regions. In this model, the responsibilities of the organizations, the factors that will increase the participation of volunteer tourists, how the benefits of the local people can be increased, what the qualities of volunteer tourism projects should be, and the things to be considered in the promotion of the activities to be carried out are grouped under five titles.

Organizations in Volunteer Tourism

There are basically three organizations that play a role in organizing volunteer tourism activities: the sending institution, the intermediary institution and the host institution. These organizations are responsible for carrying out many tasks such as developing projects, organizing them, making agreements, informing people and receiving feedback (Raymond, 2008; Tourism Research & Marketing, 2008). The main organizations that carry out these duties are local governments, non-governmental organizations (NGO) and private commercial operators (Lyons & Wearing, 2008). With the support of organizations, volunteer tourism activities have become widespread in developing countries, especially to help people in need (Vodopivec & Jaffe, 2011).

State organizations and NGOs make significant contributions to volunteer tourism activities. Although Ellis (2003 as cited in Coghlan, 2008) stated that a very small portion of the voluntary organizations in the world are state organizations, it has been observed that the grants and agreements made for volunteer tourism studies in recent years are carried out by state organizations (Tourism Research and Marketing, 2008; Mcgehee & Andereck, 2009). The aim of NGOs, which are another of the main supporters of volunteer tourism, is to carry out volunteer activities that positively affect the behavior, values and actions of volunteer tourists and local people. NGOs carry out studies on issues such as providing training to local people and entrepreneurs regarding volunteer tourism, protecting environmental values by developing volunteer projects, and contributing to regional planning and social development (Demir & Çevirgen, 2006).

The absolute cooperation to be developed and the protocols to be approved between the state, NGOs and private commercial organizations in the development and dissemination of volunteer tourism in the regions affected by the earthquakes centered in Kahramanmaraş can be described as the most important steps in correctly determining the needs of the regions and in the effective use of resources.

Volunteer Tourists

With volunteer tourism, which creates positive effects on the host community and develops a more benign form of travel, volunteers have the opportunity to think about volunteering and learn new things by participating in the experiential learning process (Wearing, 2001; Broad, 2003; Gray & Campbell, 2007). Facilitating the participation process of volunteer tourists, who see themselves as participants in "international service-learning" activities, in volunteer tourism activities forms the basis for the regulation of volunteer tourism.

When the factors that motivate volunteer tourists and facilitate the process of participating in volunteer tourism activities are examined, it is seen that projects in which the interaction between local people and volunteer tourists is high should be developed (Wearing, 2001; Brown, 2005; Coghlan, 2006; Broad & Jenkins, 2008; Benson & Seibert, 2009; Chen & Chen 2011). In addition, it is frequently emphasized in the literature that events should be organized to analyze social events and especially young people should be encouraged to participate in these events (İlbay & Acar Gürel, 2015). Also, factors such as supporting the learning process that will enable volunteer tourists to acquire new skills and contribute to their personal development also come to the fore in relevant research (Wearing, 2001; Stoddart & Rogerson, 2004; Jones, 2004; Brown, 2005; Broad & Jenkins, 2008; Sin, 2009).

Local People in Volunteer Tourism

The stakeholder most affected economically and socially by volunteer tourism is the local people. Volunteer tourism creates a positive impact on local people in terms of resource transfer, participation in social events and social change. Volunteer tourism, which provides interaction between local people and volunteer tourists, also contributes to international development (Wearing, 2001; McGehee & Andereck, 2009).

Volunteer tourism, like other types of tourism, provides many benefits to the local people, such as contributing to employment and income, preserving local culture, increasing the level of education, and ensuring the development of new sectors. In order to sustain these benefits, local people need to understand and participate in volunteer tourism activities. Educational activities can be carried out with local people on this subject. Thus, local people who can work together with volunteer tourists and share information can be made more active. The active participation of local people can ensure positive results in efforts to develop intercultural mutual understanding, which is one of the main goals of volunteer tourism (Wearing, 2001; McGehee & Andereck, 2009; Singh, 2014).

When the literature is examined, it is seen that there are studies suggesting that volunteer tourism may have negative effects on the local people. It is stated that when high-income individuals participate in volunteer tourism due to its popularity, negative effects such as the risk of deviating from the purpose of volunteer tourism, depletion of local people's resources, exceeding local capacity, commodification of local culture, underestimation of the dignity of the local community and excessive dependence on volunteer tourists may occur (Shepherd, 2002; McGehee & Andereck, 2009; Singh, 2014; Lupoli & Morse, 2015). In order to prevent these negativities from arising, what volunteer tourism is must be explained to local stakeholders and volunteer tourists through educating.

Sustainable Development Projects in Volunteer Tourism

Projects that support sustainable development, protect natural resources and benefit local people form the basis of volunteer tourism activities. Volunteer tourists allocate some of their free time and even financial income by participating in these activities (activities such as alleviating poverty, restoring structures, and participating in research on society or the environment) (Wearing, 2001).

Volunteer tourism projects, which have a great potential to contribute to sustainable development goals, are carried out on a small scale, taking into account regional characteristics and in the context of community-based conservation and development, providing greater benefits to the local community and volunteer tourists (Butcher & Smith, 2010; Lockstone-Binney & Ong, 2022). From this perspective, it can be said that volunteer tourism projects should be carried out by determining the unique needs of the provinces affected by the Kahramanmaraş earthquakes. For example, in provinces where housing destruction is common, volunteer tourists can carry out social support activities for families affected by destruction. On the other hand, in provinces where historical monuments are more damaged, they can participate in the restoration works of these monuments. As can be seen from the examples, volunteer tourism projects can be developed by determining the needs ranking of each province.

Promotion Efforts for Volunteer Tourism

National and international promotion efforts are one of the most important steps in promoting volunteer tourism to large audiences. However, it is recommended that volunteer tourism be treated differently from mass tourism and that responsible tourism be emphasized in promotion efforts (Belz & Peattie, 2012; Smith, 2015). The aim here is to eliminate the doubts of volunteer tourists and eliminate possible mistrust by emphasizing sustainability and responsible understanding in promotion efforts (Forehand & Grier, 2003).

Lyon & Maxwell (2011) emphasized that promotional messages for volunteer tourism can be created using greenwashing strategies to create a positive image and convey information about sustainable responsible tourism. However, Kreps & Monin (2011) argued that greenwashing strategies will fail because some individuals dislike those who claim to be morally superior. Researchers have stated that individuals who compare their own lives with the messages given in greenwashing strategies feel inadequate and even inferior. At this point, carrying out customer-oriented studies, making the service more accessible, giving messages that appeal to emotions rather than numerical data, introducing opportunities to work with local people and children, and making the goals clear and understandable are the prominent suggestions for the promotion of volunteer tourism (Hedlund, 2011; Smith & Font, 2014).

Results

Volunteer tourism is a type of tourism that is considered in many social, cultural and economic dimensions and provides mutual benefit to individuals and societies. The fact that individuals want to spend their holidays not only for rest and entertainment, but also for purposes such as helping societies in need, protecting the environment and contributing to sustainable development are important factors in the development of this type of tourism. From this perspective, it is obvious that the implementation of a correct volunteer tourism model will increase the motivation of those who want to become volunteer tourists.

It is thought that the implementation of the volunteer tourism model in line with community-based protection and development purposes in the provinces affected by two major earthquakes centered in Kahramanmaraş on February 6, 2023 will contribute to the reconstruction process by increasing social solidarity and awareness. In provinces affected by earthquakes, it should not be forgotten that volunteer tourism projects should be planned according to regional needs. In addition, establishing cooperation between local governments, NGOs and private commercial operators in organizing volunteer tourism activities will increase the effectiveness and sustainability of volunteer tourism projects.

Volunteer tourists, the biggest actors in the realization of volunteer tourism, contribute to both societies in need and themselves by participating in social, environmental or humanitarian projects during holiday periods. Therefore, examining the motivations of volunteer tourists is very important for the successful implementation of volunteer tourism projects. It should also be taken into consideration that projects in which the interaction between local people and volunteer tourists in the earthquake zone is high will increase the participation of volunteer tourists and that especially young people should be encouraged to participate in volunteer tourism activities.

As a result, volunteer tourism can be used as an effective tool to contribute to the reconstruction process in disaster areas and to increase social solidarity and awareness. The planning, coordination and sustainability elements required for the successful implementation of volunteer tourism projects should be taken into account and the active participation of local people should be ensured. In this way, volunteer tourism activities can contribute to sustainable development by providing mutual benefits to both the earthquake region and volunteer tourists.

References

Alexander, Z. (2012). International volunteer tourism experience in South Africa: An investigation into the impact on the tourist. *Journal of Hospitality Marketing & Management*, 21 (7), 779–799, https://doi.org/10.1080/19368623.2012.637287.

Association of Volunteer Service Organizations (n.d.). About AVSO. Retrieved May 13, 2024 from https://www.devex.com/organizations/association-of-voluntary-service-organisations-avso-45989.

Belz, F.M. & Peattie, K. (2012), *Sustainability Marketing: A Global Perspective*. Chichester: John Wiley and Sons.

Benson, A. & Seibert, N. (2009). Volunteer tourism: Motivations of German participants in South Africa. *Annals of Leisure Research*, 12 (3–4), 295–31, https://doi.org/10.1080/11745398.2009.9686826.

Broad, S. (2003). Living the Thai life – a case study of volunteer tourism at the Gibbon Rehabilitation Project, Thailand. *Tourism Recreation Research*, 28 (3), 63–72, https://doi.org/10.1080/02508281.2003.11081418.

Broad, S. & Jenkins, J. (2008). Gibbons in their midst? Conservation volunteers' motivations at the Gibbon Rehabilitation Project, Phuket, Thailand. In

K.D. Lyons & S. Wearing (Eds.), Journeys of discovery in volunteer tourism (pp. 72–85). Biddles Ltd.

Brown, S. (2005). Travelling with a purpose: Understanding the motives and benefits of volunteer vacationers. *Current Issues in Tourism*, 8 (6), 479–496, https://doi.org/10.1080/13683500508668232.

Butcher, J. & Smith, P. (2010). Making a difference': Volunteer tourism and development. *Tourism Recreation Research*, 35 (1), 27–36, https://doi.org/10.1080/02508281.2010.11081616.

Chen, L.J. & Chen, J.S. (2011). The motivations and expectations of internationalvolunteer tourists: A case study of "Chinese Village Traditions". *Tourism Management*, 32, 435–442, https://doi.org/10.1016/j.tourman.2010.01.009.

Coghlan, A. (2006). Volunteer tourism as an emerging trend or an expansion of ecotourism? A look at potential clients' perceptions of volunteer tourism organisations. *International Journal of Nonprofit and Volunteer Sector Marketing*, 11, 225–237, https://doi.org/10.1002/nvsm.35.

Coghlan, A. (2008). Exploring the Role of Expedition Staff in Volunteer Tourism. *International Journal of Tourism Research*, 10, 183–191, https://doi.org/10.1002/jtr.650.

Demir, C. & Çevirgen, A. (2006). *Ekoturizm Yönetimi*. Nobel Pub.

Easton, S., & Wise, N. (2015). Online portrayals of volunteer tourism in Nepal: Exploring the communicated disparities between promotional and user-generated content. *Worldwide Hospitality and Tourism Themes*, 7 (2), 141–158, https://doi.org/10.1108/WHATT-12-2014-0051.

Eddins, E. (2013). Bridging the gap: Volunteer tourism's role in global partnership development. In K.S. Bricker, R. Black & S. Cottrell (Eds.), Sustainable tourism and the millennium development goals: Effecting positive change (pp. 251–264). Jones & Bartlett Learning.

Forehand, M.R. & Grier, S. (2003). When is honesty the best policy? The effect of stated company intent on consumer skepticism. *Journal of Consumer Psychology*, 13 (3), 349–356, https://doi.org/10.1207/S15327663JCP1303_15.

Gray, N.J. & Campbell, L.M. (2007). A decommodified experience? Exploringaesthetic, economic and ethical values for volunteer ecotourism in Costa Rica. *Journal of Sustainable Tourism*, 15 (5), 463–482, https://doi.org/10.2167/jost725.0.

Hedlund, T. (2011). The impact of values, environmental concern, and willingness to accept economic sacrifices to protect the environment on tourists' intentions to buy ecologically sustainable tourism alternatives. *Tourism and Hospitality Research*, 11 (4), 278–288, https://doi.org/10.1177/1467358411423330.

İlbay, B. (2014). *Gönüllü ve gençlik turizmi: Eskişehir'e yönelik bir öneri* (Unpublished master's thesis). Eskişehir Anadolu University.

İlbay, B. & Acar Gürel, D. (2015). Gönüllü ve gençlik turizminin birlikte ele alınması: Eskişehir'e yönelik bir öneri. *International Journal of Human Sciences*, 12 (2), 207–234, https://doi.org/10.14687/ijhs.v12i2.3303.

Jones, A. (2004). *Review of gap year provision*. London University.

Kılıç Ekici, Ö. (2023). 6 Şubat 2023 Depremleri. *TÜBİTAK Bilim ve Teknik Dergisi*, 1–8. https://bilimteknik.tubitak.gov.tr/system/files/makale/6_subat.pdf.

Kreps, T.A. & Monin, B. (2011). Doing well by doing good? Ambivalent moral framing in organizations. *Research in Organizational Behavior*, 31 (1), 99–123, https://doi.org/10.1016/J.RIOB.2011.09.008.

Lee, S.J. (2011). *Volunteer tourists' intended participation: Using the revised theory of planned behavior* (Unpublished doctoral dissertation). Virginia Polytechnic Institute and State University.

Lin, Y., Kelemen, M. & Tresidder, R. (2018). Post-disaster tourism: Building resilience through community-led approaches in the aftermath of the 2011 disasters in Japan. *Journal of Sustainable Tourism*, 26 (10), 1766–1783, https://doi.org/10.1080/09669582.2018.1511720.

Lo, A., & Lee, C. (2011). Motivations and perceived value of volunteer tourists from Hong Kong. *Tourism Management*, 32 (2), 326–334, https://doi.org/10.1016/j.tourman.2010.03.002.

Lockstone-Binney, L. & Ong, F. (2022). The sustainable development goals: the contribution of tourism volunteering. *Journal of Sustainable Tourism*, 30 (12), 2895–2911, https://doi.org/10.1080/09669582.2021.1919686.

Lupoli, C.A. & Morse, W.C. (2015). Assessing the local impacts of volunteer tourism: Comparing two unique approaches to indicator development. *Social Indicators Research*, 120, 577–600, https://doi.org/10.1007/s11205-014-0606-x.

Lyon, T.P. & Maxwell, J.W. (2011). Greenwash: Corporate environmental disclosure under threat of audit. *Journal of Economics & Management Strategy*, 20 (1), 3–41, https://doi.org/10.1111/j.1530-9134.2010.00282.x.

Lyons, K., Hanley, J., Wearing, S. & Neil, J. (2012). Gap year volunteer tourism: Myths of global citizenship. *Annals of Tourism Research*, 39 (1), 361–378, https://doi.org/10.1016/j.annals.2011.04.016.

Lyons, K.D. & Wearing, S. (2008). Volunteer tourism as alternative tourism: Journeys beyond otherness. In K.D. Lyons & S. Wearing (Eds.), Journeys of discovery in volunteer toursim (pp. 3–11). Biddles Ltd.

McGehee, N.G. & Andereck, K. (2009). Volunteer tourism and the "voluntoured": the case of Tijuana, Mexico. *Journal of Sustainable Tourism*, 17 (1), 39–51, https://doi.org/10.1080/09669580802159693.

McGehee, N.G. & Santos, C.A. (2005). Social change, discourse and volunteer tourism. *Annals of Tourism Research*, 32 (3), 760–779, https://doi.org/10.1016/j.annals.2004.12.002.

Ministry of Interior Disaster and Emergency Management Presidency (2023). 06 Şubat 2023 Pazarcık (Kahramanmaraş) mw 7.7 Elbistan (Kahramanmaraş) mw 7.6 depremlerine ilişkin ön değerlendirme raporu. Deprem Dairesi Başkanlığı.

Raymond, E. (2008). "Make a difference!": The role of sending organisations in volunteer tourism. In K.D. Lyons & S. Wearing (Eds.), Journeys of discovery in volunteer toursim (pp. 48–60). Biddles Ltd.

Shepherd, R. (2002). Commodification, culture and tourism. *Tourist Studies*, 2 (2), 183–201, https://doi.org/10.1177/146879702761936653.

Sin, H.L. (2009). Volunteer tourism "involve me and I will learn"? *Annals of Tourism Research*, 36 (3), 480–501, https://doi.org/10.1016/j.annals.2009.03.001.

Singh, R. (2014). Volunteer tourism and host community. *International Journal of Scientific Research and Management (IJSRM)*, 2 (10), 1480–1487.

Smith, V.L. & Font, X. (2014). Volunteer tourism, greenwashing and understanding responsible marketing using market signalling theory. *Journal of Sustainable Tourism*, 22 (6), 942–963, https://doi.org/10.1080/09669582.2013.871021.

Smith, V.L. & Font, X (2015). Marketing and communication of responsibility in volunteer tourism. Worldwide Hospitality and Tourism Themes, 7 (2), 159–180, https://doi.org/10.1108/WHATT-12-2014-0050.

Stoddart H. & Rogerson, C.M. (2004). Volunteer tourism: The case of habitat for humanity South Africa. *Geojournal*, 60, 311–318, https://doi.org/10.1023/B:GEJO.0000034737.81266.a1.

Şen, S. (2023). Kahramanmaraş depremlerinin ekonomiye etkisi. *Diplomasi ve Strateji Dergisi*, 4 (1), 1–55.

Tourism Research & Marketing (2008). *Volunteer tourism: A global analysis*. ATLAS.

Vodopivec, B. & Jaffe, R. (2011). Save the world in a week: Volunteer tourism, development and difference. *European Journal of Development Research*, 23 (1), 111–128, https://doi.org/10.1057/ejdr.2010.55.

Volunteer Abroad (n.d.). Explore 1,600+ projects abroad & find your best volunteer program. Retrieved May 20, 2024 from https://www.volunteerworld.com/en?gclid=EAIaIQobChMIk5HzjMmw-AIVBn8rCh3ZGgMjEAAYASAAEgKJYvD_BwE.

Wearing, S. (2001). *Volunteer tourism: Experiences that make a difference*. CABI Pub.

Wearing, S. (2002). Re-centring the self in volunteer tourism. In G. Dann (Ed.), The Tourist as a Metaphor of the Social World (pp. 48–60). CABI Pub.

Wearing, S. (2004). Examining best practice in volunteer tourism. In R.A. Stebbins & M. Graham (Eds.), Volunteering as leisure/leisure as volunteering: An international assessment (pp. 209–224). CABI Pub.

Wearing, S., Beirman, D., & Grabowski, S. (2020). Engaging volunteer tourism in post-disaster recovery in Nepal. *Annals of Tourism Research*, 80 (2020), 1–13, https://doi.org/10.1016/j.annals.2019.102802.

Wearing, S. & McGehee, N.G. (2013). Volunteer tourism: A review. *Tourism Management*, 38 (2013), 120–130, http://dx.doi.org/10.1016/j.tourman.2013.03.002.

Wearing, S., Young, T., & Everingham, P. (2017). Evaluating volunteer tourism: Has it made a difference? *Tourism Recreation Research*, 42 (4), 512–521, https://doi.org/10.1080/02508281.2017.1345470.

Wright, D. & Sharpley, R. (2018). Local community perceptions of disaster tourism: The case of L'Aquila, Italy. *Current Issues in Tourism*, 21 (14), 1569–1585, https://doi.org/10.1080/13683500.2016.1157141.

Yea, S., Sin, H. L., & Griffiths, M. (2018). International volunteerism and development in Asia-Pacific. *Geographical Journal*, 184 (2), 110–114, https://doi.org/10.1111/geoj.12254.

Uğur Can Aykanat[1] and Gamze Şanli Ak[2]

Chapter 11 The Effects of Zero Waste and Composting Method on the Costs of Fertilizer Production from Food Waste of Hotels in Seferihisar District of Izmir Province

Introduction

The act of eating and drinking, one of the most basic phenomena that has always existed and will always exist throughout human history, is a great importance for all living things. This action, which has been going on since the existence of humanity, has gained a different dimension with the developments in many fields such as the transition to settled life, the discovery of fire, inventions, discoveries and technological advances. This phenomenon, which has evolved through many changes over time, has become a sector and has now become a science that includes concepts such as color, smell, texture and visual harmony, not only for the purpose of filling the stomach.

Today, increasing environmental responsibilities and sustainability goals on a global scale encourage many sectors to adopt environmentally friendly practices and review their waste management strategies. In this context, the tourism sector, especially in terms of accommodation facilities, plays an important role in adopting sustainable practices and effectively managing waste. While the importance of sustainable tourism is increasing, environmentally friendly projects implemented at the local level have a significant potential for regional development and economic sustainability. In this context, understanding the environmental and economic potential of the food waste utilization processes of hotels in Seferihisar district can be instructive for future tourism businesses.

The subject of this study is the food waste of 4 and 5 star hotels in Seferihisar district of Izmir province. The impact of food waste on the economy is an undeniable fact. Based on this reality, the aim of this study is to identify and evaluate food waste in line with the data provided by the businesses in the research

---

1 Istanbul Bilgi University, ugurcanaykanat@gmail.com
2 Assistant Professor, Istanbul Nişantaşı University, Faculty of Art and Design, Department of Gastronomy and Culinary Arts, gamzesanli.ak@nisantasi.edu.tr

universe. Within the scope of this evaluation, it is aimed to contribute to the concept of "clean environment" in the transformation process by using renewable energy sources.

Waste and lost products have a direct negative impact on costs. The effects of some food wastes on the environment cause environmental damage such as disruption of the ecological balance. Ecological balance must be maintained in order to sustain living life. Day by day, factors that negatively affect the ecological balance cause pollution of elements such as air, water and soil, as well as the disruption of the balance of vital factors such as nitrogen, oxygen and carbon dioxide, which is worrying for the future (Şerbet & Onursal, 2020). In recent years, the efforts of both professional kitchens and home kitchens to reduce food waste and implement food waste utilization methods have attracted attention. According to 2016 data from the Turkish Statistical Institute (TurkStat), municipalities collected 31.6 million tons of waste. Of this waste, 61.2 % was collected regularly, 29 % was disposed of irregularly, 9.3 % was processed in recycling facilities and only 0.5 % was processed using composting (Ekinci et al., 2021). Recycling and composting have been the subject of many recent studies.

The use of the composting system makes it possible to recycle waste. This recycling process not only reduces the costs of businesses, but also minimizes waste damage to the environment. Thanks to composting, some of the waste products can be used as animal food and some of them can be separated to be used as fertilizer in botanical areas such as trees, plants and flowers.

In this section, food waste amounts of 4 and 5 star hotels in Seferihisar district of Izmir province are evaluated. Zero waste and composting processes are of great importance for the sustainability of the environment and ecosystem. Meeting the energy needs of the composting machines to be used in this process entirely from renewable energy sources is of great value in terms of sustainability.

Food, Waste and Zero Waste Concepts

Food is an essential basic need for all organisms to survive. According to the Ministry of Agriculture and Forestry, food is "any processed, partially processed or unprocessed substance that is eaten, drunk or expected to be eaten or drunk by humans, excluding live animals, feed, unharvested plants, medicinal products used for therapeutic purposes, cosmetics, tobacco and tobacco products that are not offered directly for human consumption" (Ministry of Agriculture and Forestry, 2015). At the same time, food also includes substances consumed in a way that does not threaten human health. Mankind has met its daily nutritional needs by processing many edible substances since its existence. The discovery of fire undoubtedly plays an important role in the formation of the concept of cuisine. With the discovery of fire, mankind started to live around fire and started to process and consume food with the phenomenon of cooking (Arman, 2019).

Consumption has increased along with population growth worldwide. In direct proportion to consumption, the amount of waste also increases. According to the Regulation on Waste Management, waste is defined as "any substance or material that is discarded, released or required to be discarded into the environment by its producer or the real or legal person who actually possesses it" (Official Gazette, 2015). Waste can be defined simply as "the final state of matter" (Evans, 2020). The total amount of household and industrial waste generated in the European Union each year is around 3 billion tons. Of this huge volume, 100 million tons are hazardous waste (CPS, 2012).

An agreement was signed by the members of the European Union in 1975 to increase the levels of waste reduction, waste reuse and recycling (Hatipoğlu et al., 2021). This agreement is recognized as the first Waste Framework Directive (CPS, 2012). Immediately afterwards, the "Hazardous Waste Directive/Directive" entered into force in 1978 on the prevention and disposal of hazardous waste (CPS, 2012). This directive, which entered into force, basically consists of three main articles.

- Waste Prevention/Minimization: Waste prevention is based on the approach of eliminating waste during the production of the product, i.e. at the source.
- Reuse: The principle of recycling and reuse, which is one of the most important steps of waste management, includes the appropriate recovery of waste obtained after waste prevention.
- Recycling: It is a process that involves the conversion of recyclable wastes into secondary raw materials using various physical or chemical methods. This process ensures a more sustainable use of resources by integrating materials back into the production process (Ak & Genç, 2018).

According to the announcement of the Ministry of Environment and Urbanization dated 21.06.2014, "It is aimed to ensure waste minimization, which is one of the most important steps of waste management, in order to efficiently utilize wastes and to use some wastes as an alternative to raw materials used in industrial production processes in co-incineration plants established for the purpose of producing products rather than waste incineration. In this context, 'Communiqué on Waste Derived Fuel, Additional Fuel and Alternative Raw Materials' was published in the Official Gazette dated 20.06.2014 and numbered 29036 and entered into force" (Ministry of Environment and Urbanization, 2014).

The homogeneous structure of the wastes to be used will increase efficiency during the combustion process. In order to increase efficiency, wastes are subjected to various procedures such as screening, grinding, size reduction and classification during the homogenization process (Çelik, 2018). Waste-derived fuel (WDF) is widely used in industrial areas, especially in cement production. In these areas, the end products obtained by the WDF method are used for fuel substitution.

Consumption has existed as long as humanity has existed. Along with this consumption, damages to nature have also increased (Gündüz, 2021). The increase in the world population, together with the rapid increase in urbanization and population density, has led to production and consumption needs, and ultimately brought along waste problems (Bilgili, 2021). The concept of "zero waste", which we frequently encounter today, is an approach that aims to reduce, reuse and thus completely eliminate waste. According to the "Zero Waste" regulation published in 2019, it is aimed to protect the environment and human health and all resources by preventing/reducing waste generation in production, consumption and service processes, prioritizing reuse, collecting and collecting waste separately at source, and reducing the amount of waste to be sent to disposal by ensuring recycling and/or recovery. On waste, the "Zero Waste International Alliance" was established in 2022 and as a result of this establishment, a board was established to contribute to the zero waste project, provide guidance and establish a standard.

The word meaning of zero waste was updated by the Zero Waste International Alliance (ZWIA) in December 2018 and the concept of zero waste is defined as "Zero Waste: Conserving all resources by responsibly producing, consuming, reusing and recovering products, packaging and materials without incineration and without discharging them into soil, water or air in a way that threatens the environment or human health" (Zero Waste International Alliance, 2018).

Food Waste, Food Waste Reports, Waste and Loss in Food

Food waste includes all foodstuffs produced for human consumption but not consumed, left on the plate or thrown away (FAO, 2022). In this context, food waste is not only a waste of food, but also a waste of time, energy, labor, financial resources and natural resources spent in production and consumption processes (Diallo and Ünsever, 2020). Food waste poses a serious economic and environmental problem (Smith et al., 2019). In addition, food waste contributes to greenhouse gas emissions, deepening environmental problems. Considering its social, environmental and economic impacts, food waste stands out as a critical global problem that threatens our future. Reducing food waste is of great importance for sustainability. Therefore, various strategies need to be adopted to minimize food waste. These strategies include more efficient production techniques, improved storage and distribution methods, and approaches aimed at raising consumer awareness (Stuart, 2009).

In the field of gastronomy, various hazardous and non-hazardous waste materials are generated. These waste materials can occur in solid, liquid or gaseous forms. The main food wastes in the food and beverage sector are as follows:

1. Leftovers that are not consumed by consumers or overproduced during the production process,
2. Production losses as a result of errors made during the production process,
3. Food processing losses such as peels, bruises, stems, etc. generated during the food processing process.

The management and minimization of such wastes is of great importance for sustainability and environmental protection.

Although food waste has been overshadowed by other global issues such as hunger and climate change, it has become an increasingly salient issue in recent years (Republic of Turkey Ministry of Trade, 2018). By 2050, the global population is projected to reach 9.8 billion, up from 7.9 billion in 2022. Despite this, available data show that the rate of food production exceeds the rate of human population growth and that food production is currently sufficient to feed about 10 billion people (Özkan et al., 2022). According to a study conducted by FAO in 2019, 13.8 % of the food produced worldwide, or 1.3 billion tons of food, is wasted in agricultural processes. Worldwide, 30 % of food waste consists of cereal crops, 40–50 % of root crops, fruits and vegetables, and 20 % of oilseeds, meat and dairy products and fish varieties (Tekiner et al., 2021).

According to research, food waste and loss in hospitality organizations is concentrated in the kitchen, service and bar services. The main reasons for this situation include excessively large portions, customers ordering more food and beverages than they need, use of poor quality products, inadequate training of staff, cooking and storage errors, and overpurchases (US Environmental Protection Agency [EPA], 2010; Şahin & Bekâr, 2018).

According to the Republic of Turkey Ministry of Trade's 2018 Turkey Waste Report, 5.4 % of consumers throw away leftover food and 23 % of food purchased is wasted before it is consumed. According to data from the Ministry of Agriculture and Forestry's collaboration with Metro Market, one of the biggest factors in food waste is not consuming the food purchased from buffets or ordered from the menu. Preventing and reducing food waste is critical for a sustainable future.

Compost

Composting is the process by which organic waste is broken down into fertilizer by microorganisms in an oxygenated environment (Epstein, 1997). This natural process occurs spontaneously in natural environments all over the world. Humans can accelerate this process in a controlled manner to obtain a valuable fertilizer. Composting refers to the process of breaking down organic waste in an oxygenated environment and converting it into fertilizer. This method helps to reduce the amount of waste and prevent environmental pollution. Therefore, composting is an important practice in terms of both waste management and environmental protection.

In order to rebuild and protect the balance of nature and support sustainability, it is necessary to use natural composts by reducing the use of chemical pesticides and fertilizers during the production of agricultural products (Dardeniz et al., 2018). Compost provides effective utilization of organic wastes, improves soil structure, neutralizes toxic substances in the soil, aerates the soil and regulates the pH balance. It also provides nutrients to plants and supports their strengthening and growth (Rona, 2023). By adopting composting practices, businesses can reduce waste costs and contribute to increasing sustainable profitability by achieving a sustainable and competitive position in the sector (Çirişoğlu & Akoğlu, 2021). This approach provides significant environmental and economic benefits, enabling the implementation of sustainable agriculture and waste management policies.

In line with the guidelines issued by the Ministry of Agriculture and Forestry of the Republic of Turkey, cold compost production can be carried out in gardens by digging holes or creating piles. In cities or homes, it is possible to obtain natural compost using a compost bucket. Various methods have been developed, such as open composting (heap) and closed composting (silo, cell). There are two basic types of composting: cold (slow) composting and hot composting. Open composting takes place under natural conditions, while closed composting processes have many advantages. These advantages include their high capacity, their ability to produce fully mature compost in a short time, and their ability to produce continuous and controllable production without being dependent on climatic conditions. For this reason, closed systems provide more efficient and faster compost production, making them a preferred method, especially in cities and large-scale agricultural enterprises.

Process Flow Chart in Composting

- Raw material storage and processing
- Crushing/Disintegration
- Homogeneous mixing
- Ripening and processing
- Elimination
- Packaging

Result

According to the 2022 Municipally Certified Accommodation Statistics, a total of 367 establishments (excluding simple accommodation types) in the Izmir region have an annual occupancy rate of 44.76 %. The same data shows that the total number of beds in these establishments is 54,674. Based on this data, it has been determined that hotels in Seferihisar district of Izmir Province generate an average of 88,750 grams of food waste per day. However, it has also been determined that this food waste is not reused through recycling. This situation points to an important problem in terms of food waste management and sustainability in the hospitality sector. Failure to recycle food waste has negative consequences both environmentally and economically, and therefore, these wastes need to be managed effectively.

It is noteworthy that hotels in Izmir Seferihisar region generate high amounts of food waste. According to a study, it is estimated that hotels in this region produce 88,750 grams of food waste on a daily basis. Depending on the types of waste, an average of 55–65 kilograms of fertilizer with high nutritional value can be obtained from this amount of waste. The market value of the obtained fertilizers is calculated as 13,750 TL for approximately 55 kilograms of fertilizer, considering that the price of 5 kilograms of compound fertilizer varies between 250–300 TL as of February 2024. According to this calculation, if the composting machine is operated at maximum efficiency, enterprises can produce 36,500 kilograms of fertilizer on an annual basis. Thus, it is possible to obtain approximately 1.8 million TL worth of fertilizer by operating the composting machine at maximum level throughout the year. This makes it possible for hotels to both contribute to environmental sustainability and generate economic gains.

The solar-powered composting machine operates with a 15-day process and consists of four parts in total. In the first part, there are shredders that ensure the shredding and homogeneous mixing of the added waste. In the second and third sections, there are tanks that allow the compost to mature. In the last part, there is an outlet unit that takes the matured compost. The wastes added to the composting machine are kept in the first maturation tank for the first 15 days and in the second maturation tank for the next 15 days for the formation of beneficial microorganisms and the compost to be ready. The machine has a daily feeding capacity of 100 kg and with this capacity, it can turn an average of 3 tons of waste into fertilizer per month. This composter does not require large amounts of energy and has a daily energy consumption of 3.5 kW (3500 watts). This is the energy required to aerate the waste. The machine operates entirely in accordance with passive composting principles and does not emit any harmful emissions. As of February 2024, the cost of this machine is estimated at 800,000 TL.

According to Weatherspark data; there are significant changes in day length throughout the year in Seferihisar. In 2024, the shortest daylight duration is recorded as 9 hours and 30 minutes on December 21, while the longest day is expected to be 14 hours and 50 minutes on June 20. These time zones are similar to those of previous years. It has been determined that a monocrystalline solar panel can collect between 250–350 watts of energy during the hours of sunshine. Solar panels with this energy production are approximately 2 meters long and 1 meter wide. According to these calculations, 10–11 panels are required for a compost machine that needs 3.5 kW of energy. The energy from these panels will be stored in gel batteries. The stored energy will not provide enough power for the compost machine to operate. In order to convert this energy into industrial electricity, it is necessary to convert it with a 380v output three-phase inverter.

References

Ak, Ö., & Genç, A. T. (2018). A study on university students' recycling awareness: The case of Sakarya University. International Journal of Economic Studies, 4(2), 19–39.

Arman, A. (2019). Evaluation of parameters related to functionality in industrial kitchen design criteria and suggestions for design (Doctoral dissertation). Necmettin Erbakan University, Institute of Social Sciences, Konya.

Bilgili, M. Y. (2021). Origins and contemporary meaning of the zero waste approach. Istanbul Commerce University Journal of Social Sciences, 20(40), 683–703.

Çelik, S. Ö. (2018). Refuse-derived fuel: Legal framework, and the situation in Europe and Turkey. European Journal of Engineering and Applied Sciences, 1(2), 63–71.

Çirişoğlu, E., & Akoğlu, A. (2021). Food waste and its management in restaurants: The case of Istanbul. Academic Food Journal, 19(1), 38–48. https://doi.org/10.24323/akademik-gida.927664

CPS. (2012). Guide to the EU acquis on waste management. Retrieved May 24, 2023, from https://www.mess.org.tr/media/filer_public/6b/58/6b583c70-1daa-4bc5-96b5-9c988df39db1/mess_atik_yonetimi_ab_mevzuat_rehberi.pdf

Dardeniz, A., Şahin, E., Kavdır, Y., Müstüoğlu, N. M., Türkmen, C., & İlay, R. (2018). The process of making pruning residue compost and determining some of its physical and chemical properties. COMU Faculty of Agriculture Journal, 6, 19–25.

Diallo, M. L., & Ünsever, Y. S. (2020). An experimental study on the stabilization of clay soil with construction wastes and lime. Pamukkale University Journal of Engineering Sciences, 26(6), 1030–1034.

Epstein, E. (1997). The science of composting. Technomic Publishing.

Ekinci, K., Tosun, İ., & Varol, N. (2021). Compost handbook. Yalova: Atatürk Horticultural Central Research Institute.

Evans, D. (2020). We're putting the issue of waste on the table. Istanbul: Yeni İnsan Publishing.

Food and Agriculture Organization. (2022). The state of food security and nutrition in the world 2022: Repurposing food and agricultural policies to make healthy diets more affordable. FAO.

Gündüz, M. Y. (2021). Control of recyclable waste and implementation of the zero waste project: The case of Necmettin Erbakan University (Master's thesis). Necmettin Erbakan University, Institute of Science, Konya.

Hatipoğlu, A., Baran, A., Keskin, C., Baran, M. F., Aktepe, N., & Onursal, N. (2021). The concept, classification, and management of waste. Ankara: IKSA Publishing. Ministry of Agriculture and Forestry. (2015). Safe food. Kütahya: Provincial Agriculture Directorate.

Ministry of Environment and Urbanization. (2014). Communiqué on refuse-derived fuel, supplementary fuel, and alternative raw materials. Retrieved May 19, 2023, from https://cygm.csb.gov.tr/atiktan-turetilmis-yakit-ek-yakit-ve-alternatif-hammadde-tebligi-duyuru-17930

Presidency of the Republic of Turkey, Legislation Information System. (2015). Regulation on the incineration of waste. Official Gazette.

Rona, E. (2023). A compost guide for healthy soil and plants. Buğday Association for Supporting Ecological Living. Retrieved December 23, 2023, from https://www.bugday.org/blog/wp-content/uploads/2021/04/kompostreh ber_web.pdf

Smith, J. (2019). The impact of food waste on climate change. Journal of Environmental Science, 15(2), 45–62.

Stuart, T. (2009). Waste: Uncovering the global food scandal. New York, London: W. W. Norton & Company.

Şahin, S. K., & Bekar, A. (2018). A global issue, "Food waste": Its dimensions in hotel businesses. Journal of Tourism and Gastronomy Studies, 6(4), 1039–1061.

Şerbet, N., & Onursal, F. S. (2020). A systematic approach to food waste recovery processes. Turkish Journal of Agriculture-Food Science and Technology, 8(10), 2059–2067.

Tekiner, İ. H., Mercan, N. N., Kahraman, A., & Özel, M. (2021). An overview of food loss and waste in the world and Turkey. Journal of Natural Sciences of Istanbul Sabahattin Zaim University, 3(2), 123–128.

T.C. Ministry of Trade. (2018). Turkey waste report. Ankara: Directorate General for Consumer Protection and Market Surveillance.

Zero Waste International Alliance. (2018). Zero waste international alliance. Retrieved January 11, 2024, from https://zwia.org/zero-waste-definition/

Gözde Oğuzbalaban[1]

Chapter 12 Green Marketing Practices in Tourism Businesses

Introduction

Marketing, one of the business functions, has a high degree of importance in commercial life for many years (Akdemir & Akbulut, 2019). According to Kotler and Armstrong (2010), marketing is present in every aspect and every moment of our lives. Movies, TV series, advertisements, social media accounts, e-mails, smartphone applications can be given as examples. With the development of technology and industry, and the facilitation of transportation and communication in the globalizing world, businesses have started to operate and compete not only in their local markets but also in markets around the world. In this environment where the number of businesses is quite high, it is an important necessity for businesses to develop some strategies in the planning phase of their activities and while carrying out their activities.

Moreover, businesses do not operate in isolation but interact with consumers in their environment, who expect them to adopt responsible behaviors towards the environment (Berk & Celep, 2020). In the 21st century, the concept of sustainability in tourism emerged as a solution to the environmental damages caused by the industry, which relies heavily on natural resources (Weaver, 2022).

Sustainable tourism is based on planned foundations and requires the protection of natural resources for development. This ensures not only the conservation of cultural and natural assets but also promotes economic and social development in the region (Rebollo & Baidal, 2003). As the importance of environmental conservation and the understanding of sustainability in tourism increase, concepts such as "ecological marketing," "environmental marketing," and "sustainable marketing" have emerged, leading to the rise of "green marketing" practices (Chamorro & Banegil, 2006).

Worldwide, consumers' rising environmental awareness has shifted their preferences towards businesses that prioritize nature conservation in their marketing and production approaches (Ceylan & Kıpırtı, 2021). Consequently, green

---

1 Asst. Prof. Zonguldak Bülent Ecevit University, Karadeniz Ereğli Tourism Faculty, Department of Tourism Guidance, oguzbalaban@beun.edu.tr

marketing has become a crucial competitive goal for tourism businesses, driven by the need to meet the preferences and demands of environmentally conscious consumers.

This study explores the concept of green marketing, its importance, the marketing mix, and the green marketing practices implemented in tourism businesses. A literature review was conducted to utilize relevant national and international sources.

Concept of Green Marketing

The concept of green marketing, which originated from a program initiated by the American Marketing Association (AMA) in 1975 to promote environmental awareness in marketing practices, has since gained widespread recognition (Yücel & Emekçiler, 2008). The AMA defines green marketing as the activities aimed at reducing the negative environmental impacts of products, from their packaging to their potential for reprocessing (Onurlubaş & Dinçer, 2016).

According to Polonsky (1995), green marketing involves producers creating products that meet consumer needs and desires while minimizing harm to the environment. Crane (2000) defines it as incorporating environmental elements into business marketing processes. According to Karna et al. (2003), green marketing is a holistic management process responsible for predicting the needs of society and customers and finding sustainable ways to meet those needs.

Green marketing focuses on environmentally responsible production, aiming to meet customer demands and needs with products that protect the environment. It emphasizes the importance of environmentally responsible production processes (Uydacı, 2017).

Importance of Green Marketing

The recognition of environmental issues by stakeholders such as competitors, employees, consumers, and suppliers has led researchers to focus on the ecological environment. Researchers have suggested that businesses should plan their production systems to minimize negative impacts on the natural environment (Keleş et al., 2009).

Another significant factor driving businesses towards green marketing practices is the increasing environmental awareness among consumers, who tend to prefer eco-friendly products. Additionally, environmental pressures from the government, consumers, along with the competitive advantage, cost savings, and profitability offered by green marketing, have encouraged businesses to adopt these practices (Polonsky, 2008; Uydacı, 2017; Atay & Dilek, 2013). Consequently, businesses must meticulously conduct feasibility studies for environmentally friendly practices from site selection to production waste management.

Green Marketing Mix

The marketing strategies businesses must apply to continue their activities are known as the marketing mix (Kotler & Armstrong, 2010). Jerome McCarthy first introduced the concept in 1964 as the "4P's," comprising product, price, place, and promotion (Cemalcılar, 1987). These strategic elements have been redefined with environmental sensitivity to form the green marketing mix.

Green Product

A green product is defined as one produced with minimal environmental impact, reducing pollution, waste, and toxic substances while conserving energy and resources (Ottman, Stafford, & Hartman, 2006). The focus is on developing environmentally friendly production strategies and designing products that cause the least harm during their lifecycle. Green products are generally recyclable, reusable, and eco-friendly (Luttropp & Lagerstedt, 2006).

Green Price

Green price, in the context of green marketing, denotes the monetary value that consumers invest in purchasing environmentally friendly products. One of the primary concerns for businesses implementing green marketing is determining the price of green products. (Topuz, 2016). Typically, green products are more expensive than non-green products due to additional costs associated with eco-friendly practices (Varinli, 2008).

The criteria that businesses should pay attention to when making green pricing can be listed as follows (Yamak, 2007);

  i. Quality: Products should be of a quality that will satisfy consumers.
  ii. Credibility: The product should convince the consumer of its environmental benefits.
  iii. Simplicity: The green product should be easy for the consumer to understand.
  iv. Marketability: Markets should be divided into regions and appropriate strategies should be developed.
  v. Specificity: Information about the technologies developed and renewable resources should be explained to the consumer in detail.

vi. Visibility: Products should always be visible to the consumer.
vii. Concreteness: The benefits of the products to the consumer should be stated concretely.
viii. Community: The community should be informed about green products and support should be sought.

Green Distribution

Incorporating "green" attributes into distribution emphasizes the need for environmentally protective efforts at every stage of the distribution process (Simpson & Power, 2005). Measures include minimizing fuel consumption during product distribution and strategically locating sales points to save customers time. As environmentally conscious markets grow, new distribution and recycling services are expected to develop rapidly (Yücel & Ekmekçiler, 2008).

Green Promotion

Green promotion in the marketing mix helps businesses identify consumer desires and needs while achieving profitability and marketing success. The goal in tourism businesses is to create an "eco-friendly business" image in the consumer's mind through public relations, promotions, advertising campaigns, and other marketing opportunities (Uydacı, 2017). The primary challenge lies in adeptly conveying environmental information to consumers without falling into the trap of 'greenwashing,' where deceptive or exaggerated green claims can mislead consumers (Polonsky & Rosenberger, 2001).

Green Marketing Practices in Tourism Businesses

Tourism businesses have adopted green marketing principles due to consumer trends, environmental degradation, pollution, governmental pressures, and competitive forces (Atay & Dilek, 2013). This section evaluates the green marketing practices in various tourism sectors.

Green Marketing Practices in Accommodation Businesses

Environmentally conscious management in accommodation businesses focuses on minimizing or preventing environmental damage (e.g., chemical waste, natural destruction). Such businesses integrate ecological considerations into decision-making processes, redesign products, and adopt a culture of environmental protection (Nemli, 2001). Water and energy management are critical, as they are interconnected in hotel operations, affecting overall consumption (Deng & Burnett, 2002).

In Turkey, the Ministry of Culture and Tourism introduced the Green Star project in 1993, promoting environmentally friendly practices in accommodation businesses. The updated Environmental Sensitivity Classification Form in 2008 led to the issuance of the "Green Star Facility" certificate to compliant businesses. The project encourages water and energy savings, waste reduction, environmental planning, renewable energy use, ecological architecture, environmental education, and cooperation with relevant organizations (T.C. Ministry of Culture and Tourism, 2024).

Other eco-friendly practices in accommodation businesses include (Gökdeniz, 2017):

- Using non-toxic detergents in housekeeping and laundry.
- Using 100 % organic bed linens, towels, and curtains.
- Implementing smoke-free policies.
- Utilizing renewable energy sources.
- Offering organic food and beverages.
- Using recyclable materials in rooms and lobbies.
- Informing guests about the importance of not changing sheets and towels unnecessarily.
- Using energy-efficient lighting.
- Employing green transportation options.
- Implementing air exchange programs.
- Recycling water for irrigation.
- Avoiding single-use items in service areas.

Green Marketing Practices in Airlines

The aviation sector, crucial for environmental balance, places great importance on green marketing. Efforts include reducing emissions, using recyclable products on flights, testing biofuels, and implementing green airport practices (Nygren, Aleklett, & Höök, 2009). IATA's Fly Net Zero program aims for net-zero CO_2 emissions in aviation by 2050 through technological advancements, energy infrastructure, operational changes, financing, and policies (IATA, 2024a). The IATA Environmental Assessment (IEnvA) program offers a certification framework for environmental sustainability in ground operations, helping providers reduce negative environmental impacts (IATA, 2024b). Turkish Airlines has also started using sustainable aviation fuel, which does not contain harmful heavy metals (THY, 2024).

Green Marketing Practices in Travel Agencies and Tour Operators

Travel agencies and tour operators play a key role in planning and packaging travel activities, making their involvement in green marketing essential. Promoting eco-friendly hotels and airlines and creating environmentally conscious tour plans are crucial for sustainability in tourism (Şimşek, 2019). Leading online travel agency Expedia shares green travel tips on its website under "Top Tips For Eco-Friendly Travel" (Expedia, 2017). The American Society of Travel Agents (ASTA) provides a comprehensive green travel program guide for its members (Dilek, 2012). The World Travel Awards recognize green tour operators, encouraging sustainable practices (World Travel Awards, 2024).

Green Marketing Practices in Food and Beverage Businesses

With the growing awareness of the environmental impact of food and beverage businesses, particularly in developed countries, measures such as sustainable sourcing, waste reduction initiatives, and energy-efficient practices have been increasingly adopted. The Green Restaurant Association, established in the US in 1990, has guided many restaurants towards eco-friendly practices through its certification system, promoting sustainability in energy, water, waste, food, chemicals, and disposable materials (GRA, 2024). The Green Key certification, applicable to hotels, small accommodations, campsites, holiday parks, conference centers, tourist attractions, and food and beverage businesses, is awarded annually to facilities meeting 13 criteria, including staff training, environmental management, guest information, water and energy conservation, waste management, and corporate social responsibility (TÜRÇEV, 2024).

Conclusion

The advancement of technology and industrialization has diversified consumer consumption habits, leading to significant environmental pollution and threatening life. Environmental issues are now among the top global concerns. The emergence of green marketing, driven by competitive pressures and increasing awareness among consumers and producers, highlights the importance of addressing environmental issues.

Green marketing, defined as environmentally conscious marketing, aims to meet consumer needs and desires with minimal environmental impact. It is used across various sectors, including tourism. Green marketing practices are essential for tourism businesses to remain competitive.

Environmentally conscious consumers now evaluate products not only for quality and price but also for their environmental friendliness. Tourism businesses offering eco-friendly products benefit from long-term profitability and the preservation of natural resources for future generations.

In conclusion, green marketing practices provide a competitive advantage and cost savings for tourism businesses and should be adopted for sustainable success.

References

Akdemir, R. & Akbulut, O. (2019). Yeşil pazarlama stratejilerinin rekabet avantajına etkisinin incelenmesi: Muğla ilinde yer alan 4 ve 5 yıldızlı otel işletmelerine yönelik bir araştırma. *Elektronik Sosyal Bilimler Dergisi*, 18 (72), 1676–1687. https://doi.org/10.17755/esosder.530648

Atay, L. & Dilek, S. E. (2013). Konaklama işletmelerinde yeşil pazarlama uygulamaları: İbis otel örneği. *Süleyman Demirel Üniversitesi İktisadi ve İdari Bilimler Fakültesi Dergisi*, 18 (1), 203–219.

Berk, O. N. & Celep, E. (2020). Konaklama işletmelerinde yeşil pazarlama faaliyetlerinin tüketicilerin satın alma davranışlarına etkisi, konya ili örneği. *Selçuk Üniversitesi Sosyal Bilimler Enstitüsü Dergisi*, (44), 267–285.

Cemalcılar, İ. (1987). Pazarlama karması (4p) kavramında yeni gelişmeler. *Pazarlama Dünyası Dergisi*, Sayı: 4 (1), 23–24.

Ceylan, U. & Kıpırtı, F. (2021). Turizm işletmelerinde yeşil pazarlama faaliyetleri: literatür incelemesi. *Socrates Journal of Interdisciplinary Social Studies*, 8, 23–37. https://doi.org/10.51293/socrates.22

Chamorro, A. & Banegil, T. M. (2006). Green marketing philosophy: a study of spanish firms with ecolabels. *Corporate Social Responsibility and Environmental Management*, 13 (1), 11–24. https://doi.org/10.1002/csr.83

Crane, A. (2000). Facing the backlash: green marketing and strategic reorientation in the 1990s. *Journal Of Strategic Marketing*, 8 (3), 277–296. https://doi.org/10.1080/09652540050110011

Deng, S. & Burnett, J. (2002). Water use in hotels in hong kong, *International Journal of Hospitality Management*, (21), 57–66. https://doi.org/10.1016/S0278-4319(01)00015-9

Dilek, S. E. (2012). *Turizm işletmelerinde yeşil pazarlama uygulamaları: bir alan araştırması*. Çanakkale Onsekiz Mart Üniversitesi, Sosyal Bilimler Enstitüsü, Turizm İşletmeciliği Anabilim Dalı Yüksek Lisans Tezi, Çanakkale, 2012.

Expedia (2017) Top Tips For Eco-Friendly Travel. Retrieved from https://www.expedia.com/stories/top-tips-eco-friendly-travel/ (Accessed: 30.04.2024)

Gökdeniz, A. (2017). Konaklama sektöründe yeşil yönetim kavramı, eko etiket ve yeşil yönetim sertifikaları ve otellerde yeşil yönetim uygulama örnekleri. *International Journal of Social and Economic Sciences*, 7(2), 70–77.

GRA (2024). Mission: Green the Restaurant Industry. Retrieved from http://www.dinegreen.com/about (Accessed: 30.04.2024).

IATA (2024a) Net Zero Roadmaps. Retrieved from https://www.iata.org/en/programs/environment/roadmaps/ (Accessed:10.05.2024).

IATA (2024b) Environmental Assessment (IEnvA). Retrieved from https://www.iata.org/en/services/certification/ienva/environmental-assessment/ (Accessed:10.05.2024).

Karna, J. Hansen, E. N. & Juslin, H. (2003). Social responsibility in environmental marketing planning. *European Journal of Marketing*, 37 (5/6), 848–871. https://doi.org/10.1108/03090560310465170

Keleş, R. Hamamcı, C. & Çoban, A. (2009). *Çevre politikası*. Ankara: İmge Kitabevi Yayınları.

Kotler, P. & Armstrong, G. (2010). *Principles of Marketing*. Harlow: Pearson Education Limited.

Luttropp, C.& Lagerstedt, J. (2006). Ecodesign and the ten golden rules: generic advice for merging environmental aspects into product development. *Journal of cleaner production*, 14(15–16), 1396–1408. https://doi.org/10.1016/j.jclepro.2005.11.022

Nemli, E. (2001). Çevreye duyarlı yönetim anlayışı. *İstanbul üniversitesi siyasal bilgiler fakültesi dergisi*. (23–24), 211–224.

Nygren, E., Aleklett, K. & Höök, M. (2009). Aviation fuel and future oil production scenarios. *Energy Policy*, 37 (10), 4003–4010. https://doi.org/10.1016/j.enpol.2009.04.048

Onurlubaş, E. & Dinçer, D. (2016). *Yeşil pazarlama tüketici algısı üzerine bir araştırma*. İstanbul: Beta Yayın.

Ottman, J. A., Stafford, E. R.& Hartman, C. L. (2006). Avoiding gren marketing myopia: ways to improve consumer appeal for environmentally preferable products. *Environment: science and policy for sustainable development*, 48 (5), 22–36. https://doi.org/10.3200/ENVT.48.5.22-36

Polonsky, M. J. (1995). A stakeholder theory approach to designing environmental marketing strategy. *Journal of Business & Industrial Marketing*, 10 (3), 29–46. https://doi.org/10.1108/08858629510096201

Polonsky, M. J. (2008). An ıntroduction to green marketing. *Global environment:problems and policies*, 2(1), 1–10.

Polonsky, M.J. Rosenberger III, P.J. (2001). Reevaulating green marketing: a strategic approach. *Business Horizons*, 44(5), 21–30.

Rebollo, J. F. V. & Baidal, J. A. I. (2003). Measuring sustainability in a mass tourist destination: pressures, perceptions and policy responses in Torrevieja, Spain. *Journal of sustainable tourism*, 11(2–3), 181–203.

Simpson, D. F. & Power, D. (2005). Use the supply relationship to develop lean and green suppliers. *Supply Chain Management: An International Journal*. 10 (1), 60–68.https://doi.org/10.1108/13598540510578388

Şimşek, Ç. (2019). *Konaklama işletmelerinde yeşil pazarlama uygulamaları: otel yöneticileri üzerinde bir alan araştırması*. Necmettin Erbakan Üniversitesi, Sosyal Bilimler Enstitüsü, Turizm İşletmeciliği Anabilim Dalı Yüksek Lisans Tezi, Konya, 2019.

T.C. Ministry of Culture and Tourism (2024). Yatırım ve İşletmeler Genel Müdürlüğü: Çevreye Duyarlılık Kampanyası (Yeşil Yıldız). Retrieved from https://yigm.ktb.gov.tr/TR-11596/cevreye-duyarlilik-kampanyasi-yesil-yildiz.html (Accessed:07.05.2024).

THY (2024) Geleceğe Doğru Türk Havayolları ve Sürdürülebilir Turizm, Retrieved from https://blog.turkishairlines.com/tr/gelecegi-dogru-surdurulebilir-turizm-ve-turk-hava-yollari/ (Accessed: 30.04.2024).

Topuz, S. (2016). Yeşil pazarlama ve üretici işletmelerin yeşil pazarlama faaliyetlerine ilişkin bir araştırma. Beykent Üniversitesi, Sosyal Bilimler Enstitüsü, İşletme Anabilim Dalı, Yüksek Lisans Tezi, İstanbul, 2016.

TÜRÇEV (2024). Yeşil Anahtar. Retrieved from https://turcev.org.tr/v2/icerikDetay.aspx?icerik_id=15 (Accessed: 30.04.2024).

Uydacı, M. (2017). *Yeşil pazarlama*. İstanbul: Türkmen Kitabevi.

Varinli, İ. (2008). *Pazarlamada yeni yaklaşımlar*. Ankara: Detay Yayıncılık.

Weaver, D. B. (2022). *Sustainable tourism*. Buhalis D. (ed.) In Encyclopedia of Tourism Management and Marketing (pp. 317–321). Edward Elgar Publishing. https://doi.org/10.4337/9781800377486.sustainable.tourism

World Travel Awards (2024) World Travel Awards Winners. Retrieved from https://www.worldtravelawards.com/winners/2024 (Accessed: 30.04.2024)

Yamak, O. (2007). *Üretim yönetimi*. İstanbul: Türkmen Kitabevi.

Yücel, M. & Ekmekçiler, Ü. S. (2008). Çevre dostu ürün kavramına bütünsel yaklaşım; temiz üretim sistemi, eko-etiket, yeşil pazarlama. *Elektronik Sosyal Bilimler Dergisi*, 7 (26), 320–333.

Neslihan Onur[1] and Ayşen Ertaş Sabanci[2]

Chapter 13 Gastrotourists' Approaches towards Destinations

Introduction

Today, living conditions are changing, the pace of business life is gradually increasing and people's needs, expectations and tendencies are differentiating. People are trying to evaluate their work and non-work lives in a more efficient, enjoyable and different way by shaping their needs in line with their wishes (Kozak et al. 2017). In this direction, activities such as rest, entertainment and eating and drinking have diversified, and the understanding of vacation has gradually changed and personalized. This situation has led the tourism sector, destinations and tourism businesses to adopt an approach to provide a better tourism experience to their customers with various and different opportunities (Kivela & Crotts, 2006; Kozak et al. 2017).

In addition to the desires, curiosity and expectations of the tourists that cause the increase and diversification of tourism mobility, the destinations that meet or have the potential to meet them also come to the fore (Haven-Tang & Jones 2005; Fox 2007). Tourists looking for an alternative to the classic sea, sand, sun and nature holidays focus on food and beverage, in other words gastronomy. Gastro tourists turn to destinations for reasons such as spending time in gastronomy-based destinations, social interaction, personal development and recognizing what is new and different in their lives (Barkat & Vermignon 2006).

The relationship between tourists and destinations is explained by various theories in terms of tourism (Ambrož & Ovsenik 2011). The relationship between gastro tourists and destinations is based on push-pull factors in tourism literature. Gastro-tourist's intrinsic approaches belonging to the gastro-tourist and the gastronomy-based facilities and services belonging to the destination and offered in the destination are extrinsic approaches (Quan & Wang 2004).

---

1 Assoc. Prof. Dr., Akdeniz University, Manavgat Faculty of Tourism, Department of Gastronomy and Culinary Arts, neslihanonur@akdeniz.edu.tr
2 Research Assistant, Recep Tayyip Erdoğan University, Ardeşen Tourism Faculty, Department of Gastronomy and Culinary Arts, aysen.ertas@erdogan.edu.tr

Gastronomy, Gastro Tourism, Gastro Tourist Concepts

Gastronomy, defined as the science and art of eating and drinking, has been handled as a term in dictionaries, in poems, newspaper articles, as a subject in books and as a department in schools for centuries, as a result of which the scope of the concept has developed and this concept has been included in the consciousness of the public (Scarpato 2002; Kivela & Crotts 2005).

People travel and stay temporarily to fulfill their needs such as resting, having fun, and engaging in cultural activities outside the place where they live. This explains the multifaceted and functional tourism that these people have been practicing for many years (Doğan 2004). However, nowadays tourists also want to taste foods and beverages that are unique to a region. Understanding and recognizing the different and new has always been an attractive idea for people. *Gastronomy tourism* or *gastro tourism* is an important type of tourism that capitalizes on the ideas of curiosity and discovery (Quan & Wang 2004; Kivela & Crotts 2006). It is the main and only reason for many tourists to visit a region for purposes such as getting to know the food culture of that region, seeing and experiencing the preparation and production stages of the food, as well as tasting the flavors specific to the region (Mak et al. 2012). In other words, the factor that encourages people to travel is food and beverage and culinary culture. Gastrotourism is different from the conventional tourism concept and is based on food and beverage, taste and habits, and getting to know the culture (Wolf 2006).

Gastro tourists are those who participate in gastro tourism, traveling for food and beverage beyond the need to eat and drink. They travel for the purpose of recognizing the culinary culture in destinations and discovering the differences of food by experiencing foods specific to different cultures (Mitchell & Hall 2003; Long 2004). Every nation, country and society has a food culture shaped according to its geography, socio-cultural and economic structure, historical identity, agricultural production and structure, traditions, eating behaviors and habits, and taste (Kim & Choe 2019).

Understanding gastrotourist behavior is of great importance. Because they are tourists who can afford to travel for kilometers to taste the food of the destination and introduce their food culture, are inquisitive, eager to try local products, and eager for culinary experiences. In order to attract gastro tourists to destinations, the first of the tourism marketing steps should be to focus on research to understand the tourists who are already visiting or have the potential to visit the destination. In this direction, it is essential to identify and understand the motivations of gastro tourists and the factors that affect the choice of destination.

Factors Affecting Destination Choices of Gastro Tourists

In the tourism sector, destinations are geographical regions that offer touristic values and services with natural and attractive beauties for tourists (Baydeniz et al. 2023). Tourists travel to destinations for intrinsic factors such as escaping from everyday life and experiencing the new and different, social and cultural interaction, gaining prestige, adventuring, spending time with other people and nature (Hjalager 2004; Kivela & Crotts 2006; Mak et al. 2012; Hjalager & Johansen 2013).

Since tourists come from different cultures, their wants and needs vary. Tourists choose destinations that will meet their wants and needs. Understanding the motivational factors, wants, needs and behaviors of tourists is very important for both businesses and other tourism stakeholders in the destination (Nisari & Sakin Yılmazer 2018).

Factors such as local culture, places to visit, sun, sea, prestige, cheap transportation facilities of the destination are effective in tourists' choice of destination (Tellioğlu 2021; Kurt & Arslan Ayazlar 2021.) In addition, destinations with a variety and number of tourism facilities, tourism institutions and organizations are preferred. This is an important factor in revisiting destinations and extending the length of stay (Smith & Xiao 2008). The lifestyle of the destination, local food and beverages and agricultural products grown in the region, which are among the local-original values among the destination attraction elements, are very important in motivating gastro tourists externally in choosing a destination (Fields 2002). As a gastro tourist, it is possible to take part in gastronomy tourism with intrinsic motivations such as physical, cultural, social and prestige. The food experience satisfies the desire for experience by appealing to all five senses and becomes the most remembered element during the vacation (Quan & Wang 2004; Nisari & Sakin-Yılmazer 2018).

Gastronomy is one of the reasons why tourists, especially gastrotourists, prefer a destination because it reflects the cultural richness and diversity of a destination (Okumuş et al. 2007). In addition, the distance of the destination and prices in the region are among the factors that cannot be ignored for gastro tourists (Safarov et al. 2022). The preferred destination is a unique tourism activity that contributes to the personal development of gastro tourists such as relaxing both psychologically and physically, getting to know a culture and satisfying their curiosity (Bucak & Aracı 2013; Ayaz & Yalı 2017). Gastro tourists, who act with cultural motivation, act with motives such as learning about a certain culture, tasting dishes specific to this culture, and visiting restaurants and festivals where these dishes are served (Guzman & Canizares 2011). From this point of view, local cuisines are a motivational tool that offers original and unique cultural experiences and creates opportunities for those seeking diversity and innovation in the food experience (Üzülmez 2021; Smith & Xiao, 2008).

Gastro tourists are addressed in various ways in the literature due to their different attitudes towards destination choice. *Food-oriented gastro tourists* prefer to do food and beverage-oriented shopping from markets and shopping centers in the destination, especially during the holiday period. *Organic gastro tourists* prefer to buy and taste products produced by farmers in the destination they visit. *Innovative gastro tourists* prefer to experience different, new flavors instead of ordinary eating and drinking habits, and to participate in festivals and events related to destination-specific food and beverages. Gastro tourists who aim to learn prefer to follow gastronomy-related publications, watch programs and participate in trainings so that they can assimilate the food culture specific to the destination. *Local-oriented gastro tourists,* on the other hand, prefer to experience the local food and beverages of the destination, observe the preparation processes of these products, learn their stories and purchase these products (Şimşek & Selçuk 2018; Arıcı & Bayram 2021).

According to another classification of gastrotourists, tourists seeking locality primarily want to experience local foods in the restaurant in the destination. On the other hand, innovation-seeking gastro tourists prefer to experience new flavors in the destination they visit. Regular gastro tourists prioritize pricing in service. For the authentic tourist, food and beverage alone is not an effective factor and other cultural elements in the destination are more effective (Kurt & Arslan Ayazlar 2021).

Gastro tourists' experience components; product-based experiences (quality, price, cleanliness, recognition, story, naturalness, diversity, originality, distinctiveness, presentation), relationship-based experiences (relations with local people, relations with the staff working in the area, guide, local managers, other tourists), activity-based experiences(educational-tutorial activities, entertaining activities, healthy activities, relaxing-relaxing activities) and experiences based on the gastronomic environment (authentic gastronomic environments, local gastronomic environments, historical gastronomic environments, nostalgic gastronomic environments, fun gastronomic environments, scenic gastronomic environments) (Akyürek & Kutukız 2020).

According to Wolf (2006), the factors that influence the choice of destination for gastro tourists are; watching a famous chef's cooking performance, having a dining experience in a famous restaurant or participating in a special program in the restaurant, participating in food and beverage festivals and gastronomic events held in the destination, researching local food ingredients, going to wine production regions, watching the process and attending cooking courses.

Gastronomy tours, which positively affect the destination choices of gastro tourists, increase product diversity in destinations. Gastronomy tours include gastronomy-based events and activities (festivals, fairs, etc.), visits to production and harvest areas, museum visits, conversations about gastronomy history, product, harvest, etc., competitions and shows (Gövce et al., 2018; Çağlı, 2012). A wide range of activities ranging from experiencing products produced in the field, vineyard or garden, eating in prestigious restaurants or receiving special culinary training from famous chefs attracts interest.

Destination-specific, local dishes and flavors have a positive effect on attracting tourists to the destination and on gastronomy image (Stepchenkova & Mills, 2010; Üzülmez 2021). Likewise, festivals and events affect the image of the destination. It is indirectly effective in increasing the number of gastro tourists (Baydeniz et al. 2023). It also affects the length of stay in the destination, the satisfaction levels of the visits and the intention to visit the destination again (Çakır Keleş & Özkaya, 2022). This is not the case for gastro tourists who have a negative attitude towards local food for neophobia or other reasons.

The presentation of culinary culture with its traditional features in the menus of hotels and restaurants in the destination, the knowledge, attitudes and behaviors of tour guides about the local food and beverages of the destination, organizations and activities that will allow gastro tourists to try the food of that region are beneficial for the development of gastronomy tourism and play a meaningful role in destination choices (Şen & Aktaş, 2017; Kivela & Crotts 2006).

Local food is an attraction factor that increases the motivation to visit for gastro tourists by serving and consuming the gastronomic product obtained by using locally specific materials, equipment and techniques in a locally specific way (Nisari & Sakin Yılmazer, 2018).

Food-centered activities in a destination include traditional or high quality restaurants, food and wine festivals, cooking schools, wine education, local markets, culinary and food competitions or events, wineries and vineyards, vegetable and fruit picking areas, food and wine routes, food and wine routes, hawkers and farms (Kivela & Crotts, 2005; 2006; Karim & Chi 2010).

Food and beverage, even food, is the most important part of the tourism sector. Its impact on tourists is huge. It causes competition between countries, cities and even small holiday regions. Because its characteristics such as being valuable, rare, non-substitutable and inimitable have made local gastronomy and food a strategic resource for destinations (Yılmaz, 2017). Destinations' gastronomic image, gastronomic diversity/richness, local cuisine, gastronomic events (festivals, culinary museums, cooking courses, etc.), gastronomic products and food and beverage establishments increase the preference of the region.

Conclusion and Evaluation

As a tool that increases tourism movements, food and beverages have an important place in the marketing and development plans of most destinations (Getz 2008). Today, food and beverages have become an integral and important part of tourism development and marketing strategies. Gastro tourists prefer to travel to a destination in order to taste the food that has become famous and integrated with the destination, and to observe the production sites and stages. Tourists who want to have different experiences and gain prestige and status are on the move in the tourism sector, which serves their expectations (intrinsic motivation factors or push factors). The tourism sector, on the other hand, tries to make destinations more attractive (extrinsic motivation factors or pull factors) for reasons such as product differentiation and making them more attractive than others. However, in order to attract gastro tourists to destinations or to create a potential for gastro tourists, destinations need to recognize tourist motivations well.

The gastronomic image, gastronomic diversity/richness, local cuisine, gastronomic events (festivals, culinary museums, cooking courses, etc.), gastronomic products and food and beverage establishments increase the preferability of the region. The gastronomy tourism resources and attraction elements of the destination vary depending on the destination. A tourism destination in a rural area may have many different agricultural products, local foods or farms, while in a big city, restaurants, museums or different gastronomic activities may be more common (Gözce et al. 2018). Therefore, when analyzing the approaches of gastro tourists, the fact that each destination is unique should not be ignored.

References

Akyürek, S. & Kutukız, D. (2020). Gastro Turistlerin Deneyimleri: Gastronomi Turları Kapsamında Nitel Bir Araştırma. *Journal of Tourism and Gastronomy Studies, 8*(4), 3319–3346.

Ambrož, M. & Ovsenik, R. (2011). Tourist Origin and Spiritual Motives. *Management,* 16 (2), 71–86.

Arıcı, S. & Bayram, Ü. (2021). Gastronomi Turizmi ve İlişkili Kavramlar. *Gastronomi Turizmi Kavramlar, İlkeler ve Uygulamalar* İçinde (pp. 25–38). Detay Yayıncılık, Ankara.

Ayaz, N. & Yalı, S. (2017). Kültürel Turistlerin Seyahat Tercihleri ve Yiyecek-İçecek Beklentileri: Safranbolu Örneği, *Türk Turizm Araştırmaları Dergisi*, 1(1), 43–61.

Barkat, S. M. & Vermignon, V. (2006). Gastronomy tourism: A comparative study of two French regions: Brittany and la martinique. Paper presented at Sustainable Tourism with Special Reference to Islands and Small States Conference, Malta.

Baydeniz, E., Kılıcı, L. & Çelik, S., (2023). Yerel Mutfak Algısı Gastro Aktivite ve Gastro Deneyimin Destinasyon Marka İmajına Etkisi: Afyonkarahisar Örneği. Safran Kültür ve Turizm Araştırmaları Dergisi, 6 (1): 133–153.

Bucak T., & Aracı E. (2013). Türkiye'de Gastronomi Turizmi Üzerine Genel Bir Değerlendirme, *Balikesir University The Journal of Social Sciences Institute*, 16(30), 203–216.

Çakır Keleş, M. & Özkaya, F. (2022). Gaziantep'i Ziyaret Eden Yerli Turistlerin Yiyecek Neofobisi ve Çeşitlilik Arayışı Eğilimlerinin Yerel Yemek Tüketimine Etkisi. Afyon Kocatepe Üniversitesi Sosyal Bilimler Dergisi, 24(1), 365–381.

Çağlı, I. B. (2012). Türkiye'de Yerel Kültürün Turizm Odaklı Kalkınmadaki Rolü: Gastronomi Turizmi Örneği. Master Thesis. İstanbul Teknik Üniversitesi, Fen Bilimleri Enstitüsü, İstanbul.

Doğan, H. Z. (2004). Turizmin Sosyo-kültürel Temelleri. Detay Yayıncılık. Ankara.

Fields, K. (2002). *Demand For The Gastronomy Tourism Product: Motivational Factors*. In A. M. Hjalanger & G. Rithards (Eds.), Tourism and Gastronomy (pp. 36–51). London: Routledge.

Fox, R. (2007). Reinventing the Gastronomic Identity of Croatian Tourist Destinations. International Journal of Hospitality Management, 26(3), 546–559.

Getz, D. (2008). Event Tourism: Definition, Evolution, and Research, *Tourism Manangement*, 29: 403–428.

Gövce, M., Özdoğan, O. N. & Şimşek, O. U. (2018). Destinasyon Pazarlamasında Gastronominin Rolü: Bibliyometrik Bir Analiz. International Gastronomy Tourism Studies Congress, Kocaeli.

Guzman, T. L. & Canizares, S. S. (2011). Gastronomy, Tourism and Destination Differantiation: A Case Study in Spain. Review of Economics & Finance, 1, 63–72.

Haven-Tang, C. & Jones, E. (2005). Using Local Food and Drink to Differentiate Tourism Destinations Through a Sense of Place. Journal of Culinary Science & Technology, 4(4), 69–86.

Hjalager, A. M. (2004). What Do Tourists Eat and Why? Towards A Sociology of Gastronomy and Tourism. *Tourism*, 52(2), 195–201.

Hjalager, A. M. ve Johansen, P. H. (2013). Food Tourism in Protected Areas- Sustainability for Producers, The Environment and Tourism? *Journal of Sustainable Tourism*, 21 (3), 417–433.

Karim, S. A. & Chi, C. G. Q. (2010). Culinary Tourism as A Destination Attraction: An Empirical Examination of Destinations' Food İmage. *Journal of Hospitality Marketing & Management*, 19, 531–555.

Kim, S., & Choe, J.Y. (2019). Testing an Attribute-Benefit-Value-Intention (ABVI) Model of Local Food Consumption as Perceived by Foreign Tourists. *International Journal of Contemporary Hospitality Management, 31*(1), 123–140.

Kivela, J., & Crotts, C. J. (2006). Tourism and Gastronomy: Gastronomy's Influence on How Tourists Experience a Destination. Journal of Hospitality & Tourism Research, 30(3), 354–377.

Kivela, J., & Crotts, J.C. (2005). Gastronomy tourism. *Journal of Culinary Science & Technology*, 4(2–3), 39–55.

Kozak, N., Kozak, M. A. & Kozak, M., (2017). Genel Turizm, Detay Yayıncılık, Ankara.

Kurt, G. & Arslan Ayazlar, R. (2021). Gastronomi Turist Taksonomisi, Türk Turizm Araştırmaları Dergisi, 5(2): 1280–1298.

Long, L. (Ed.) (2004). *Culinary Tourism*. Lexington: The University Press of Kentucky.

Mak, A. H., Lumbers, M., Eves, A. & Chang, R. C. (2012). Factors Influencing Tourist Food Consumption. *International Journal of Hospitality Management,*31(3), 928–936.

Mitchell, R. & Hall, M., (2003). Tourism As A Force For Gastronomic Globalization And Localization. In Tourism and Gastronomy (pp. 85–102). Routledge.

Nisari, M. A. & Sakin-Yılmazer, M. (2018). Yerel Yemeklerin Ziyaretçi Motivasyonuna Etkisi. *International Journal of Contemporary Tourism Research* 1,68–77.

Okumuş, B., Okumuş, F., & McKercher, B. (2007). Incorporating Local and International Cuisines in The Marketing Of Tourism Destinations: The Cases Of Hong Kong and Turkey. Tourism Management, 28(1), 253–261.

Quan, S. & Wang, N. (2004). Towards a Structural Model of the Tourist Experience: An Illustration from Food Experiences in Tourism. *Tourism Management, 25*(3), 297–305.

Safarov, B., Mirzaev, K., Janzakov, B. & Ruzibayev, O. (2022). A Study on the Impact of Distance and Income on Potential Gastrotourists' Decision-Making Process. African Journal of Hospitality, Tourism and Leisure, 11(6), 2052–2062.

Scarpato, R. (2002). Gastronomy studies in search of hospitality. in CAUTHE 2002: *Tourism and Hospitality on The Edge; Proceedings of The 2002 CAUTHE Conference*, 546. Edith Cowan Universty Press.

Smith, S. L. & Xiao, H. (2008). Culinary Tourism Supply Chains: A Preliminary Examination. *Journal of Travel Research*, 46, 289–299.

Stepchenkova, S., & Mills, J. E. (2010). Destination Image: A Meta-Analysis Of 2000–2007 Research. *Journal Of Hospitality Marketing & Management*, 19(6), 575–609.

Şen, A. & Aktaş, N. (2017). Tüketicilerin Seyahatleri Sırasında Besin Seçimleri, Yöresel Gastronomi Davranışları ve Destinasyon Seçiminde Gastronomi Unsurlarının Rolü: Konya-Karaman Örneği. *Karamanoğlu Mehmet Bey Üniversitesi Sosyal ve Ekonomik Araştırmalar Dergisi* 19(32): 65–72.

Şimşek, A. & Selçuk, G. (2018). Gastro Turistlerin Tipolojisinin Belirlenmesi: Gaziantep Ölçeğinde Bir Uygulama. *Uluslararası Türk Dünyası Turizm Araştırmaları Dergisi*, 3(1), 28–43.

Tellioğlu, S. (2021). Türk ve Alman Turistleri Tatile İten ve Çeken Faktörlerin Analizi, Alanya Akademik Bakış Dergisi, 5(1): 287–299.

Üzülmez, M. (2021). Yöresel Mutfak ile Destinasyon Ve Gastronomi Turizmi Arasındaki İlişkiye Yönelik Bir İnceleme. *Journal of Hospitality and Tourism Issues*, 3(1), 23–36.

Wolf, E. (2006). Culinary Tourism The Hidden Harvest. Lowa: Kendall/Hunt Publishing Company.

Yılmaz, G. (2017). Gastronomi ve Turizm İlişkisi Üzerine Bir Değerlendirme. *Seyahat ve Otel İşletmeciliği Dergisi/ Journal of Travel and Hospitality Management*,14(2), 171–191.

Olca Sezen Dogancili[1] and Ramazan Guzel[2]

Chapter 14 Artificial Intelligence and Post-Luddism in Tourism Industry

Introduction

Artificial intelligence offers online/offline interaction by focusing on both emotions and intelligence of people (Kazak, Chetyrbok, & Oleinikov, 2020). It is thought that artificial intelligence, whose presence in the tourism sector has increased with the development of innovative technologies, will contribute to customer satisfaction by improving efficiency and productivity in the tourism sector. At the same time, it is predicted that some jobs will be automated and made easier by providing tourism businesses with the opportunity to take a more active role in business processes (Samala, Katkam, Bellamkonda, & Rodriguez, 2022).

The simplification of tasks that people see as laborious will help people to fulfill their tasks more efficiently as they become simpler with technological innovations. In the tourism sector, artificial intelligence applications have found their place in various areas such as wireless network technologies, souvenir design in museums (Ding, 2022), space tourism (Kim, Hall, Kwon, & Sohn, 2024), providing personalization in services, making recommendations, facilitating operations with robot systems, creating speech systems, developing smart travel agencies, language translation applications, and natural language processing technology (Bulchand-Gidumal, 2020).

Artificial Intelligence in Tourism Industry

The use of artificial intelligence in the tourism industry will enhance service quality, customer satisfaction, and customer loyalty. Artificial intelligence applications that enable the determination of the criteria for tourists to assess service quality (Prentice, Lopes, & Wang, 2020) are based on targeting tourists to provide personalized experiences, providing support on various issues, and

---

1 Assoc. Prof. Dr., Sinop University, Faculty of Tourism, Department of Tourism Guidance, o.dogancili@sinop.edu.tr
2 Lecturer, Sinop University, School of Foreign Languages, rguzel@sinop.edu.tr

collecting data on tourist behaviors and preferences during this process (Pisoni, Díaz-Rodríguez, Gijlers, & Tonolli, 2021). This data provides information on substantial matters such as analyzing tourists, analyzing trends in the customer market, and contributing content recommendations for future planning (Sabili, Khalisa, & Dewi, 2024).

The obtained data enables the analysis of tourists, analysis of trends in the customer market, and providing content recommendations for future planning(Sabili, Khalisa, & Dewi, 2024). This information is crucial in understanding and analyzing tourists, identifying customer preferences and behaviors, and making informed decisions for future marketing strategies. By utilizing artificial intelligence applications, businesses can gain valuable insights from the data collected, allowing them to tailor their offerings and enhance the overall customer experience in the tourism sector.

Applications of artificial intelligence in tourism are categorized into five types: Search/Booking Engines, AR/VR Devices, Kiosks/Self-service Screens, Robots/Autonomous Vehicles and Virtual Assistants/Chatbots (Huang, Chao, Velasco, Bilgihan, & Wei, 2022).

Technological elements used in place of human workforce or to assist them in work processes (Kim, Kim, Badu-Baiden, Giroux, & Choi, 2021) such as preventive machine learning for work accidents, personal customer service, and chat and messaging robots have found their place in the tourism industry (Citak, Owo, & Weichbroth, 2021). In addition to the service robots that have emerged in the forefront in the last 10 years, technologies such as AI-supported self-service kiosks, chatbots, metaversal tourism technologies, ML, and NLP contribute value to the tourism sector (Solakis, Katsoni, Mahmoud, & Grigoriou, 2022).

There are different perspectives on the use of artificial intelligence in the tourism sector. While some people have positive attitudes towards receiving travel recommendations from artificial intelligence and implementing these routes (Martin, et al., 2020), it should be stated that there are others who argue that artificial intelligence will negatively affect the career competency of the human workforce, reduce organizational commitment, and adversely impact job burnout (Kong, et al., 2021). There are also concerns about the negative outcomes of using artificial intelligence, such as the loss of job and autonomy due to robotics, reliability and privacy concerns (Limna, 2023), and the potential social isolation (Grundner & Neuhofer, 2021).

In hospitality businesses, the replacement or support of service providers such as chefs and bartenders with artificial intelligence attracts customers' attention, acting as an element encouraging tourists to make purchases (Kashem, Shamsuddoha, Nasir, & Chowdhury, 2022). The distinct use of artificial intelligence in terms of tourists' intention to revisit the businesses and the creation of customer satisfaction makes it more feasible to reach excellent levels of service (Mariani & Borghi, 2021; Limna, 2023). In this context, in the study by Gupta et al. (2022), artificial intelligence robot activities are categorized under automation, data collection, personalization, and uninterrupted service offering in the context of experiences in smart city applications.

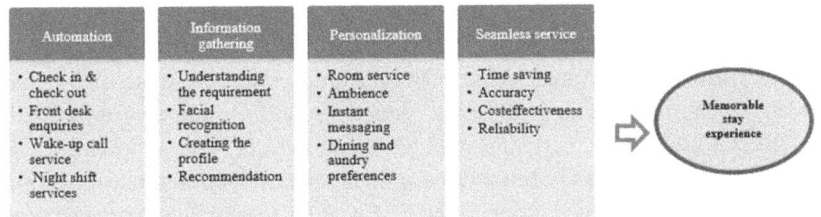

Figure 14.1. Examples of Artificial Intelligence Activities in Smart City Applications.
Source: Gupta, Modgil, Lee, Cho, & Park, 2022.

Starting from reservation processes in hospitality businesses, it can be seen that technology takes place in many processes, from shaping pre-accommodation services (Buhalis & Moldavska, 2022) to the use of generative artificial intelligence like Chat GPT (Dwivedi, Pandey, Currie, & Micu, 2024). This integration of technology enriches the fulfillment of human roles by leveraging the power of artificial intelligence and automation, while also storing the elements of tourists' desires and requests. For example, the Hilton Hotels utilize an AI-powered robot named "Connie" to interact with tourists and provide the necessary information. At Henna Hotel in Japan, service robots are also used to interact with tourists in tasks such as reception procedures and room assistance (Solakis, Katsoni, Mahmoud, & Grigoriou, 2022). Another example is the use of the Aloft application, which allows individuals to control various aspects of their rooms through voice commands, such as adjusting the lighting color, playing a video from YouTube, watching a movie from Netflix, and setting the room temperature. These are examples of artificial intelligence applications used in hospitality establishments (Bisoi, Roy, & Samal, 2020).

The use of artificial intelligence technology in the food and beverage industry is seen as an increase in productivity and a decrease in repair costs and labor requirements. Particularly in the food and beverage production stage, neural networks (NNs) and machine learning (ML) technologies have made it possible to optimize time, labor, and expertise (Oliveira, Coppola, & Vermesan, 2021).

Innovative methods involving artificial intelligence are also employed in processes such as packaging and delivery of the produced goods, in addition to food processing (Dani, Rawal, Bagchi, & Khan, 2022). Artificial intelligence applications have also found their place in ensuring food safety and establishing waste management systems, contributing to more acceptable and healthy production environments for both customers and employees (Kumar, Rawat, Mohd, & Husain, 2021)

Artificial intelligence applications such as QR, facial recognition applications, self-ordering kiosks, robot chefs (Wei & Simay, 2023), ordering and payment systems with voice-activated technology (Cheong, Seah, Loh, & Loh, 2021) are available to increase cooking efficiency, reduce service time and ease of payment.

Artificial intelligence activities applied in the field of food and drinks have many advantages such as product development, ensuring operational efficiency, increasing customer satisfaction (Wu & Ku, 2024), enabling more accurate forecasting in planning, saving staff from repetitive tasks (Groene & Zakharov, 2024), minimizing waste, ensuring quality control (Wahab & Nor, 2023), and increasing productivity in marketing texts (Kuang, Lim, Tan, Ho, & Husaini, 2024). However, there are also disadvantageous aspects such as the high cost of artificial intelligence applications, difficulties in providing qualified personnel to operate and maintain the technology, and experiencing barriers in direct communication with customers (Abass, Zohry, & Soliman, 2022).

It is seen that artificial intelligence has achieved wide social impacts by providing personalization in tourists' experiences in destination trips and taking place in many areas of tourism (Islam, 2023). Smart tourism technologies also provide services to museum visitors and these applications should be considered as a new type of service in museums rather than supporting existing services (Yang & Zhang, 2022). For example, with the interactive guide robot named "Doris", applications in various modules were presented to tourists by providing a certain level of human-machine interaction (Vásquez & Matía, 2020).

In travel agencies, the use of artificial intelligence with the PNR (Passenger Name Record) database allows for the automation of reservation processes and enables the use of information derived from created reservations and cancellation requests as a resource (Sáncheza, Sánchez-Medinaa, & Pellejero, 2020).

Post-Luddism in the Tourism Sector

Post-Luddism is defined as "the fear of losing one's job related to artificial intelligence and even the tendency to harm artificial intelligence" (Anadolu Agency

Website). Since artificial intelligence applications, which are frequently preferred in many areas in the tourism sector such as reception services, food and beverage management, are seen to be indistinguishable from the online representative interview of customers (Rauf, Zurcher, Pantelidis, & Winbladh, 2022), the literature on post-Luddism in the tourism sector is discussed

It is known that the use of artificial intelligence in the tourism sector has a direct and indirect impact on job insecurity, job commitment, and intentions to leave among the working human force (Koo, Curtis, & Ryan, 2021). However, it is also stated that determining and meeting not only the needs and expectations of the customers but also the employees by using artificial intelligence technologies will also bring increases in job performance (Limna, 2023).

While it is acknowledged that there is a risk of artificial intelligence technology replacing many jobs - especially low skilled service tasks- in the tourism industry (Li, Bonn, & Ye, 2019), it is also mentioned that the goal is not to completely replace the human workforce in the tourism sector, but rather to take over their monotonous tasks, create opportunities to enhance employee productivity, and increase job enrichment by increasing job responsibilities (Dwivedi, Pandey, Currie, & Micu, 2024).

Results

Although the concept of post-luddism has not yet been included in the literature since it has recently been introduced to the academic literature, existing researches talk about the proper and correct use of artificial intelligence technology to support the working manpower and undertake monotonous jobs. Even though it is stated that the loading of standard jobs to artificial intelligence has a positive effect on the productivity and work performance of the human power, it also creates job insecurity in people due to the fact that it undertakes some jobs. In this context, firstly, in-depth interviews should be conducted with employees regarding the use of artificial intelligence in different areas of the tourism sector, then, based on the data obtained, some managerial suggestions should be presented to business managers.

References

Abass, M. N., Zohry, M. A.-F., & Soliman, S. G. (2022). The possibility of using robot as one of the artificial intelligence techniques in the food and beverage department in five-star hotels: Managers' perspective. *Research Journal of the Faculty of Tourism and Hotels Mansoura University, 11*(2), 857–907.

Anadolu Ajansı Web Sitesi (2024). URL: https://www.aa.com.tr/tr/bilim-teknol oji/yapay-zeka-nedeniyle-isini-kaybetme-korkusunu-post-luddism-olarak-adlandirdilar/3174169 (Erişim Tarihi: 16.04.2024).

Bisoi, S., Roy, M., & Samal, A. (2020). Impact of artificial intelligence in the hospitality industry. *International Journal of Advanced Science and Technology, 29*(5), 4265–4276.

Buhalis, D., & Moldavska, I. (2022). Voice assistants in hospitality: using artificial intelligencefor customer service. *Journal of Hospitality and Tourism Technology, 13*(3), 386–403. doi:10.1108/JHTT-03-2021-0104.

Bulchand-Gidumal, J. (2020). Impact of artificial intelligence in travel, tourism and hospitality. Z. Xiang, M. Fuchs, U. Gretzel, & W. Höpken içinde, *Handbook of e-Tourism* (s. 1943–1962). SpringerLink.doi:https://doi.org/10.1007/978-3-030-05324-6_110-1.

Cheong, Y. S., Seah, C. S., Loh, Y. X., & Loh, L. (2021). Artificial Intelligence (Ai) In the food and beverage industry: Improves the customer experience. *2nd International Conference on Artificial Intelligence and Data Sciences (AiDAS)*, (s. 1–6). IPOH, Malaysia. doi:10.1109/AiDAS53897.2021.9574261.

Citak, J., Owo, M. L., & Weichbroth, P. (2021). A note on the applications of artificial intelligence in the hospitality industry: Preliminary results of a survey. *Procedia Computer Science*(192), 4552–4559. doi:10.1016/j.procs.2021.09.233.

Dani, R., Rawal, Y. S., Bagchi, P., & Khan, M. (2022). Opportunities and challenges in implementation of artificial intelligence in food & beverage service industry. *International Conference on Advancements in Engineering and Sciences (ICAES2021).*

Ding, L. (2022). Research on innovation in the design of museum tourist souvenirs based on artificial intelligence. *International Journal of Intelligent Systems and Applications in Engineering, 10*(2s), 54–58.

Dwivedi, Y. K., Pandey, N., Currie, W., & Micu, A. (2024). Leveraging ChatGPT and other generative artificial intelligence (AI)-based applications in the hospitality and tourism industry: Practices, challenges and research agenda. *International Journal of Contemporary Hospitality Management, 36*(1), 1–12. doi:10.1108/IJCHM-05-2023-0686.

Groene, N., & Zakharov, S. (2024). Introduction of AI-based sales forecasting: How to drive digital transformation in food and beverage outlets. *Discover Artificial Intelligence, 4*(1), 1–17. doi:https://doi.org/10.1007/s44 163-023-00097-x.

Grundner, L., & Neuhofer, B. (2021). The bright and dark sides of artificial intelligence: A futures perspective on tourist destination experiences. *Journal of Destination Marketing & Management*(19). doi:https://doi.org/10.1016/j.jdmm.2020.100511.

Gupta, S., Modgil, S., Lee, C.-K., Cho, M., & Park, Y. (2022). Artificial intelligence enabled robots for stay experience in the hospitality industry in a smart city. *Industrial Management & Data Systems, 122*(10), 2331–2350. doi:10.1108/IMDS-10-2021-0621.

Huang, A., Chao, Y., Velasco, E. d., Bilgihan, A., & Wei, W. (2022). When artificial intelligence meets the hospitality and tourism industry: An assessment framework to inform theory and management. *Journal of Hospitality and Tourism Insights, 5*(5), 1080–1100. doi:10.1108/JHTI-01-2021-0021.

Islam, M. S. (2023). Decoding tourist experiences in the digital age: An introductory guide to conducting interviews in smart tourism research. *Smart Tourism, 4*(2), 1–13. doi:10.54517/st.v4i2.2533.

Kashem, M. A., Shamsuddoha, M., Nasir, T., & Chowdhury, A. A. (2022). The role of artificial intelligence and blockchain technologies in sustainable tourism in the Middle East. *Worldwide Hospitality and Tourism Themes, 14*, 178–191. doi:0.1108/WHATT-10-2022-0116.

Kazak, A. N., Chetyrbok, P. V., & Oleinikov, N. N. (2020). Artificial intelligence in the tourism sphere. *IOP Conf. Series: Earth and Environmental Science*(421), 1–6. doi:10.1088/1755-1315/421/4/042020.

Kim, M. J., Hall, C. M., Kwon, O., & Sohn, K. (2024). Space tourism: Value-attitude-behavior theory, artificial intelligence, and sustainability. *Journal of Retailing and Consumer Services*(77), 1–12.

Kim, S., Kim, J., Badu-Baiden, F., Giroux, M., & Choi, Y. (2021). Preference for robot service or human service in hotels? Impacts of the COVID-19 pandemic. *International Journal of Hospitality Management*(93),1–12. doi:https://doi.org/10.1016/j.ijhm.2020.102 795.

Kong, H., Yuan, Y., Baruch, Y., Bu, N., Jiang, X., & Wang, K. (2021). Influences of artificial intelligence (AI) awareness on career competency and job burnout. *International Journal of Contemporary Hospitality Management, 33*(2), 717–734. doi:10.1108/IJCHM-07-2020-0789.

Koo, B., Curtis, C., & Ryan, B. (2021). Examining the impact of artificial intelligence on hotel employees through job insecurity perspectives. *International Journal of Hospitality Management*(95), 1–12. doi:https://doi.org/10.1016/j.ijhm.2020.102763.

Kuang, A. C., Lim, T. M., Tan, C. W., Ho, C. F., & Husaini, N. A. (2024). AI Ads: Practicability of text generation for F&B Marketing. *Journal of Logistics, Informatics and Service Science, 11*(2), 324–345. doi:10.33168/JLISS.2024.0220.

Kumar, I., Rawat, J., Mohd, N., & Husain, S. (2021). Opportunities of artificial intelligence and machine learning in the food industry. *Hindawi Journal of Food Quality*, 1–10. doi:https://doi.org/10.1155/2021/4535567.

Li, J., Bonn, M., & Ye, B. (2019). Hotel employee's artificial intelligence and robotics awareness and its impact on turnover intention: The moderating roles of perceived organizational support and competitive psychological climate. *Tourism Management*(73), 172–181. doi:https://doi.org/10.1016/j.tourman.2019.02.006.

Limna, P. (2023). Artificial Intelligence (AI) in the hospitality industry: A review article. *International Journal of Computing Sciences Research*(7), 1306–1317. doi:https://doi.org/10.25147/ijcsr.2017. 001.1.103.

Mariani, M., & Borghi, M. (2021). Customers' evaluation of mechanical artificial intelligence in hospitality services: A study using online reviews analytics. *International Journal of Contemporary Hospitality Management, 33*(11), 3956–3976. doi:10.1108/IJCHM-06-2020-0622.

Martin, B. A., Jin, H. S., Wang, D., Nguyen, H., Zhan, K., & Wang, Y. X. (2020). The influence of consumer anthropomorphism on attitudes towards artificial intelligence trip advisors. *Journal of Hospitality and Tourism Management*, 108–111. doi:https://doi.org/10.1016/j.jhtm.2020.06.004.

Oliveira, R. O., Coppola, M., & Vermesan, O. (2021). AI in Food and Beverage Industry. O. Vermesan içinde, *Artificial Intelligence for Digitising Industry – Applications* (s. 251–259). New York: River Publishers.

Pisoni, G., Díaz-Rodríguez, N., Gijlers, H., & Tonolli, L. (2021). Human-centered artificial intelligence for designing accessible cultural heritage. *Applied Sciences, 11*(2), 1–30. doi:https://doi.org/10.3390/app11020870.

Prentice, C., Lopes, S. D., & Wang, X. (2020). The impact of artificial intelligence and employee service quality on customer satisfaction and loyalty. *Journal of Hospitality Marketing & Management, 29*(7), 739–756. doi:https://doi.org/10.1080/19368623.2020. 1722304.

Rauf, A., Zurcher, M., Pantelidis, I., & Winbladh, J. (2022). Millennials' perceptions of artificial intelligence in hotel service encounters. *Consumer Behavior in Tourism and Hospitality, 17*(1), 3–16. doi:10.1108/CBTH-04-2021-0104.

Sabili, A. A., Khalisa, A., & Dewi, R. K. (2024). The effect of applying artificial intelligence in brand marketing strategies to improve company effectiveness and efficiency (case study: j.co coffee & donuts Indonesia). *Jurnal Scientia, 13*(1), 317–330. doi:10.58471/ scientia.v13i01.

Samala, N., Katkam, B. S., Bellamkonda, R. S., & Rodriguez, R. V. (2022). Impact of AI and robotics in the tourism sector: A critical insight. *Journal of Tourism Futures, 8*(1), 73–87. doi:10.1108/JTF-07-2019-0065.

Sáncheza, E. C., Sánchez-Medinaa, A. J., & Pellejero, M. (2020). Identifying critical hotel cancellations using artificial intelligence. *Tourism Management Perspectives*(35), 1–8. doi:https://doi.org/10.1016/j.tmp.2020.100718.

Solakis, K., Katsoni, V., Mahmoud, A. B., & Grigoriou, N. (2022). Factors affecting value co-creation through artificial intelligence in tourism: A general literature review. *Journal of Tourism Futures*, 1–15. doi:https://doi.org/10.1108/JTF-06-2021-0157.

Vásquez, B. P., & Matía, F. (2020). A tour-guide robot: Moving towards interaction with humans. *Engineering Applications of Artificial Intelligence*(88), 1–17. doi:https://doi.org/10.1016/j.engappai.2019. 103356.

Wahab, N. A., & Nor, R. B. (2023). The role of artificial intelligence in waste reduction in the beverage industry: A comprehensive strategy for enhanced sustainability and efficiency. *AI, IoT and the Fourth Industrial Revolution Review, 13*(11), 1–8.

Wei, Y., & Simay, A. E. (2023). AI adoption in the Chinese food and beverage industry: An exploratory study. *FIRM Journal of Management Studies, 8*(2), 145–160. doi:10.33021/firm.v8i2.4412.

Wu, S.-H., & Ku, E. C. (2024). Aligning restaurants and artificial intelligence computing of food delivery service with product development. *Journal of Hospitality and Tourism Technology*, 1–18. doi:10.1108/JHTT-10-2023-0322.

Yang, X., & Zhang, L. (2022). Smart tourism technologies towards memorable experiences for museum visitors. *Tourism Review, 77*(4), 1009–1023. doi:10.1108/TR-02-2022-0060.

Duygu Doğan[1] and Hasan Köşker[2]

Chapter 15 The Role of Effective Communication in Crisis Management within the Tourism Sector

Introduction

The tourism sector is inherently vulnerable to a range of external effects, which can occasionally lead to crises within the industry. These crises manifest in various forms, such as natural disasters, pandemics, and political instabilities, and can deeply impact the operations of tourism businesses and destinations. Effective crisis management is, therefore, an essential management discipline within the tourism sector. However, the success of crisis management is largely dependent on effective communication strategies.

A crisis is defined as:

> an unpredictable event that threatens important expectancies of stakeholders related to health, safety, environmental, and economic issues, which can seriously impact an organisation's performance and generate negative comments (Coombs, 2007, p. 2–3).

A crisis represents a significant event that can have negative outcomes for an organization, affecting its public image, products, services, or reputation. It disrupts normal business operations and can endanger an organization's survival. Crises that countries face can arise from various causes, including economic, political, and health-related issues (Ateş & Baran, 2020). According to Doern et al., (2019), crises are often difficult to control and foresee, requiring organizations to make quick and effective decisions. Fearn (2007) stated that "A crisis can be a strike, terrorism, a fire, a boycott, product tampering, a product failure, or numerous other things". If not managed properly, crises can cause severe damage to organizations and lead to adverse outcomes (Özcan, 2021).

This section addresses the nature of crises in the tourism sector, the role of communication in crisis situations, and how communication should be managed

---

1   Lecturer, Zonguldak Bülent Ecevit University, Karadeniz Ereğli Tourism Faculty, duyguttnc@beun.edu.tr
2   Ass. Prof., Zonguldak Bülent Ecevit University, Karadeniz Ereğli Tourism Faculty, Department of Tourism Management, hasankosker@beun.edu.tr

during and after crises. It also explores how communication can act as a savior during crisis moments and serve as a strategic tool in the post-crisis recovery process. Understanding the role of communication in crisis management is not only a theoretical necessity but also a practical requirement for tourism professionals.

Crisis in the Tourism Sector

A tourism crisis refers to any event that poses a threat to the regular functioning of tourism-related businesses. It can harm the overall reputation of a tourist destination in terms of safety, attractiveness, and comfort, leading to negative perceptions among visitors. Consequently, this can result in a decline in the local travel and tourism economy, disrupting business operations for the industry due to a decrease in tourist arrivals and spending (Sönmez & Allen, 1994, p. 22). Köşker (2017) described a tourism crisis as a chaotic situation that disrupts the normal course of the tourism sector, raises concerns about the safety and security of tourists in the area, and leads to a decrease in tourism demand and revenues.

The tourism sector can be subjected to various crises, including natural disasters, terrorist incidents, socio-cultural tensions, legal restrictions, political and economic instabilities, internal management errors and financial issues (Uyar et al., 2020). These crises can be natural or human-made (Pforr & Hosie, 2008). The global nature of the tourism sector makes it sensitive to crises (Ritchie, 2004; Zhong et al., 2021; Maxim & Morrison, 2022). A crisis can impact not only the country or region where it occurs but also the entire world. Consequently, the extent of the impact on tourism demand varies based on the magnitude of the crisis.

In the past 20 years, events such as the September 11 attacks in 2001, the SARS outbreak in 2003, the MERS outbreak in 2015, and the COVID-19 pandemic in 2019 have demonstrated how vulnerable the tourism sector is to crises worldwide (Hidayat et al., 2023). The fact that a crisis emerging in any part of the globalized world can pose a threat to the tourism sector necessitates the continuous preparedness of the industry for crises (Pforr & Hosie, 2008). Crises, such as travel restrictions, natural disasters, or health-related issues, directly affect the tourism industry, deeply shaking tourist flows, investments, and the overall economic situation (Henderson, 2007). These threats, which include not only infectious diseases but also social events (Zhong et al., 2021), disrupt the operations of tourism-related businesses and significantly impact the economic, social, environmental, and cultural structures of destinations (Pongsakornrungsilp et al., 2021).

Crisis Management

According to Çamdereli (2004), crisis management can be defined as a conscious effort to protect institutional identity and reputation and to avoid negative impacts on communication with the target audience in the face of unexpected situations that have the potential to produce adverse outcomes. This emphasizes the importance of strategic and measured actions by organizations during crises. The scope of crisis management includes crisis prevention, crisis preparedness, crisis response, and crisis revision (Hosie & Smith, 2004). Crisis management also encompasses risk management, as a crisis occurs when risks are not managed properly and effectively. For example, if tourism providers do not pay attention to risk management, they can endanger tourists' lives (Wut et al., 2021). Effective crisis management is essential for handling these situations.

Crisis management has two fundamental aspects: preventing the crisis from occurring and managing it positively for the organization when it does occur (Canöz & Öndoğan, 2015). The role of leading managers is critical in crisis management. These leaders are responsible for making correct decisions during a crisis, guiding teams, and turning adverse situations into opportunities. As Erol (2022) pointed out, the mission of leading managers in this process is to minimize the impacts of the crisis and ensure that the organization emerges stronger. How crises are handled directly affects whether organizations become stronger or fail. Therefore, organizations must manage these processes effectively and use their success to improve their positions and experiences (Bal, 2023). Although crises are often perceived as events that produce negative outcomes, each crisis situation can also present opportunities. As Tüz (2001) noted, crises not only create threats and problems but also offer significant opportunities for organizational development. The nature of the crisis necessitates effective communication strategies to manage perceptions and minimize damage. This highlights the critical importance of selecting the right communication tools based on the crisis's scope, target audience, and message sensitivity (Noel, 1987).

Good crisis management allows for identifying weaknesses and making improvements in these areas. The extraordinary effort required to resolve a crisis can reveal different capabilities, leading to change and transformation and creating a learning opportunity (Bal, 2023). A crisis can have serious consequences for a company's credibility and operations due to its potential impacts, creating a significant public interest situation for the industry or organization (Noel, 1987). During a crisis, survival is the primary concern for all businesses. Many businesses learned valuable lessons from the 2002 SARS outbreak and recognized the vital importance of crisis management, including developing a recovery plan (Soemodinoto et al., 2001; Hiemstra & Wong, 2002).

Crisis management is divided into three stages: pre-crisis, crisis moment (with three sub-stages: emergency; interim; relief), and post-crisis (with two sub-stages: recovery; learning process) (Pira & Sohodol, 2004; Filiz, 2007; Pongsakornrungsilp et al., 2021). The pre-crisis stage focuses on planning and preparation to prevent a crisis before it occurs, based on lessons learned from past events and being ready for potential future crises. The crisis stage involves actively managing and mitigating the crisis's effects through immediate action. This stage is characterized by three sub-stages: emergency, when initial support is needed; interim, when the situation is managed; and relief, when efforts are made to normalize. The post-crisis stage, called recovery and learning, involves focusing on strategies to regain losses, conducting post-action research to manage information and resources better, and improving future crisis management.

This process involves dealing with situations that could jeopardize an organization's values, vision, and mission. Crises vary depending on the organization's characteristics and the field in which it operates, so a standard solution is not possible. Each crisis requires knowledge and experience from different disciplines, such as law, sociology, psychology, and engineering. Additionally, because crisis management during crises requires skills and knowledge different from daily operations, it is considered a special management form. During crises, instead of working longer hours, more intelligent and efficient work methods should be adopted (Filiz, 2007). An effective crisis management process offers the opportunity to see and evaluate opportunities within the crisis. Taking the right steps during a crisis not only minimizes damage but also protects and strengthens the organization's image, re-establishing its reputation in the eyes of the public (Coombs, 1999).

The Role of Communication in Crisis Management in the Tourism Sector

The term "communication" in the context of crisis management refers to proactive communication with stakeholders with the goal of minimising the degree of damage done to the organization's image or reputation and reducing the impact on those who are affected (Dhanesh & Sriramesh, 2018, p. 207). In today's world, factors such as globalization and the widespread use of information technologies have made modern crises more visible. International companies must effectively communicate not only with stakeholders and employees during crises but also with the general public. Therefore, businesses need not only crisis response plans but also effective communication plans that they can use during crises (Çakmak, 2019, p. 47). Crisis communication aims not only to prevent negative events but also to mitigate their impact, playing a crucial role in managing and resolving crises effectively. This process includes strengthening organizations, educating people on managing crises, helping society understand risks, improving response actions, sharing important messages, building trust, and changing perspectives on handling crises. It also involves businesses sharing their values and rules, building trust, and finding new business partners (Wong et al., 2021).

Dhanesh and Sriramesh (2018, p. 204–205) discussed two important approaches to crisis management: SCCT (Situational Crisis Communication Theory) and Image Restoration Theory. These theories emphasize the importance of communication with stakeholders, aiming to minimize the impact on affected parties and the damage to the organization's image. They also consider macro-environmental variables.

Regarding the role of the media in crisis communication, messages in the media, whether print or new media, can act as triggers in a crisis, intensifying it. These messages, stored and not deleted on the internet, can have negative impacts on the reputation of institutions, organizations, and businesses even after the crisis is over (Dhanesh & Sriramesh, 2018, p. 207). Senić & Marinković (2016) emphasized that during a crisis, an organization must respond proactively to maintain control over the information sent to the general public; otherwise, the information vacuum can be filled with misinformation, causing lasting damage to the organization's or destination's image. Therefore, fast and clear communication is crucial during the post-crisis reconstruction process, ensuring that stakeholders interact and work in harmony (Mair et al., 2016; Bremser et al., 2018; Ramkissoon, 2023).

Publications are expected to meet people's information needs during crises, and television, often considered a reliable information source by the wider community, emerges as an important communication tool with both positive and negative aspects (Koluman et al., 2024). The primary goal of crisis communication is to navigate the crisis effectively, mitigate negative perceptions, and foster a positive image for the institution, organization, or business involved. Effective crisis communication strategies play a crucial role in maintaining institutions' continuity and reputation, giving them an advantage over competitors (Aydın, 2020). Effective communication during the post-crisis period is a critical tool for destination development and image restoration (Xie et al., 2022) and is important for informing people far from the crisis center and potential tourists about the destination's reliability and the crisis (Oliveira & Huertas, 2019). As Liu et al., (2011) stated, messages used in crisis communication should provide information and contain emotional content. To achieve this, destination management offices (DMOs) should not only inform the public but also emotionally support them through effective communication methods and evoke empathy. This will ensure effective communication and restore the destination's lost reputation.

Established in 1967 in Bangkok, The Association of Southeast Asian Nations, or ASEAN, listed some basic principles for effective and responsible crisis communications in its 2015 "ASEAN Tourism Crisis Communications Manual":

- Do not assume communications responsibility if your organisation is not the most appropriate source.
- Establish credibility with audiences by identifying the relevant channels in advance.
- Respond in the same medium.
- Stay on message.
- Disclosure.
- Accuracy.
- Transparency.
- Honesty.
- Accessibility.
- Know your audience.
- Respond quickly.
- Update frequently.
- Cooperation.
- Stay cool.

Effective crisis communication forms an integral part of comprehensive crisis management strategies. One of the primary goals of crisis management is for institutions, organizations, and businesses to apply predetermined strategies and respond systematically, planned, and methodically to crises they frequently encounter. While the communication tools used in crisis communication vary, the internet is generally considered the most preferred source by society (Seltzer & Mitrook, 2007). This is due to the high visibility and accessibility of messages shared through the media (Ateş & Baran, 2020). The "blog crisis communication model (BMCC)" developed by Jin and Liu in 2010 examined the role of blogs in crisis communication management and how to use blogs effectively during crises (Jin & Liu, 2010). This model played a significant role in forming the SMCC model (Figure 15.1) as a framework for crisis communication management in the changing media environment (Austin et al., 2012).

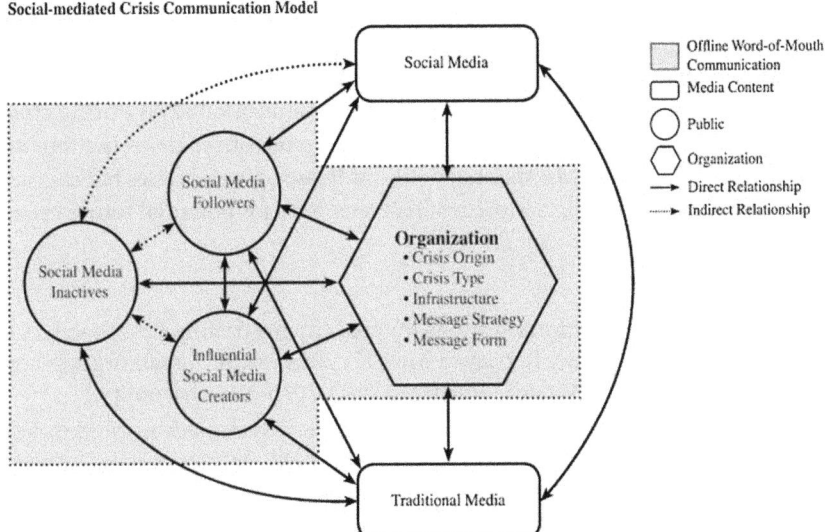

Figure 15.1: Social Mediated Crisis Communication Model.
Reference: Austin et al., (2012).

The Social-Mediated Crisis Communication (SMCC) model, developed by Jin and Liu (2010), is a framework for understanding strategies and interactions for using social media during crises. This model adapts traditional crisis communication theories, such as Situational Crisis Communication Theory (SCCT) by Timothy Coombs and Image Restoration Theory (Benoit, 2015), to the modern social media context, analyzing how organizations communicate during crises, how the public and stakeholders respond, and the role of social media in this process (Austin et al., 2012). For example, research by Wang (2016) highlights how strategic and effective use of social media platforms can significantly enhance reputation and rebuild trust following crises. Maintaining communication with the public can reinforce a sense of security and better prepare the community to support the tourism industry's recovery with coping strategies (Schimmenti et al., 2020; Ramkissoon, 2023).

Results

In conclusion, implementing effective communication strategies in managing crises in the tourism sector is vital. Communication established at the right place and time can reduce uncertainty during crises, prevent adverse situations from worsening, and ensure public and tourist trust. Transparent and clear communication can minimize potential damage from crises. Therefore, investing in education and resources to develop effective communication skills during crises is of great importance for the sustainability and competitiveness of the tourism sector. Effective crisis management not only addresses current crises but also lays the foundation for building a more resilient structure for potential future crises.

References

Asean Tourism Crisis Communications Manual (Incorporating Best Practices of Pata and Unwto) (2015). Retrieved May 17, 2024. https://asean.org/wp-content/uploads/2012/05/ASEAN-Tourism-Crisis-2015-New-Layout.pdf

Ateş, N. B., & Baran, S. (2020). Kriz iletişiminde sosyal medyanın etkin kullanımı: Covid-19 (koronavirüs) salgınına yönelik twitter analizi. *Kocaeli Üniversitesi İletişim Fakültesi Araştırma Dergisi*, 16, 66–99.

Austin, L., Fisher Liu, B., & Jin, Y. (2012). How audiences seek out crisis information: Exploring the social-mediated crisis communication model. *Journal of Applied Communication Research*, 40(2), 188–207.

Aydın, G. (2020). Sosyal medya ve kriz iletişimi. *Selçuk İletişim*, 13(3), 1202–1230.

Bal, F. (2023). Turning the crisis into an opportunity. In Seyhan, M. (Ed.), Studies on Management, Organization, and Strategy (pp. 269–299). Gaziantep: Özgür Publications. https://doi.org/10.58830/ozgur.pub66.c228

Benoit, W. L. (2015). Image restoration theory. Wiley online library. https://doi.org/10.1002/9781405186407.wbieci009.pub2

Bremser, K., Alonso-Almeida, M. D. M., & Llach, J. (2018). Reducing costs or increasing marketing: Strategic suggestions for tourism firms in crisis situations. *Journal of Customer Behaviour*, 17(3), 211–228.

Çakmak, T. F. (2019). Turizm endüstrisinde bütüncül yaklaşımla kriz yönetimi ve örnek olaylar, 1.Baskı. Ankara: Detay Yayıncılık.

Çamdereli, M. (2004). Ana çizgileriyle halkla ilişkiler, İstanbul: Salyangoz Yayınları.

Canöz, K., & Öndoğan, A. (2015). Kriz yönetiminde dönüşümcü liderin rolü. *Gümüşhane Üniversitesi İletişim Fakültesi Elektronik Dergisi*, 3(1), 36–61.

Coombs, W. T. (1999). Information and compassion in crisis responses: A test of their effects. *Journal of Public Relations Research*, 11(2), 125–142.

Coombs, W. T. (2007). *Ongoing crisis communication: Planning. managing and responding*. 4th ed. London: Sage.

Dhanesh, G. S., & Sriramesh, K. (2018). Culture and crisis communication: Nestle India's maggi noodles case. *Journal of International Management*, 24(3), 204–214.

Doern, R., Williams, N., & Vorley, T. (2019). Special issue on entrepreneurship and crises: Business as usual? An introduction and review of the literature. *Entrepreneurship & Regional Development*, 31(5–6), 400–412.

Erol, V. (2022). Kamu yönetimi alanında kriz yönetiminde lider yöneticinin rolü. *Dicle Üniversitesi Sosyal Bilimler Enstitüsü Dergisi*, 30, 344–366.

Fearn-Banks, K. (2007). *Crisis communications: A casebook approach.* New York: Routledge.

Filiz, E. (2007). Türk *kamu yönetiminde kriz yönetimi*. İstanbul: Alfa Akademi Basım Yayım.

Henderson, J. (2007). *Tourism crisis: Causes, consequences and management.* Oxford: Butterwaorth- Heinemann.

Hidayat, M., Pinariya, J. M., Yulianti, W., & Maarif, S. (2023). Crisis communication management in building tourism resilience in Bali Province during the pandemic of Covid-19. *International Journal of Professional Business Review: Int. J. Prof. Bus. Rev.*, 8(7), 29.

Hiemstra S., Wong K. K. (2002). Factors affecting demand for tourism in Hong Kong. *Journal of Travel and Tourism Marketing*, 13(1/2), 41–60.

Hosie, P., & Smith, C. (2004). Preparing for crises with online security management education. *Research and Practice in HRM*, 12(2), 90–127.

Jin, Y., & Liu, B. F. (2010). The Blog-Mediated Crisis Communication Model: Recommendations for Responding to Influential External Blogs. Journal of Public Relations Research, 22(4), 429–455. https://doi.org/10.1080/1062726 1003801420

Koluman, H., Özşirin, S., & Aslan, P. (2024). Kriz anlarında kitle iletişim araçları: Afet dönemlerinde televizyon yayıncılığı üzerine bir araştırma. *Erciyes İletişim Dergisi*, 11(1), 205–225.

Köşker, H. (2017). Krizlerin turizm sektörüne etkileri üzerine bir araştırma: 2016 yılı Türkiye örneği. *Akademik Bakış Uluslararası Hakemli Sosyal Bilimler Dergisi*, 62, 216–230.

Liu, B. F., Austin, L., & Jin, Y. (2011). How publics respond to crisis communication strategies: The interplay of information form and source. *Public Relations Review*, 37(4), 345–353.

Mair, J., Ritchie, B. W., & Walters, G. (2016). Towards a research agenda for post-disaster and post-crisis recovery strategies for tourist destinations: A narrative review. *Current Issues in Tourism*, 19(1), 1–26.

Maxim, C., & Morrison, A. M. (2022). Crisis in the city? A systematic literature review of crises and tourism cities. *Tourism Recreation Research*, 47, 1–13. https://doi.org/10.1080/02508281.2022.2078063

Noel, V. (1987). Communicating in a crisis - choosing the right vehicle. *Industrial Crisis Quarterly*, 1(2), 27–37. https://doi.org/10.1177/108602668700100205

Oliveira, A., & Huertas, A. (2019). How do destinations use twitter to recover their images after a terrorist attack?. *Journal of Destination Marketing & Management*, 12, 46–54.

Özcan, Ş. (2021). *Etkili kriz yönetimi*. Ankara: İksad Yayınevi.

Pforr, C., & Hosie, P. J. (2008). Crisis management in tourism: Preparing for recovery. *Journal of Travel & Tourism Marketing*, 23(2–4), 249–264.

Pira, A., & Sohodol, Ç. (2004). Kriz yönetimi halkla ilişkiler açısından bir değerlendirme. İstanbul: İletişim Yayınları.

Pongsakornrungsilp, S., Pongsakornrungsilp, P., Kumar, V., & Maswongssa, B. (2021). The art of survival: Tourism businesses in Thailand recovering from COVID-19 through brand management. *Sustainability*, 13(12), 6690.

Ramkissoon, H. (2023). Perceived social impacts of tourism and quality-of-life: A new conceptual model. *Journal of Sustainable Tourism*, 31(2), 442–459.

Ritchie, B. W. (2004). Chaos, crises and disasters: a strategic approach to crisis management in the tourism industry. *Tourism Management*, 25(6), 669–683.

Schimmenti, A., Billieux, J., & Starcevic, V. (2020). The four horsemen of fear: An integrated model of understanding fear experiences during the COVID-19 pandemic. *Clinical Neuropsychiatry*, 17(2), 41.

Seltzer, T., & Mitrook, M. A. (2007). The dialogic potential of weblogs in relationship building. *Public Relations Review*, 33(2), 227–229.

Senić, V., & Marinković, V. (2016). Crisis communication in tourism. *Tourism International Scientific Conference Vrnjačka Banja - TISC*, 1(2), 275–290. http://www.tisc.rs/proceedings/index.php/hitmc/article/view/186

Soemodinoto, A., Wong, P. P., & Saleh, M. (2001). Effect of prolonged political unrest on tourism. *Annals of Tourism Research*, 28(4), 1056–1060.

Sönmez, S. F., & Allen, L. R. (1994). *Managing tourism crises: A guidebook*. Department of Parks, Recreation and Tourism Management, Clemson University.

Tüz, M.V. (2001). *Kriz ve işletme yönetimi*. Alfa Yayınları, İstanbul.

Uyar, H, Solmaz, A, S, & Kasapoğlu, C. (2020). Turizm endüstrisinde pazarlama iletişimi in Solmaz S. A. (Ed),Turizm işletmelerinde halkla ilişkiler. Ankara: Nobel Yayıncılık, pp. 143–217

Wang, Y. (2016). Brand crisis communication through social media: A dialogue between brand competitors on Sina Weibo. *Corporate Communications: An International Journal*, 21(1), 56–72.

Wong, I. A., Ou, J., & Wilson, A. (2021). Evolution of hoteliers' organizational crisis communication in the time of mega disruption. *Tourism Management*, 84, 104257.

Wut, T. M., Xu, J. B., & Wong, S. M. (2021). Crisis management research (1985–2020) in the hospitality and tourism industry: A review and research agenda. *Tourism Management*, 85, 104307.

Xie, C., Zhang, J., Huang, Q., Chen, Y., & Morrison, A. M. (2022). An analysis of user-generated crisis frames: Online public responses to a tourism crisis. *Tourism Management Perspectives*, 41, 100931.

Zhong, L., Sun, S., Law, R., & Li, X. (2021). Tourism crisis management: Evidence from COVID-19. *Current Issues in Tourism*, 24(19), 2671–2682.

Haldun Demirel[1], Muhammed Demiralp[2] and Evren Güçer[3]

Chapter 16 Tourism in Turkey after 1980

Introduction

The term post-1980 refers to an important milestone, especially in the Turkish literature. The process that started with the January 24, 1980 Economic Stability Program was completed with the return to multi-party political life in 1983 after the military coup of September 12, 1980 and The Anavatan Party's coming to power alone in the elections held in 1983 and started a comprehensive economic, political and social transformation in Turkey.

The January 24 Economic Stabilization Program, which will be known as the January 24 Decisions, was based on an export-oriented approach by abandoning the import-substitution industrialization strategy that was started to be implemented in 1963 in a planned manner. In this context, in order to overcome the foreign exchange crisis, it has a perspective of structural adjustment, carried out by international capital through the World Bank, which is oriented towards two strategic goals: internal and external market liberalization and the empowerment of international and domestic capital against labor (Boratav, 2006; Yeldan, 2001).

The post-1980 transformation has in fact been an extension of two important transformations on a global scale: Neoliberal economic policies and Globalization. The inflation and economic stagnation of the developed economies in the early 1970s, i.e. the stagflation crisis, the increase in costs and exchange rate fluctuations have initiated a transformation in the international economy since the mid-1970s. In 1979, with the election of Margaret Thatcher as Prime Minister of the United Kingdom and Roland Reagan as President of the United States in 1981, the new economic model turned into a global wave. The new economic model is based on the free market, advocates that the state should not be involved in the economy as an entrepreneur and sees the international liberalization of capital as the basic rule. Since the early 1990s, the new economic model, which would come to be known as neoliberalism, was presented

---

1 The Republic of Türkiye Privatization Administration, hdemirel@oib.gov.tr
2 The Republic of Türkiye Privatization Administration, mdemiralp@oib.gov.tr
3 Ankara Hacı Bayram Veli University, Faculty of Tourism, evren.gucer@hbv.edu.tr

to developed and developing economies through international organizations as a prescription for emerging from the crisis.

As a result of neoliberalism's determination of the free movement of capital as a precondition, economic globalization began in the 1980s, and since the early 1990s, globalization has materialized in its social and cultural dimensions.

After a period of relative income growth and development, especially in developing countries such as Turkey, as a result of the neoliberal economic policies of the 1980s, the 1990s was a period in which the social costs of the 1980s were questioned with the impact of globalization. The 2000s have been a period of economic, social and cultural restructuring with the advances in technology making itself felt at every point of the globalized world, and shaped on the axis of sustainability, especially after 2010.

The aim of this study is to examine the developments in the field of tourism in Turkey in the post-1980 period, which can be considered as the beginning of the most turbulent period of the historical process. In addition, this study aims to evaluate the 1980s as the foundation period of tourism, the 1990s as the period of competition of Turkish tourism in the globalizing world, and the 2000s in terms of the sustainability of Turkish tourism.

1980s

In the early 1980s, neoliberalism was used in a very different way to describe a wave of market liberalization, privatization and the withdrawal of the welfare state in developed, developing and underdeveloped economies. Subsequently, neoliberalism has continued to expand as a concept to refer not only to an economic model but to a broader political, ideological, cultural and spatial phenomenon. By the early 1990s, neoliberalism had risen to the level of a major phenomenon (Venugopal, 2015). The neoliberal approach redefines the role of the state in the economy, since it is essentially based on the principle of market determinism. Economist John Williams has called the model of neoliberalism as a prescription the "Washington Consensus" and lists the consensus points as privatization, competitive exchange rates, facilitating foreign direct investment, and removing rules that impede market entry and competition (Williamson, 1990).

The Turkish economy experienced a period of economic and political crisis throughout the 1970s, with economic instability and foreign exchange bottlenecks making the economy unmanageable while political conflicts continued. The January 24 Decisions were taken mainly to fulfill the criteria put forward by the IMF and the World Bank in order to get out of the economic crisis. These decisions meant the acceptance of the new economic model.

Developing countries such as Turkey had to rely on policies that would provide the necessary foreign credits to overcome the foreign exchange crisis in the short term and increase foreign exchange earnings in the medium term. In terms of medium-term targets, it was a state policy in the early 1980s to concentrate on the service sector and to achieve this through privatization or facilitating foreign direct investment. The full realization of this policy would be realized in 1983 with the return to multi-party life and the single-handed rule of the Anavatan Party after the general elections.

In the post-1980 period, the first step towards the transformation of tourism policies was taken with the January 24 Economic Decisions. With the January 24 Decisions, a 32.7 % devaluation was announced, daily exchange rates were announced, and it was decided to make the necessary arrangements to facilitate foreign capital investments. In this context, tourism was prioritized in the development phase and this approach was legalized with the Tourism Incentive Law Numbered 2634 in 1982 (Çeken 2003: 134)

The development policies of developing country economies are based on the development of industry and foreign trade. However, in order to provide the necessary financing for this, foreign exchange earning activities need to be strengthened. Tourism has an important role in terms of financing the industry with its foreign currency earning aspect (Çakır ve Bostan, 2000: 35). Tourism not only boosts the economy through tourism revenues, but also through its effect of increasing investments for tourism purposes (Tutar, 1999: 126). Due to these characteristics of tourism, tourism started to be evaluated in the context of sectors of special importance in development in 1985 following the Tourism Incentive Law Numbered 2634 (Zengin, 2010).

Tourism Incentive Law Numbered 2634 aims to establish tourism infrastructure in Turkey and encourage private sector fixed capital investments. Within the framework of the new economic policies, the private sector was encouraged to invest in tourism in order to ensure sectoral development under the leadership of the private sector. Not only legal arrangements were made, but the Prime Minister himself advised businessmen to invest in tourism. In 1985, a businessman with textile investments stated that the Prime Minister of the time said, "Look, you are exporting, let's invest in tourism." After the businessman hesitated, the Minister of Tourism and the relevant General Manager called him and advised him to invest in tourism (https://www.hurriyet.com.tr/turizm-ozali-ariyor-148877).

By the 1980s, tourism was limited to domestic tourism, foreigners were considered "guests" and the places visited by foreign tourists were mostly historical and cultural excursions. 1From the beginning of the planned period in 1963 until 1983, the period in which the state worked for the development of tourism, and the period after 1983 is seen as the "liberalization" period (DPT, 2007: 6). In this period, the tourism policy was based on making a positive contribution to the current account deficit through tourism revenues and providing vacation opportunities for citizens (DPT, 2007: 6). In 1955, Hilton Istanbul, the first modern hotel built from scratch in post-World War II Europe, was opened. The hotel was built by Emek Construction Company, owned by the Pension Fund. Emek Construction Company also built the Hilton, Grand Tarabya and Maçka Hotels in Istanbul, Bursa Çelik Palace, Ankara Stadium, Grand Ankara and Grand Ephesus Hotels in Izmir. 20 million square meters of public land was allocated for tourism investments by the end of the 1990s with the implementation of the Tourism Incentive Law Numbered 2634, which was initiated in 1983 (Afşar, 2002: 151). While the allocation of public land for tourism investments is under the authority of the Ministry of Culture and Tourism in accordance with the Tourism Incentive Law Numbered 2634, there are also allocations made by the Ministry of Finance and the General Directorate of Forestry and Water Affairs. As of 2017, 39 public lands were allocated for tourism purposes by the Ministry of Finance, 90 by the General Directorate of Forestry and Water Affairs, and 264 by the Ministry of Culture and Tourism (Çayır, 2017).

The period between 1980 and 1990 was the period when Turkey experienced the fastest development in the field of tourism. During this period, the number of beds increased from 56,000 to 173,000 and the number of tourists visiting the country rose from 1.2 million to 5.3 million (Kan ve Kuleyin, 2017).

In 1983, within the framework of the transition to the free market model, the state withdrew from superstructure investments and privatization practices were initiated and free market conditions were taken as the basis for prices in touristic facilities. A new incentive system was introduced in accordance with the Tourism Incentive Law Numbered 2634 and related legislation. In this context, incentives such as allocation of public lands, tax exemptions, low-interest, long-term incentive loans and employment supports are conditional upon obtaining an incentive certificate from the State Planning Organization and an investment certificate from the Ministry of Tourism (DPT, 2007). Within the framework of this system, 18 allocation lists were published between 1983 and 1997, and as a result, 297 investors received permission to build tourist facilities in 139 tourism development areas and centers with a total capacity of 95,178 beds (DPT, 2007).

Although the post-1980 policies achieved success beyond the target in terms of foreign exchange earnings, "*When the positive and negative results of the neoliberal policy implementations that started in the 1980s are evaluated, it is seen that tourism gained a significant momentum with the regulations brought to the sector by the Tourism Incentive Law that entered into force in 1982; however, some of the practices introduced by the said law are in contradiction with the concepts of spatial planning, conservation, sustainability, etc."* (DPT, 2014).

The 1980s was a period when, on the one hand, incentive policies were implemented to encourage the private sector to invest in tourism, and on the other hand, preparations for the privatization of state-owned tourism investments began. Within the scope of Law Numbered 3291, the banking activities of the Republic of Turkey Tourism Bank were transferred to the Development Bank of Turkey and its enterprises were transferred to TURBAN, which was established with the decision of the High Planning Council dated 27.09.1988 and numbered 88/9, and whose capital is wholly owned by the Prime Ministry Public Partnership and Public Housing Administration. Thus, TURBAN was included in the scope of privatization with one travel agency and 23 accommodation establishments. The privatization of TURBAN's enterprises was intended to be completed in the 1990s. The 1980s represent the period when tourism was shaped as an area where the private sector operates within the framework of free market rules with the encouragement and guidance of the state.

1990s

The 1990s were the years in which globalization in economic, cultural and political dimensions materialized as a result of neo-liberal policy practices on a global scale. Globalization is defined as a complex network of social processes that intensifies and expands economic, cultural, political and technological exchanges and linkages around the world, thus its impact and dimensions are more clearly understood (Campbell vd., 2010).

As the concept of "Global Village" is becoming more and more a reality as a result of great breakthroughs in transportation and communication thanks to technological developments, tourism has also been positively affected by these developments. Developments such as the reunification of Germany, the return to democracy in the former USSR and other Eastern European countries, and the peaceful end of apartheid in South Africa in the early 90s have opened up new markets for international tourism and increased competition.

In addition to positive developments, there have also been negative developments such as the 1991 Gulf Crisis and the 1994 Financial Crisis, and these events have caused the expectation of security to gain importance in terms of tourism. A new process that started in the 1990s and will be carried into the 2000s; individualization and differentiation efforts in tourism demand with the effect of globalization, the social structure in which leisure time and self-actualization are social needs have gained a determining character (Kozak vd., 2013).

The 1990s was a period in which the growth trend that started in the world tourism industry after the 1980s increased, and accordingly, competition intensified. During these years, there was an annual increase of 4 % in the number of tourists and 7.2 % in tourism revenues worldwide (DPT, 2001). Turkey, on the other hand, accounted for a share between 1.5 % and 2 % in terms of the number of tourists and tourism revenues compared to global market data (DPT, 2001).

As of 1991, tourism in Turkey can be examined in two periods; during the period up to 1997, due to the impact of the First Gulf War, tourism businesses reduced prices to compete with rival countries. In this period, unlike the 1980s, domestic tourism was seen as an important market, and tourism businesses engaged in activities aimed at domestic tourism. After 1998, the business approach and environmental sensitivity became more decisive in guiding investments. While in the 1980s, companies predominantly operating in the construction sector invested in tourism, there was a change in the investor profile after 1998. Although the 'leave everything to the state' period ended in tourism during this process, a proper role distribution between the state and the private sector could not be achieved (DPT, 2007).

Tourism activities carried out between 1980 and 1998 showed an increasing trend in terms of foreign exchange earnings, except for the year 1998. Compared to 1980, there was an increase of nearly 8 billion dollars in foreign exchange earnings from tourism in 1997. However, in 1998, due to the economic crisis in Russia and the Far East, foreign exchange earnings from tourism declined (Şanlıoğlu ve Özcan, 2017).

During the period between 1986 and 1992, the increase in the pace of tourism investments was influenced by incentives provided under the name 'Resource Utilization Support Premium,' which allowed the portion of investment expenditures covered by equity to be paid to businesses in cash, effectively as grants. Although these incentives accounted for only 13 % of the total investment amount, their impact on tourism investments was much higher. Since 1992, the 'Resource Utilization Support Premium' system was abolished and replaced with the Fund-Sourced Credit scheme. This loan, which had a two-year grace period and a five-year term, was not suitable for the nature of tourism investments due to its short duration. Since the expected positive impact was not achieved, the fund-sourced credit scheme was abolished. Consequently, tourism investments first stagnated and then declined between 1995 and 1999 (DPT, 2001).

The privatization efforts initiated at the end of the 1980s culminated in the establishment of the Privatization Administration in 1994. TURBAN Enterprises, transferred to the Prime Ministry Public Housing Administration in 1988, were included in the privatization scope, and privatization efforts commencedIn 1995, TURBAN's Kemer Marina and Çeşme Hotel, Elmadağ Mountain House and Ilıca Motel in 1996, and Akçay Holiday Village in 1997 were privatized. TURBAN's eight properties in Istinye, Istanbul were transferred to the Istanbul Stock Exchange for a fee.

The change in tourism demand, which would be more tangibly felt in the 2000s, began to manifest itself towards the end of the 1990s. In addition to the traditional tourism trio of sea, sand, and sun, people's expectations such as cultural, health, congress, and yachting started to emerge. As a result of this trend, Turkey will need to develop tourism concepts and programming that will spread tourism throughout the year

2000s

The 1980s, based on the establishment of a state-supported investment environment, was a period of great advancement due to the allocation of public lands, incentives, and grant support. The 90s are characterized by the struggle of the existing fixed capital to maintain growth and not to fall behind rival countries under free market conditions. The 2000s are evaluated within the framework of sustainability as a volatile period shaped by neo-liberal policies and globalization.

International tourist movements worldwide surpassed the one billion tourist milestone for the first time in 2012 (UNWTO, 2013). In parallel, the United Nations World Tourism Organization (UNWTO) announced that the number of tourists requiring a visa before traveling reached its lowest level in 2015 (UNWTO, 2016). The advancements in transportation, paralleling the rapid progress of technology in the 2000s, have enabled travel plans to be made to every corner of the World.

Especially the period after the 1980s has been a growth period for international tourism, and factors such as intense competition and climate crisis have brought up the sustainability of countries' tourism potentials ndeed, UNWTO declared 2017 as the 'International Year of Sustainable Tourism Development' and aimed to draw the attention of tourism stakeholders to sustainable development goals in tourism with the 'Chengdu Declaration' published in 2017 (UNWTO, 2017).

Sustainability contributes to the preservation and enhancement of natural beauty, historical heritage, and cultural assets. Thus, tourism supply resources are preserved, and the range of tourist products is diversified. Tourism does not possess a finite characteristic like underground and aboveground riches. The most fundamental factor that would turn tourism into a depletable resource is irreversible environmental damage. A forward-looking perspective towards the future is indispensable for sustainability in realizing the existing potential. While mass tourism may seem an attractive option under free market conditions, its nature entails significant risks in terms of sustainability in Tourism.

Within the framework of the transformation in international tourism, a new type of tourism based on nature and ecology and carrying sustainability elements has started to be preferred instead of traditional tourism consisting of the sea, sand and sun trilogy, which is the main focus of mass tourism. In parallel with this change, awareness has developed towards the green lifestyle and increasing ecological problems, referred to as the 3E (Entertainment-Excitement-Education) trio - Enjoyment-Excitement-Education. As a tangible example of this change, the tourism concept presented as 4L (Landscape-Leisure-Learning-Limit) in the Alps can be given (SBB, 2018).

The 2000s were a period in Turkey where alternative tourism initiatives increased both in terms of setting tourism policies and due to sustainability concerns as well as political and economic crises. Especially towards the end of the 1990s and the beginning of the 2000s, the implementation of the all-inclusive accommodation concept to increase the number of tourists and promote mass tourism, along with the high demand from the Russian Federation and Europe, had an impact on even the smallest accommodation facilities (Gülbahar, 2008: 165).

In the 2000s, tourism continued to grow. The number of foreign visitors, which was 1,228,000 in 1980, increased by 17.6 times to reach 21,122,798 in 2005. Similarly, tourism revenue, which was 400 million dollars in 1980, increased by 45 times to reach 18 billion dollars in 2005 (DPT, 2007). ourism revenues peaked in 2014, reaching $34.3 billion, while the number of incoming tourists reached 41.5 million. However, due to developments in the vicinity, situations such as terrorism and the refugee crisis, a decline was experienced between 2015 and 2017. Although a recovery process began in 2017, the global Covid-19 pandemic in 2020 and the Russia-Ukraine War negatively affected international tourism, consequently impacting Turkish tourism as well.

Turkey also has significant potential in terms of alternative tourism. Throughout the 2000s, goals for alternative tourism were set in development plans and strategic plans; however, due to national and international political and economic crises, it was not possible to achieve these goals. In this regard, the ineffective and inefficient management of public and private sector relations also plays a role in the inability to achieve these goals.

Conclusion

The study aimed to analyze the post-1980 development of tourism in Turkey, considering its natural beauty, geographical location, historical and cultural heritage. The study evaluated the establishment of the tourism sector under private sector leadership with government incentives in the 1980s, the sector's development under global competitive conditions in the 1990s, and its sustainability in the 2000s.

Efforts to overcome the economic hardships post-1980 resulted in tourism being seen as a lifeline in Turkey, parallel to the development of international tourism. Tourism, which has connections with hotels, restaurants, transportation, travel agencies, insurance companies, food, and health industries, not only has an impact on national development but also provides income for the local population. Starting from the second half of the 1980s, investors who had previously accumulated capital in sectors such as textiles and construction took part as tourism investors for the first time in tourism investments in the Aegean and Mediterranean regions. Although these investments yielded significant financial returns, especially from the early 2000s onwards, the magnitude of its sociocultural cost was clearly seen, emphasizing the necessity of division of labor between the public and private sectors and the importance of a holistic approach for the sustainability of Tourism.

In terms of public-private sector relations, local governments in tourism regions play a crucial role. In order to achieve cultural and ecological sustainability goals, local governments need to demonstrate a socially responsible attitude in issues such as zoning plans, urban planning, and infrastructure investments.

In the historical process discussed in this study, tourism has experienced global transformations, and as a result, the upward trend that started as mass tourism has transformed into alternative forms of tourism shaped by individual passions and expectations. The rate of increase in technological developments has also led to an increase in individual expectations. Demands that could have been achieved through comprehensive national campaigns in the 80s can now be achieved with a single social media post.

The current situation not only facilitates but also has the potential to lead to destructive effects. Therefore, tourism has ceased to be an area that can only be sustained with natural resources and will continue to grow with the highest level of standardization and implementation in terms of accommodation, transportation, security, and concepts.

In order to realize its potential, Turkey, which has a brand value in terms of international tourism, should adopt a tourism approach determined in cooperation with stakeholders at national and local levels, taking into account global trends.

References

Afşar, A. (2002). Türkiye Turizm Sektöründe Konaklama İşletmelerinin Finansal Analizi. Anadolu Üniversitesi İktisadi ve İdari Bilimler Fakültesi Dergisi, 18(1), 147–166.

Boratav, K. (2006). Türkiye iktisat tarihi, 1908–2009. (147–148).

Campbell, P. J., MacKinnon, A., & Stevens, C. R. (2010). An introduction to global studies. John Wiley & Sons.

Çakır, M., & Bostan, A. (2000). Turizm Sektörünün Ekonominin Diğer Sektörleri ile Bağlantılarının Girdi-Çıktı Analizi ile Değerlendirilmesi. Anatolia: Turizm Araştırmaları Dergisi, 11(2), 35–44.

Çayır, A.S. (2017) Tahsis Sürelerinin Sonlarına Yaklaşan Turizm Amaçlı Kamu Taşınmazları, 3. Turizm Şurası 1–3 Kasım 2017 Tebliğler Kitabı, (602–610)

Çeken, H. (2003). Türk Turizminde Yabancı Sermaye ve Yabancı Sermaye Ortamının İyileştirilmesine Yönelik Öneriler. Balıkesir Üniversitesi Sosyal Bilimler Enstitüsü Dergisi, 6(10), 25–45.

DPT, (2014) 10. Kalkınma Planı, Turizm Özel İhtisas Komisyonu Raporu, Ankara, 2014, Erişim Tarihi: 30 Nisan 2024

DPT, (2007) Dokuzuncu Kalkınma Planı Turizm Özel İhtisas Komisyonu Raporu, Erişim Tarihi: 29 Nisan 2024

DPT, (2001) Sekizinci Kalkınma Planı Turizm Özel İhtisas Komisyonu Raporu, Erişim Tarihi: 26 Nisan 2024

Gülbahar, O. (2008). Turizmin Türkiye'de 1980 Sonrası Dönemde Cari İşlemler Dengesine Etkisi. Journal of Qafqaz University, (24).

Kan, N., & Kuleyin, B. (2017). Kalkınma Planları Çerçevesinde Türkiye'nin Deniz Turizmi Stratejilerinin Tarihsel Gelişimi. Dokuz Eylül Üniversitesi Denizcilik Fakültesi Dergisi, 51–64.

Kozak, M. A., Evren, S., & Çakır, O. (2013). Tarihsel süreç içinde turizm paradigması. Anatolia: Turizm Araştırmaları Dergisi, 24(1), 7–22.

Strateji ve Bütçe Başkanlığı, https://sbb.gov.tr/wp-content /uploads/ 2018/11/Sekt%C3%B6rler-%C4%B0tibar%C4%B1yla-Sabit-Sermaye Yat%C4%B1r%C4%B1mlar%C4%B1 - Toplam.pdf Erişim Tarihi: 29 Nisan 2024

Şanlıoğlu, Ö., & Özcan, E. Ö. (2017). Türkiye'de Uygulanan Turizm Teşvik Politikaları Ve Sonuçları Üzerine Bir Değerlendirme. Kırıkkale Üniversitesi Sosyal Bilimler Dergisi, 7(2), 97–118.

Tutar, F. (1999). Türk turizm sektöründeki gelişmelerin cari işlemler dengesine muhtemel etkileri: Akdeniz ülkeleriyle karşılaştırmalı bir analiz (Doctoral dissertation, Anadolu University (Turkey)).

UNWTO, (2013). World Tourism Barometer (PDF). UNWTO World Tourism Barometer. 11 (1). January 2013. Archived from the original (PDF) on 28 February 2013. Erişim Tarihi: 30 Nisan 2024

UNWTO, (2016). World Tourism Barometer (PDF). UNWTO World Tourism Barometer. 11 (1). Erişim Tarihi: 30 Nisan 2024

UNWTO, (2017). Chengdu Declaration on 'Tourism and the Sustainable Development Goals'. Erişim Tarihi: 30 Nisan 2024. https://www.e-unwto.org/doi/abs/10.18111/unwtogad.2017.1.g51w645001604506

Venugopal, R. (2015). Neoliberalism as concept. Economy and society, 44(2), 165–187.

Williamson, J. (1990). Latin American adjustment: How much has happened? (Vol. 4). Washington, DC: Institute for International Economics.

Yeldan, E. (2001). Küreselleşme sürecinde Türkiye ekonomisi. İstanbul: İletişim Yayınları, 20–23.

Zengin, B. (2010). Turizm sektörünün Türkiye ekonomisine reel ve moneter etkileri. Akademik İncelemeler Dergisi, 5(1), 102–126.

Çağrı Erdoğan[1] and Zeynep Yamaç Erdoğan[2]

Chapter 17 Tourism Product Generation Function of Wars and the Reflections on Tourism

Introduction

Prominently characterized by the destructiveness throughout human history, wars exert a suffocating effect not only in military but also in civilian sphere. As a matter of fact, the shaping effect of wars on a global scale is felt in all areas of life. There is no doubt that one of these areas is tourism (Smith, 1998), and while at first sight, the contrast between tourism –the representative of "peace"– and war stands out at opposite poles (Mirisaee & Ahmad, 2018), nevertheless, the complex relationship between them is analysed from a broader perspective over time and contributions made to be found a more realistic basis (Butler & Suntikul, 2013; Suntikul, 2019).

In the direction of making the mentioned realistic ground visible, the demand shrinkage caused by wars and demand shifting towards alternative destinations are scrutinised in this study, considering that security sensitivity is one of the most important factors shaping demand and the elasticity of tourism demand is high in general. Furthermore, the study addresses –despite all the destruction it brings and being a niche market – tourist pulling power of the wars. Subsequently the tourism products formed by the fundamental nature of wars leaving lasting imprints in history becomes the focus of the study.

In order to support a holistic perspective, the subject is discussed on a global scale with examples. Finally, the study is concluded by mentioning, beyond being an activity that takes place in peacetime, the value of tourism as an activity that strengthens and spreads peace.

---

1 Ph.D., Sakarya University of Applied Sciences, Faculty of Tourism, Department of Tourism Management, cagrie@sakarya.edu.tr
2 Ph.D., Bilecik Şeyh Edebali University, Faculty of Applied Sciences, Department of Tourism Guidance, zeynep.erdogan@bilecik.edu.tr

The Shrinking and Shifting Demand

The first effect coming to mind in the war-tourism relationship are the shrinking tourist traffic in the region where the war is effective (Hamadeh & Bassil, 2017), the significant decline in tourism revenues in related destinations (Novelli, Morgan & Nibigira 2012), and the shifting demand to alternative regions where the perception of risk is low (Seabra, Dolnicar, Abrantes & Kastenholz, 2013). In order to concretise these effects, the issue will be clarified through the discussion of the ongoing Russia-Ukraine War.

The Russia-Ukraine War

Russia has developed policies to control the countries in its immediate vicinity in military, political and economic terms, driven by security needs and energy dominance considerations due to its position. Specifically, Ukraine's position between Europe and Russia, its potential NATO/*The North Atlantic Treaty Organization* membership, and its moves towards integration with the West have alarmed Russia and increased its interventions in Ukraine (Keskin, 2015 p. 59; Özsoy, 2022 p. 556). Years of ongoing disputes made military intervention inevitable, leading to a large-scale military operation launched by Russia on 24 February 2022, targeting Ukraine's political unity, sovereignty, and territorial integrity (T.C./*Türkiye Cumuriyeti/Republic of Türkiye* Dışişleri Bakanlığı, 2024). In addition to the multifaceted general impact of the war extending beyond the borders of the two countries, there have been significant global economic repercussions. In this context—despite the varying effects on different countries—the intensification of the war's negative impact on global tourism has become unavoidable. As emphasised by UN/*United Nations* Tourism (2024), Russia's military assault on Ukraine has escalated the overall risk burden for tourism, further increasing already high oil prices and transportation costs, exacerbating uncertainty, and disrupting travel in Eastern Europe. It is evident that the war has caused billions of dollars in damage to the tourism economy and that military attacks pose a risk to the recovery of confidence in global travel.

With its profound social impacts and long-term consequences, war (Smith, 1998, p. 202) stands out as one of the primary factors creating a crisis environment within the tourism industry. As a result of the war's negative effects across psychological, social, and economic dimensions, individuals experiencing reduced welfare levels and increased financial hardship often prioritize cutting tourism expenditures at the top of their discretionary spending reductions (Johnson & Ashworth, 1990; Alegre & Pou, 2004). To provide a more concrete illustration of the issue, it would be appropriate to examine the impact of the war on tourism demand in Türkiye, a significant country with which both Russia and Ukraine have demonstrated substantial economic and political cooperation.

The war between Russia and Ukraine has generated a shock effect on Türkiye's tourism expectations for 2020 regarding these markets (Kızıl Erol, 2022 p. 99). Russia has held a significant position in Türkiye's tourism industry due to its large population, geographical proximity, and close political and economic ties (Dalkıran & Bayrak, 2020: pp. 221–222). As can be seen from Table 17.1, the arrival of over 7 million visitors from Russia to Türkiye in 2019 was a crucial factor in enhancing these tourism expectations. Moreover, as highlighted by the data in Table 17.1, Russia consistently ranked as the top country sending the most visitors to Türkiye over the five-year period from 2019 to 2023, except for 2022, when it ranked second.

Table 17.1. Number of Visitors from Russia and Ukraine to Türkiye (2019–2023)

Country	2019	2020	2021	2022	2023
Russia	7.017.657	2.128.758	4.694.422	5.232.611	6.313.675
Ukraine	1.547.996	997.652	1.153.092	675.467	839.729
Ranking*	R: 1. U: 7.	R: 1. U: 4.	R: 1. U: 3.	R: 2. U: 16.	R: 1. U: 14.

* Ranking of countries sending the most visitors to Türkiye, R: Russia, U: Ukraine.
Source: (T.C. Kültür ve Turizm Bakanlığı, 2023)

On the other side of the conflict, Ukraine has progressively secured a significant position within the Turkish tourism market over time, particularly after 2015 (T.C. Kültür ve Turizm Bakanlığı, 2023). Indeed, Ukraine's ranking among the countries sending the most visitors to Türkiye showed a significant increase from 2019 until the onset of the war in 2022 (see Table 17.1), following which there was a natural and rapid decline in its position.

In order to evaluate the rate of increase in visitor arrivals, the year 2020, characterized by a notable decline in global visitor numbers due to COVID-19/ *2019 novel coronavirus/coronavirus disease*, is disregarded. Instead, the analysis focuses on data from 2019, representing the pre-pandemic period, and on the year 2022, which marks the onset of the conflict. The increase rate in the number of visitors from Russia to Türkiye in 2019 was determined to be 17.65 % compared to the previous year, while this rate declined to 11.46 % in 2022 compared to the previous year. For Ukraine, these figures recorded an increase of 11.61 % in 2019, but a decrease of 67.21 % in 2022 (T.C. Kültür ve Turizm Bakanlığı,

2023). As Özsoy (2022, p. 571) has also highlighted, the decline in tourism revenues, directly correlated with the number of visitors from Russia and Ukraine, has had a detrimental impact on the tourism industry of Türkiye.

Transitioning from a Türkiye-centric perspective to a global scale, it becomes apparent that Moldova (-69 %), Slovenia (-42 %), Latvia (-38 %), Finland (-36 %), Czech Republic (-35 %), and Sweden (-34 %) emerge as the destinations most significantly impacted by the proportional decline in flights to European destinations (excluding Russia and Ukraine) between February 24 and May 11, 2022, compared to 2019 figures (UN Tourism, 2024). Looking from a broader viewpoint, beyond these specific instances and involved nations, Figure 17.1 distinctly illustrates the overarching repercussions of the conflict in Ukraine on the entirety of the Northern Hemisphere and underscores the potential adverse effects on the tourism industry, thus emphasizing the war's fundamental role.

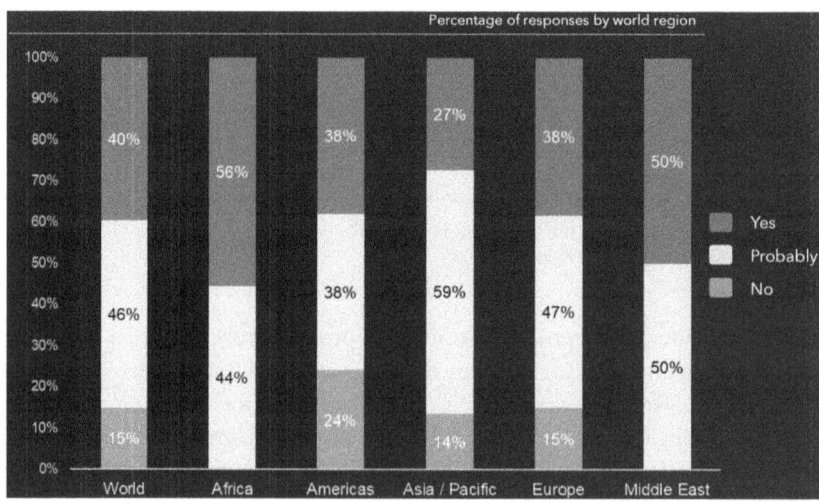

Figure 17.1. Will the Russia-Ukraine War Affect Northern Hemisphere 2022 Summer Season?
Source: (UNWTO/*United Nations World Tourism Organisation*, 2022, p. 23)

The atmosphere of despair and uncertainty in tourism destinations accompanying the war, travel restrictions, and human rights violations increase the concerns of tourism stakeholders, investors, and potential tourists (Tan & Cheng, 2024, p. 11). The Russia-Ukraine War has brought several factors that may lead to changes in consumer behaviour to the forefront, notably the "fear of traveling near conflict zones," "preference for destinations perceived as safe," "adopting a wait-and-see approach," "inclination towards domestic tourism," "preference for nearby travel," and "lower confidence" (UNWTO, 2022). These factors vividly depict the tourism industry's sensitivity and vulnerability in terms of demand dynamics. The swift transition of tourist preferences towards destinations perceived as secure not only disturbs the balance of global tourism dynamics but also adversely impacts the relevant stakeholders in the tourism industry, who typically devise their strategies based on expectations grounded in more stable periods.

In the wake of the global pandemic's impact (Hall & Cooper, 2021), the tourism industry endeavors to navigate through the reluctance of tourists to travel and address the shock of the year 2020, which is deemed as a complete loss (Fotiadis, Polyzos & Huan 2021). In this context, the concept of war carries a significantly more ominous connotation for the tourism industry as it strives to heal the wounds of the subsequent years (Pandey & Kumar, 2023, p. 692). Even upon the conclusion of hostilities, the destructive effects of war on tourism may necessitate an additional decade or so for normalization (Bandara, 1997; Grzinic, 2010; Okafor & Khalid, 2020), thus providing a more realistic portrayal of the extent of devastation.

"Danger" as a Tourism Product Element

While the sensitivity of tourism demand elasticity to events threatening personal safety, such as terrorism and conflict, is well acknowledged, this section has also been elucidated with concrete examples. Particularly, the impact of man-made disasters like war on tourism tends to be greater than that of natural disasters such as earthquakes (Ma, Chiu, Tian, Zhang & Guo, 2020). Nonetheless, tourism products, despite containing elements of 'danger,' can still attract demand (Lisle, 2016). Ultimately, it is a contradictory reality that war and conflict-ridden environments, which most individuals seek to avoid, can also appeal to tourists (Cochrane, 2015).

In their examination of the dimensions of war tourism, Williams, Wassler, and Fedeli (2023) focused on visiting areas where wars have occurred in the past as well as regions with ongoing conflicts, within the frameworks of voluntarism and voyeurism. As a result, the authors have proposed four categories: hot war travel, combat volunteering, non-combat volunteering, and virtual war travel, aiming to present a more comprehensive view of war tourism.

Going beyond the issues that seem to be at the centre of war tourism, the author touches on the points of real risk being lower than perceived risk and achieving exclusive authenticity, Mahrouse (2016) focused on the War Zone Tours company, which turns war, conflict and destruction zones into a tourism product. The privilege of purchasing safety in conditions where others are struggling for survival, and the normalisation of these man-made disasters, are points opened up for discussion by the author. Thus, it is made more apparent that the potential for wars to attract tourists is not limited to areas such as tourism type, tourism product components, and tourist motivation.

Tourism Product Generation Function of Wars: War-Based Tourism Products

The interplay between war and tourism presents a sophisticated relationship (Butler & Suntikul, 2013). The profound impacts of war on the tourism industry can lead to the decline of renowned tourist spots, coupled with economic and social repercussions (Mansfeld, 1999, p. 36; Tan & Cheng, 2024). Conversely, there is a discourse suggesting that war may serve as a tourism asset, with active conflict zones potentially attracting tourists seeking unconventional experiences (Timothy, 2013, p. 13). In recent years, alongside the visits to regions that have witnessed conflicts in the past, "hot" war tourism has gained attraction even in regions actively experiencing warfare (Williams et al., 2023). In this context, the tourism industry is witnessing a growing trend towards the marketing of destinations and products linked to conflict zones and the legacies of war (Henderson, 2000). Such offerings typically fall under the umbrella of dark tourism or thanatourism, defined as "travel to a location wholly, or partially, motivated by the desire for actual or symbolic encounters with death…" (Seaton, 1996, p. 240).

Many examples exist worldwide of areas directly affected by war being opened up for tourism purposes. For instance, Vietnam offers visitors sites of historical significance related to its resistance against the United States and its struggle for national liberation (Henderson, 2000, p. 273). An exploration of tourism offerings and elements within the framework of thanatourism, which expands beyond the boundaries of war tourism, is presented by Seaton (1996, pp. 240-242; 1999, p. 131). He outlines five fundamental categories, examining historical and global examples. The first category includes travelling to witness public deaths, including Roman gladiatorial fights, political and public executions. The second category encompasses travels to sites where mass or individual fatalities occurred, including Pompeii, obliterated by a volcanic eruption and excavated since 1748, the Colosseum in Rome, the Auschwitz genocide camp, the book depository in Dallas, where Kennedy was assassinated, and locations like Graceland, linked to the death of Elvis Presley. The third category includes journeys to places of detention ending in death, graves, underground tombs, memorial graves, and war memorials. The fourth category encompasses travels to places containing evidence or symbolic representations of death events. Examples include the Revolution Museum in Cuba, where weapons, torture devices, and items belonging to victims are displayed, as well as Madame Tussauds in London, which exhibits wax figures of infamous killers. The fifth category comprises the tourist attraction created by the reenactment of deaths and associated events, predominantly featuring religious figures until the 20th century. In the context of the diversity it presents over time by incorporating various products, the comprehensive coherence of thanatourism also facilitates the establishment of a connection with war tourism.

Examples of the transformation and utilization of war zones for tourism abound. Notable instances include the Somme/Picardie region in France, which holds significant importance in the context of the Western Front during World War I, and the Gettysburg Battlefield, a pivotal site in the American Civil War (Atay & Yeşildağ, 2010, p. 69). Additionally, visits by Anzacs to Gallipoli to pay homage to their ancestors' graves, journeys to Bosnia and Herzegovina, and Serbia for protected areas of ruin and remains of war, and trips by Turks to South Korea, where their forebears fought and are interred, can be cited as further examples (Kızıl Erol, 2022, p. 100). Among the war sites located directly within Türkiye that attract visitors are the Gelibolu Yarımadası Tarihi Milli Parkı/*Gallipoli Peninsula Historical National Park*, the Başkomutan Milli Parkı/*Commander-in-Chief National Park*, and the Sakarya Meydan Muharebesi Tarihi Milli Parkı/*Sakarya Battlefield Historical National Park*. The house used as the 12th Group Headquarters during the Battle of Sakarya, and later stayed in by the famous author Halide Edip Adıvar while she was serving on the Yunan Mezalimini Araştırma Komisyonu/*Commission for Investigating Greek Atrocities*, was opened as the Halide Edip Adıvar ve Kadın Kahramanlar Müzesi/*Halid Edip Adıvar and Women Heroes Museum*. This transformation serves as an example of converting battle sites and their memories into tourism products (Yamaç Erdoğan, 2023, p. 690). Atatürk ve Kurtuluş Savaşı Müzesi/*Atatürk and the War of Liberation Museum*, structured into four sections, is among the most visited museums in Türkiye. It showcases panoramas of the War of Liberation alongside personal belongings, memorabilia, and war artefacts of Mustafa Kemal Atatürk.

Conclusion

A decline in tourism demand may result from both a reduction in the number of tourists visiting destinations affected by war and a decrease in these countries' capacity to generate tourists. The reduction in tourist production in warring nations is accompanied by negative repercussions that extend beyond national boundaries, such as escalating inflation proportional to the magnitude of the conflict and security-based travel restrictions, which can further influence tourism demand. Alternatively, in the absence of such circumstances, and assuming a stable tourist production in countries not involved in conflicts; the appeal of destinations immersed in war, conflict, and compromised safety will significantly wane. Consequently, the number of tourist magnets will diminish, thus a demand shift will occur due to the decrease in supply diversity while demand is constant.

The Israel-Palestine conflict serves as an example, highlighting how the detachment of involved parties from events with profound implications for conscience and legitimacy, such as war crimes and crimes against humanity, can severely damage the overall reputation of nations and, consequently, their destination image. In a broader sense, the wanton destruction attributed to Israel has generated a profound accumulation of moral discomfort within the international community (TÜBA/*Türkiye Bilimler Akademisi/Turkish Academy of Science*, 2023; UN General Assembly 2024). Consequently, the enduring traces of this unease persist even after the conflict, posing a challenge to Israel's attractiveness as a destination.

Alongside considerations regarding demand, it is important to recognize that active war zones possess a degree of tourist allure, albeit in a niche dimension. Despite being noticeably weaker compared to the overall tourist appeal generated by non-conflict tourism products, this allure requires investigation and comprehension due to the nuances it presents. Examining the interplay between wars, active conflict zones, and tourism products, as opposed to merely assessing their potential to attract tourists, reveals a dual effect; while wars cause the destruction of existing tourism infrastructure, they also have a share in the production of new tourism products. Indeed, the historical context and narrative construction of wars exert a profound influence on societies, and it is imperative to acknowledge that this influence extends to the sphere of tourism without exception. The reflections of war, as witnessed in sites such as memorials, monuments, and ruins, alongside the war-centric cultural artifacts like anthems, songs, elegies, paintings, museums, novels, theatrical productions, and other forms of cultural and artistic expression, have the potential to either become standalone tourism products or serve as integral components or motivational factors within the realm of tourism. These elements can function independently as tourism attractions or significantly contribute to tourism experiences as essential components or motivational influences.

Tourism is predominantly an activity undertaken during times of peace, which are perceived as periods devoid of war (Litvin, 1998). These activities are likely to contribute significantly to the understanding of the historical context of wars that have been either forgotten or whose importance has been diminished, as well as the devastation and inhumane conditions they brought about. Furthermore, they are expected to shed light on the profound impact of warfare in shaping the essence of societies and its wide-ranging implications on all aspects of life. The realization of the cognitive acquisition potential inherent in tourism experiences (Erdoğan & Kıngır, 2019) in this regard holds promise for fostering pathways towards goodness. This could serve as a constructive departure from the futile destruction often associated with wars, potentially leading towards positive outcomes and averting senseless violence. In the manner that D'Amore (1988, 2009) highlighted the issue; There is a desire for tourism – as a value that prevents the direct labeling of difference and the other as "enemies" – to strengthen its position in extending and fortifying peace.

References

Alegre, J., & Pou, L. (2004). Micro-economic determinants of the probability of tourism consumption. *Tourism Economics, 10*(2), 125–144.

Atay, L., & Yeşildağ, B. (2010). Savaş alanları ve turizmi [Battlefields and Tourism]. *Aksaray Üniversitesi İktisadi ve İdari Bilimler Fakültesi Dergisi/ Aksaray University Journal of Faculty of Economics and Administrative Sciences, 2*(2), 65–72.

Bandara, J. S. (1997). The impact of the civil war on tourism and the regional economy. *South Asia: Journal of South Asian Studies, 20*(s1), 269–279.

Butler, R., & Suntikul, W. (Eds.). (2013). *Tourism and war* (1st ed.). Routledge.

Cochrane, F. (2015). The paradox of conflict tourism: The commodification of war or conflict transformation in practice. *Brown Journal of World Affairs, 22*, 1.

D'Amore, L. (1988). Tourism-the world's peace industry. *Journal of Travel Research, 27*(1), 35–40.

D'Amore, L. (2009). Peace through tourism: The birthing of a new socio-economic order. *Journal of Business Ethics, 89*(4), 559–568.

Dalkıran, G. B., & Bayrak, Ö. A. (2020). Pandemi dönemi turizmde Rusya pazarı ve Türkiye'ye yönelik seyahat planlarında sağlık turizmi boyutu [Russian market in pandemic period tourism and dimension of health tourism plans for Turkey]. *Balkan ve Yakın Doğu Sosyal Bilimler Dergisi/Balkan and Near Eastern Journal of Social Sciences, 6*, 221–228.

Erdoğan, Ç. & Kıngır, S. (2019). Turistik deneyimlerin bilişsel edinim bağlamında değerlendirilmesi [Evaluation of touristic experiences in the context of cognitive acquisition]. *Social Sciences Studies Journal, 5*(48), 6074–6084.

Fotiadis, A., Polyzos, S., & Huan, T. C. T. C. (2021). The good, the bad and the ugly on COVID-19 tourism recovery. *Annals of Tourism Research, 87*, 103117.

Grzinic, J. (2010). Croatian tourism offer in the war and after war period. *Journal of Administrative Sciences, 8*(2), 121–136.

Hall, C. M., & Cooper, C. (2021). Introduction to current issues in Asia: COVID-19 and beyond. In C. M. Hall & C. Cooper (Eds.), *Current issues in Asian tourism: Volume II* (pp. 3–8). Routledge.

Hamadeh, M., & Bassil, C. (2017). Terrorism, war, and volatility in tourist arrivals: The case of Lebanon. *Tourism Analysis, 22*(4), 537–550.

Henderson, J. C. (2000). War as a tourist attraction: The case of Vietnam. *International Journal of Tourism Research, 2*(4), 269–280.

Johnson, P., & Ashworth, J. (1990). Modelling tourism demand: A summary review. *Leisure Studies, 9*(2), 145–161.

Keskin, M. (2015). Yakın çevre doktrini bağlamında Rus dış politikası: Ukrayna müdahalesi [Russian foreign policy in the context of the near periphery doctrine: Ukraine intervention]. *Barış Araştırmaları ve Çatışma Çözümleri Dergisi/Journal of Peace Research and Conflict Resolution, 3*(2), 45–62.

Kızıl Erol, K. (2022). Turizmde karanlık bir kavram: Savaş turizmi ve Rusya'nın taraf olduğu iki olay [A dark concept in tourism: War tourism and two incidents with Russia]. *Journal of Applied Tourism Research, 3*(1), 95–104.

Lisle, D. (2016). *Holidays in the danger zone: Entanglements of war and tourism.* University of Minnesota Press.

Litvin, S. W. (1998). Tourism: The world's peace industry?. *Journal of Travel Research, 37*(1), 63–66.

Ma, H., Chiu, Y. H., Tian, X., Zhang, J., & Guo, Q. (2020). Safety or travel: Which is more important? The impact of disaster events on tourism. *Sustainability, 12*(7), 3038.

Mahrouse, G. (2016). War-zone tourism: Thinking beyond voyeurism and danger. *ACME: An International Journal for Critical Geographies, 15*(2), 330–345.

Mansfeld Y. (1999). Cycles of war, terror, and peace: Determinants and management of crisis and recovery of the Israeli tourism industry. *Journal of Travel Research, 38*(1), 30–36.

Mirisaee, S. M., & Ahmad, Y. (2018). Post-war tourism as an urban reconstruction strategy case study: Khorramshahr. *International Journal of Tourism Cities*, 4(1), 81–97.

Novelli, M., Morgan, N., & Nibigira, C. (2012). Tourism in a post-conflict situation of fragility. *Annals of Tourism Research*, 39(3), 1446–1469.

Okafor, L. E., & Khalid, U. (2020). Regaining international tourism attractiveness after an armed conflict: The role of security spending. *Current Issues in Tourism*, 24(3), 385–402.

Özsoy, B. (2022). Rusya-Ukrayna Savaşı ve Türk Devletleri Teşkilatı Ülkeleri [Russia-Ukraine War and the Organization of Turkic States Countries]. *Türk Dünyası İncelemeleri Dergisi/Journal of Turkish World Studies*, 22(2), 549–576.

Pandey, D. K., & Kumar, R. (2023). Russia-Ukraine War and the global tourism sector: A 13-day tale. *Current Issues in Tourism*, 26(5), 692–700.

Seabra, C., Dolnicar, S., Abrantes, J. L., & Kastenholz, E. (2013). Heterogeneity in risk and safety perceptions of international tourists. *Tourism Management*, 36, 502–510.

Seaton, A. V. (1996). Guided by the dark: From thanatopsis to thanatourism. *International Journal of Heritage Studies*, 2(4), 234–244.

Seaton, A. V. (1999). War and thanatourism: Waterloo 1815–1914. *Annals of tourism Research*, 26(1), 130–158.

Smith, V. L. (1998). War and tourism: An American ethnography. *Annals of Tourism Research*, 25(1), 202–227.

Suntikul, W. (2019). Tourism and war: Global perspectives. In D. J. Timothy (Ed.) *Handbook of globalisation and tourism* (pp. 139–148). Edward Elgar Publishing.

T.C. Dışişleri Bakanlığı. (2024). Rusya Federasyonu Tarafından Ukrayna'ya Yönelik Başlatılan Askeri Operasyon Hk (No: 62). Retrieved from: https://www.mfa.gov.tr/no_-62_-rusya-federasyonu-tarafindan-ukrayna-ya-yonelik-baslatilan-askeri-operasyon-hk.tr.mfa

T.C. Kültür ve Turizm Bakanlığı. (2023). Yıllık Bülten-2023 Yılı Sınır Giriş-Çıkış İstatistikleri. Retrieved from: https://yigm.ktb.gov.tr/TR-249709/yillik-bultenler.html

Tan, J., & Cheng, M. (2024). Tourism, war, and media: The Russia-Ukraine War narrative. *Journal of Travel Research*, 1–15.

Timothy D. J. (2013). Tourism, war, and political instability: territorial and religious perspectives. In R. Butler & W. Suntikul (Eds.), *Tourism and war* (pp. 12–25) Routledge.

TÜBA. (2023). *TÜBA Filistin – İsrail Savaşı raporu [TÜBA report on the Palestinian – Israeli War] (Report no: 53)*. Retrieved from: https://tuba.gov.tr/tr/yayinlar/suresiz-yayinlar/raporlar/tuba-filistin-israil-savasi-raporu

UN General Assembly. (2024). *Illegal Israeli actions in Occupied East Jerusalem and the rest Occupied Palestinian Territory: Admission of new Members to the United Nations (A/ES-10/L.30/Rev.1)*. Retrieved from: https://documents.un.org/doc/undoc/ltd/n24/129/97/pdf/n2412997.pdf?token=3ONo9KoHU6SmS0Ci9I&fe=true

UN Tourism, 2024. Impact of the Russian offensive in Ukraine on international tourism. Retrieved from: https://www.unwto.org/impact-russian-offensive-in-ukraine-on-tourism

UNWTO. (2022). Impact of the Russian offensive in Ukraine on international tourism. UNWTO Tourism Market Intelligence and Competitiveness Department, Issue: 4, 16 May 2022.

Williams, N. L., Wassler, P., & Fedeli, G. (2023). Social representations of war tourism: A case of Ukraine. *Journal of Travel Research*, 62(4), 926–932.

Yamaç Erdoğan, Z. (2023). Kurtuluş Savaşı'ndaki kadın kahramanların Cumhuriyet Dönemi'ndeki tematik bir müze bağlamında incelenmesi [Analysing the female heroes in the War of Liberation in the context of a thematic museum in the Republic Period]. *Gaziantep University Journal of Social Sciences, 22*(Cumhuriyet'in 100. Yılı Özel Sayısı), 679–699.

Yasin Ozaslan[1]

Chapter 18 Service Robots and Artificial Intelligence in the Hospitality Industry: A Literature Review

Introduction

Nowadays, innovations in information and communication technologies are leading to changes in individuals' needs and expectations (Gomber et al., 2017). With these developments, individuals want to expand their comfort zones in their work and social lives. Businesses, on the other hand, aim to attract individuals' attention and gain a competitive advantage over rival companies by utilizing developing technological applications (Kao & Huang, 2023). Since the tourism sector encompasses many different industries, businesses operating in this field must adapt to innovations to succeed in a highly competitive environment.

Melissen et al. (2019) state that the hospitality sector is labor-intensive and dependent on humans, but it has begun to embrace technological innovations. Service robots offer advantages such as automating work processes, reducing costs, and attracting customer interest (Allman, 2014; Harari, 2017; Bowen & Morosan, 2018). The tourism and hospitality sector are witnessing the rapid development of technologies such as online reservations, check-in, mobile payments, and virtual reality (Wu et al., 2015; Shen et al., 2016; Tussyadiah et al., 2018; Liu & Mattila, 2019). These technologies aim not to replace humans in service delivery but to facilitate processes. According to Le et al. (2022), the integration of service robots is anticipated to enhance productivity, improve service quality, and boost operational efficiency. The execution of service tasks encompasses a diverse array of activities necessitating varying degrees of intelligence and expertise, thereby underscoring the significance of human-robot collaboration in the current hospitality industry. This collaboration is critical because both employees and robots possess unique advantages compared to each other (Huang & Rust, 2018). Tourism businesses aware of technological advancements have started using service robots since 2015. The first robot hotel, Henn na Hotel, opened in 2015 at Japan's Huis Ten Bosch amusement park (Rajesh, 2015). At

---

1 Ph.D., Yalova University, Yalova Vocational High School, Department of Tourism Management, yasinozaslan@yalova.edu.tr

Amsterdam Schiphol Airport, the android robot named Spencer guided passengers, while KLM Royal Dutch Airlines introduced the autonomous Care-E robot for luggage assistance in 2018 (KLM, 2016). Gradually, the use of service robots has increased, and applications such as dinosaur-shaped receptionists, cloakroom robots, robot porters, and in-room personal assistants have become widespread in hotel operations. Although fully robotic hotels are still limited, many hotels are implementing smart automation applications such as room delivery robots, autonomous check-in, and virtual personal assistants. The increasing use of service robots and artificial intelligence in the tourism sector has also attracted the attention of academics, leading to a rise in the number of publications since 2018. Studies have focused on the perceptions of employees and customers regarding AI-supported service robots, as well as on issues of risk, security, privacy, and ethics. However, most of the existing literature consists of literature reviews. This is due to the limited prevalence of robots in the industry and the high cost of conducting experiments. In the future, as the adoption of robots increases, it is expected that the number of conceptual and descriptive papers will decrease, and empirical research will increase. Given that this field is new, and the application area is limited, this study also conducted a literature review. A search conducted on Web of Science (WOS) on May 3, 2024, using the keywords "Artificial Intelligence" "Service Robot" and "Tourism" yielded 40 studies (including 33 articles, 5 review articles, and 2 conference papers). The reviewed studies aimed to explain the perceptions and attitudes of both customers and employees regarding AI and service robots in tourism businesses. Additionally, suggestions for future research were provided.

Artificial Intelligence and Service Robots

In recent years, technological advancements have enabled the use of artificial intelligence across a wide range of areas, from production to marketing and from education to the service sector, leading to significant transformations in various industries. Artificial intelligence (AI) can be defined as human intelligence technology created using computer programs (Tsaih & Hsu, 2018). Shukla Shubhendu and Vijay (2013) define AI as decision-making systems that imitate human intelligence by synthesizing large databases with complex hardware and software. The fourth industrial revolution is characterized by technological advancements such as robotics, nanotechnology, artificial intelligence, quantum computing, the internet of things, and fully autonomous vehicles (Schwab, 2017). While the third industrial revolution was characterized by the implementation of intelligent and autonomous systems supported by data analytics and machine learning, the fourth industrial revolution is distinguished by machines that communicate with one another and autonomously make decisions without human intervention. AI enables machines to produce human-like solutions by applying human characteristics through computer algorithms. This allows businesses to offer personalized suggestions based on customers' previous needs and preferences, and to conduct human-like interactions without human intervention. Huang and Rust (2018) identified four types of intelligence that AI can use to perform service tasks: mechanical, analytical, intuitive, and empathetic intelligence. AI is increasingly being used in the context of customer services and is becoming an important source of service innovation. These interactions can take place virtually or face-to-face. AI-based applications include query-based response systems like Siri on iPhone, Alexa on Amazon's Echo, Cortana on Microsoft, and Macy's on Call developed for retailers. Additionally, AI service robots are becoming widespread in various service sectors, both as online virtual bots (such as chatbots) and as physical entities (Huang & Rust, 2018).

Current studies on the role of AI-based ideas and related devices in the business world indicate that service robots are becoming an increasing reality and are likely to replace humans providing services in various environments (Loureiro et al., 2021). Wirtz et al. (2018) note that service robots encompass a variety of design features, including non-humanoid forms like iRobot's Roomba and virtual AI software such as Google Assistant. However, humanoid robots, exemplified by models like Pepper, stand out as some of the most significant innovations in the business world. For example, Forbes (2019) states that humanoid robots generate positive interest for companies. This can be explained by the fact that humanoid robots are found to be more approachable by users due to their ability to exhibit human-like gestures and emotions. Humanoid service robots distinctly differ from traditional machines in their capacity to engage in socially meaningful interactions with consumers (Brengman et al., 2021; Wirtz et al., 2018). These capabilities allow robots to establish deeper and more empathetic communication with users, enriching the customer experience and helping businesses enhance customer satisfaction.

In the current tourism industry, numerous hotels are at the forefront of adopting service robots, significantly transforming the guest experience with cutting-edge technology. For instance, the Hilton McLean Tysons Corner hotel in Virginia, USA, features a robot concierge named Connie. Connie assists guests with recommendations for local attractions, dining options, and directions within the hotel, providing a personalized touch to their stay (Gagliordi, 2016; Trejos, 2016). This integration of AI and robotics not only enhances guest convenience but also sets a new standard for hospitality services. In Nagasaki, Japan, the Henn na Hotel is notable for being the first fully automated hotel that utilizes service robots, setting a precedent for future hospitality innovations (Osawa et al., 2017). The hotel's use of robots for check-in, room service, and even luggage storage exemplifies the potential for automation in reducing operational costs and improving efficiency. Guests experience a futuristic stay with minimal human interaction, which can be particularly appealing during health crises like the COVID-19 pandemic. Aloft Hotels have introduced robot bellboys, streamlining luggage handling and guest services, while Marriott Hotels employ a robot concierge named Mario, who enhances guest interactions by providing valuable information and a unique, engaging experience. These robots are designed to assist with repetitive tasks, allowing human staff to focus on more personalized and complex guest needs, thereby improving overall service quality. At Hotel EMC2 in Chicago, two robots named Cleo and Leo are employed to deliver items such as food or towels to guests' rooms and take orders, significantly improving room service efficiency and guest satisfaction (Chestler, 2020). These robots can navigate the hotel autonomously, ensuring timely deliveries without the need for human intervention. This not only speeds up service but also ensures consistency and reliability. The FlyZoo Hotel in Hangzhou, China, which opened in 2018, offers a unique lodging experience characterized by minimal human interaction. It leverages service robots and advanced technologies such as facial recognition check-in and voice control, showcasing the potential for futuristic hotel stays (Song et al., 2022). The hotel's innovative approach includes using robots for room service, cleaning, and even entertainment, providing guests with a seamless, tech-driven experience. The Crowne Plaza San Jose-Silicon Valley Hotel features a robot named Dash that delivers items to guest rooms, enhancing the convenience and speed of room service (Hornyak, 2020). Dash can interact with elevators, navigate hallways, and deliver items directly to guests' doors, ensuring that requests are fulfilled promptly. This level of service is particularly beneficial in high-occupancy scenarios where human staff might be overwhelmed. Additionally, in California, the integration of robots into the food service industry is exemplified by a burger robot that can fulfill up to 120 orders per hour and Café X, which uses robot baristas capable of preparing three drinks in forty seconds. These advancements highlight the efficiency and precision that robotics can bring to hospitality and food services (Troitino, 2018; Canales, 2018). By reducing wait times and improving consistency, these robots enhance customer satisfaction and operational efficiency. These examples demonstrate how service robots are evolving in the hospitality industry and enhancing the customer experience. Through these innovative technologies, hotels can provide faster, more efficient, and personalized services to their guests while also increasing their operational efficiency. As technology continues to advance, the role of service robots in the hospitality industry is likely to expand, offering even more opportunities to improve guest satisfaction and streamline hotel operations.

Service Robots from the Customers' Perspective

The adoption of robot technology in the tourism sector has occurred relatively late compared to other industries. This is likely due to the complex responses required to meet customer needs in the services provided. For example, while robots in a car factory typically take on repetitive and predictable tasks, complex tasks such as ensuring guest satisfaction and providing personalized service gain more importance in the hospitality sector. Therefore, even by the mid-1990s, while many car factories extensively utilized robots, a predominantly robot-operated hotel only opened in 2015. This indicates that the adoption process of technology in the hospitality industry has been slower than in the automotive sector (Ivanov et al., 2019). However, tourism businesses are increasingly emphasizing technologies that will enhance customer experiences and meet customer needs (Gretzel, 2021).

The increasing popularity of service robots in the hospitality sector has led many researchers to show interest in this area. Numerous studies have explored various aspects of this phenomenon, including customer perceptions of service robots, their experiences (Tung and Au, 2018), service satisfaction levels (Prentice et al., 2020), technology acceptance (Zhang et al., 2021), and intention (Yang et al., 2021). Moreover, empirical research has examined the impact of service robots on consumers' perceived value (de Kervenoael et al., 2020), their propensity for positive word-of-mouth, and their intentions to interact with robots (Lu et al., 2021).

Companies planning to use service robots face a specific challenge, known in the literature as anthropomorphism. Anthropomorphism refers to the attribution of human characteristics, behaviors, or qualities to non-human entities or objects (Koda and Maes, 1996). The creation of human-like devices, known as anthropomorphism, is a crucial aspect of robotics (Mende et al., 2019). Despite its importance, consumers may experience discomfort when interacting with these robots (Blut et al., 2021). Marketing professionals have identified that anthropomorphism can enhance affection for brands and products (Sheehan et al., 2020). Nonetheless, the effects of service robots' anthropomorphic characteristics on consumer responses and the reasons behind these effects are still not fully understood. Some scholars suggest that anthropomorphic features in service robots can increase consumer engagement, as these human-like attributes align with the fundamental principles and expectations people apply in social interactions with robots (Duffy, 2003). Conversely, other researchers, such as Mende et al. (2019), express skepticism, arguing that heightened perceived anthropomorphism may lead to consumer discomfort, particularly due to feelings of eeriness and a perceived threat to human identity.

Mori's (1970) Uncanny Valley Theory posits that while anthropomorphism initially enhances positive impressions of service robots, excessively high levels of anthropomorphism can elicit negative reactions. A service robot with human-like features can induce greater unease in consumers compared to one that appears more mechanical. Robots exhibiting moderately high levels of anthropomorphism may diminish consumers' attitudes and influence their judgments of warmth (Kim et al., 2019). However, excessive anthropomorphism can lead to positive perceptions because of the U-shaped relationship that exists between individuals and highly human-like robots (Blut et al., 2021). This type of relationship can be described as one where, initially, an increase in one variable leads to a decrease in the other variable (or vice versa), and after a certain point, this trend reverses.

In a study analyzing online reviews of service robots, Yu (2020) found that individuals generally hold negative perceptions of human-like robots in hotel settings. Despite these adverse reactions to anthropomorphic robots, Van Pinxteren et al. (2019) discovered that anthropomorphism can enhance customers' trust in and their intention to use AI robots. It has been shown that anthropomorphic designs also influence consumer behaviors and perceptions (De Visser et al., 2016). Additionally, other researchers (Blut et al., 2021) indicate that customers are more likely to trust a human-like robot because such a robot is generally assumed to perform tasks more effectively. As emphasized in the consumer behavior literature, customers who perceive employees as attractive and charming tend to attribute higher service value to them. This perception increases customers' willingness to tip, spend more money, and purchase more expensive products (Jacob & Guéguen, 2014). According to Mori (1970), as robots become more human-like, the perceived affinity towards them increases. For example, industrial robots in factories that lack faces and legs elicit almost no sense of closeness from people due to their lack of human resemblance. Conversely, as robots acquire human-like appearances and features, people can develop a sense of attachment towards them (Mori et al., 2012).

Mende et al. (2019), in their research on customer interactions with humanoid service robots in the hospitality industry, found that the acceptance of service robots is influenced by the context in which they are utilized and the customer's familiarity and comfort with technology. Furthermore, Lin et al. (2020)

noted that customers' acceptance of service robots varies based on whether they are staying at a full-service hotel or a limited-service hotel. There are also studies that measure perceptions of service robots based on gender and age. For example, studies on service robots in the hospitality sector have shown that young people are more supportive of the introduction of service robots, and women are more skeptical towards them (Ivanov et al., 2018). It has been found that men rate female robots more positively than women do (Yu & Ngan, 2019), indicating that the design of service robots can influence perceptions. Additionally, children found interactions with service robots to be fun, innovative, and playful (Lee et al., 2017). However, parents were not very keen on service robots babysitting their children (Ivanov & Webster, 2020). Li et al. (2010) discovered that Germans exhibited lower levels of sympathy, engagement, trust, and satisfaction in interactions with anthropomorphic robots compared to their Chinese and Korean counterparts. Existing studies indicate that the acceptance of service robots in hospitality businesses depends on factors such as the purpose of using the robots, the customer's attitude towards technology, age, and gender; for instance, young people and men tend to have a more positive approach towards service robots, while women and parents may be more skeptical.

Service Robots from the Employees' Perspective

The deployment of service robots in the hospitality industry is rapidly expanding to reduce costs and enhance efficiency. In China, waiter robots operate at a cost lower than the average annual salary of hospitality workers and can serve 50 % to 100 % more than human employees (Hospitality and Marketing News, 2019). This highlights the significant impact of service robots on the workforce. The introduction of these robots will considerably alter workflows, job design, and the organizational environment, thereby profoundly influencing employees' perceptions and job performance (Song et al., 2022). Numerous studies have investigated the factors that influence employees' willingness to work with robots and the reasons behind their resistance (Pizam et al., 2022; Fu et al., 2022). Some employees may find service robots enjoyable and engaging, thus fostering a hedonic motivation to work with them (Ali et al., 2023). Conversely, the integration of service robots into the workforce can be perceived as a threat by some employees, resulting in resistance to their use (Li et al., 2019).

Theoretically, service robots are intended to serve as a job resource to assist hospitality employees. However, as technology is not without flaws, service failures are inevitable (Hertzfeld, 2019). Individual readiness and acceptance of technology vary, and while employees may acknowledge the benefits of robots, they may also view them as a threat (Paluch et al., 2022). Consequently, the use of service robots may impose an additional burden on employees' roles, making it essential to determine whether this is perceived as a resource that can enhance job performance, particularly in the face of physically and emotionally demanding tasks. The integration of service robots also brings the fear of replacing humans in the workplace (Decker et al., 2017). However, research on the adoption of service robots indicates that more information is needed on how they affect employees' levels of stress and anxiety (Wirtz et al., 2018). It remains uncertain whether service robots will reduce the workload or enhance employee performance (Qiu et al., 2020). Employees may adopt robots with the assumption that they will help them perform their jobs better (Xiao and Kumar, 2019). However, the level of interaction employees has with robots can influence the formation of acceptance and adaptation (Gretzel & Murphy, 2021).

Although human-to-human bonding is well-documented in the workplace, the extent of reciprocity, bonding, and trust that can develop between employees and service robots remains largely unexplored (Qiu et al., 2020). Additionally, discomfort in working with robots may elevate an employee's intention to resign (Li et al., 2019). The implementation of smart automation also brings risks such as job loss, especially the loss of "low-tech" jobs, and loss of control due to robot autonomy (Huang and Rust, 2018). Employees' impressions of robots can vary across different roles and levels in the hospitality sector (Li et al., 2019). For example, research shows strong evidence of robot fear among women, nonwhites, and less educated individuals (McClure, 2018).

In conclusion, the use of service robots in the hospitality sector can significantly affect how employees perceive and perform their jobs. Whether these changes will increase the emotional labor requirements of employees is still a topic of research, and further examination of employee experiences during this process is necessary.

Results

Service robots offer various advantages to businesses with their innovations (Bowen & Morosan, 2018). However, the effects of these innovations on employees and customers in the sector should be carefully evaluated. From the employees' perspective, the integration of service robots can change workflows and job design, which may lead to a perceived threat for some employees (Li et al., 2019). On the other hand, some employees may welcome these innovations positively and see working with robots as a source of hedonic motivation (Ali et al., 2023).

From the customers' perspective, interactions with service robots can lead to both positive and negative perceptions. While anthropomorphic robots have the potential to establish more meaningful social interactions with consumers, a high level of anthropomorphism can sometimes cause discomfort (Mori, 1970; Blut et al., 2021). Therefore, it is important to consider customers' perceptions and reactions when designing and using service robots. The existing literature on the integration of service robots into the sector is generally limited to conceptual and descriptive studies, with a noticeable lack of empirical research. The limited use of robots in the industry or the high cost of laboratory experiments has led to a scarcity of empirical studies in this area. However, as the adoption of service robots increases, it is expected that more research will be conducted in this field, and more detailed data on the functionality of robots will be obtained.

In the future, some recommendations can be made for the successful implementation of service robots in the tourism sector. Firstly, comprehensive training programs should be organized, and psychological support provided to ensure that employees can work harmoniously with service robots. Secondly, customers should be informed about the advantages and use of service robots to alleviate their concerns. Additionally, ethical and security standards for service robots should be established, and continuous feedback mechanisms should be implemented.

In conclusion, the use of service robots in the tourism sector can lead to significant changes for both employees and customers. In this process, carefully managing the adaptation and acceptance of technology is critical for increasing operational efficiency and ensuring customer satisfaction. Future research will provide more comprehensive and in-depth data on the role and impact of service robots in the sector, contributing to the more effective and efficient use of these technologies.

References

Allman, T., (2014). China restaurant introduces robot waiters. BBC News. https://www.bbc.com/news/av/world-asia-30460737/china-restaurant-introduces-robot-waiters. Accessed 29 May 2024.

Blut, M., Wang, C., Wünderlich, N. V., & Brock, C. (2021). Understanding anthropomorphism in service provision: a meta-analysis of physical robots, chatbots, and other AI. Journal of the Academy of Marketing Science, 49, 632–658.

Bowen, J., & Morosan, C. (2018). Beware hospitality industry: the robots are coming. Worldwide Hospitality and Tourism Themes, 10(6), 726–733.

Brengman, M., De Gauquier, L., Willems, K., & Vanderborght, B. (2021). From stopping to shopping: An observational study comparing a humanoid service robot with a tablet service kiosk to attract and convert shoppers. Journal of Business Research, 134, 263–274.

Canales, K. (2018). This futuristic café relies on robots to take your order and make your coffee—No human interaction required. Business Insider.

Chestler, D. (2020). The future is now: How robots are storming the travel industry. SiteMinder.

de Kervenoael, R., Hasan, R., Schwob, A., & Goh, E. (2020). Leveraging human-robot interaction in hospitality services: Incorporating the role of perceived value, empathy, and information sharing into visitors' intentions to use social robots. Tourism Management, 78, 104042.

De Visser, E. J., Monfort, S. S., McKendrick, R., Smith, M. A., McKnight, P. E., Krueger, F., & Parasuraman, R. (2016). Almost human: Anthropomorphism increases trust resilience in cognitive agents. Journal of Experimental Psychology: Applied, 22(3), 331.

Decker, M., Fischer, M., & Ott, I. (2017). Service Robotics and Human Labor: A first technology assessment of substitution and cooperation. Robotics and Autonomous Systems, 87, 348–354.

Duffy, B. R. (2003). Anthropomorphism and the social robot. Robotics and Autonomous Systems, 42(3–4), 177–190.

Forbes (2019), Artificial intelligence in humanoid robots. www.forbes.com/sites/cognitiveworld/2019/02/25/artificial-intelligence-in-humanoid-robots/?sh=2127796424c7 Accessed 30 May 2024.

Fu, S., Zheng, X., & Wong, I. A. (2022). The perils of hotel technology: The robot usage resistance model. International Journal of Hospitality Management, 102, 103174.

Gagliordi, N. (2016). This Watson-Powered robot concierge is rethinking the hotel industry. ZDNET. https://www.zdnet.com/article/this-watson-powered-robot-concierge-is-rethinking-the-hotel-industry/. Accessed 30 May 2024.

Gomber, P., Koch, J. A., & Siering, M. (2017). Digital Finance and FinTech: current research and future research directions. Journal of Business Economics, 87, 537–580.

Gretzel, U. (2021). Conceptualizing the smart tourism mindset: Fostering utopian thinking in smart tourism development. Journal of Smart Tourism, 1(1), 3–8.

Gretzel, U., & Murphy, J. (2019). Making sense of robots: Consumer discourse on robots in tourism and hospitality service settings. In Robots, artificial intelligence, and service automation in travel, tourism and hospitality (pp. 93–104). Emerald Publishing Limited.

Harari, Y. N. (2017). Reboot for the AI revolution. Nature, 550(7676), 324–327.

Hertzfeld, E. (2019). Japan's Henn na Hotel fires half its robot workforce. Hotel Management, 31.

Hornyak, T. (2020). Meet the robots that may be coming to an airport near you. CNBC. https://www.cnbc.com/2020/01/10/meet-the-robots-that-may-be-coming-to-an-airport-near-you.html Accessed 30 May 2024.

Hospitality and Marketing News (2019). Robot Waiters, it's happening now and coming to a restaurant near you soon. https://www.hospitalityandcateringnews.com/2019/09/robot-waiters-happening-now-coming-restaurant-near-soon/. Accessed 01 June 2024.

Huang, M. H., & Rust, R. T. (2018). Artificial intelligence in service. Journal of Service Research, 21(2), 155–172.

Ivanov, S., & Webster, C. (2020). Robots in tourism: A research agenda for tourism economics. Tourism Economics, 26(7), 1065–1085.

Ivanov, S., Webster, C., & Garenko, A. (2018). Young Russian adults' attitudes towards the potential use of robots in hotels. Technology in Society, 55, 24–32.

Ivanov, S., Gretzel, U., Berezina, K., Sigala, M., & Webster, C. (2019). Progress on robotics in hospitality and tourism: a review of the literature. Journal of Hospitality and Tourism Technology, 10(4), 489–521.

Jacob, C., & Guéguen, N. (2014). The effect of employees' clothing appearance on tipping. Journal of Foodservice Business Research, 17(5), 483–486.

Kao, W. K., & Huang, Y. S. S. (2023). Service robots in full-and limited-service restaurants: Extending technology acceptance model. Journal of Hospitality and Tourism Management, 54, 10–21.

Kim, S. Y., Schmitt, B. H., & Thalmann, N. M. (2019). Eliza in the uncanny valley: Anthropomorphizing consumer robots increases their perceived warmth but decreases liking. Marketing Letters, 30, 1–12.

KLM (2016). Spencer robot completed tests guiding KLM passengers at Schiphol https://news.klm.com/spencer-robot-completed-tests-guiding-klm-passengers-at-schiphol/ Accessed 29 May 2024.

Koda, T., & Maes, P. (1996, November). Agents with faces: The effect of personification. In Proceedings 5th IEEE international workshop on robot and human communication. RO-MAN'96 TSUKUBA (pp. 189–194). IEEE.

Le, K. B. Q., Sajtos, L., & Fernandez, K. V. (2022). Employee-(ro)bot collaboration in service: an interdependence perspective. Journal of Service Management, 34(2), 176–207.

Lee, J. J., Kim, D. W., & Kang, B. Y. (2017). Esthetic interaction model of robot with human to develop social affinity. International Journal of Advanced Robotic Systems, 14(4).

Li, D., Rau, P. P., & Li, Y. (2010). A cross-cultural study: Effect of robot appearance and task. International Journal of Social Robotics, 2, 175–186.

Li, J. J., Bonn, M. A., & Ye, B. H. (2019). Hotel employee's artificial intelligence and robotics awareness and its impact on turnover intention: The moderating roles of perceived organizational support and competitive psychological climate. Tourism Management, 73, 172–181.

Lin, H., Chi, O. H., & Gursoy, D. (2020). Antecedents of customers' acceptance of artificially intelligent robotic device use in hospitality services. Journal of Hospitality Marketing & Management, 29(5), 530–549.

Liu, S. Q., & Mattila, A. S. (2019). Apple Pay: Coolness and embarrassment in the service encounter. International Journal of Hospitality Management, 78, 268–275.

Loureiro, S. M. C., Guerreiro, J., & Tussyadiah, I. (2021). Artificial intelligence in business: State of the art and future research agenda. Journal of Business Research, 129, 911–926.

Lu, L., Zhang, P., & Zhang, T. C. (2021). Leveraging "human-likeness" of robotic service at restaurants. International Journal of Hospitality Management, 94, 102823.

McClure, P. K. (2018). "You're fired," says the robot: The rise of automation in the workplace, technophobes, and fears of unemployment. Social Science Computer Review, 36(2), 139–156.

Melissen, F., Van der Rest, J. P., Josephi, S., & Blomme, R. (2019). Hospitality Experience. Routledge.

Mende, M., Scott, M. L., Van Doorn, J., Grewal, D., & Shanks, I. (2019). Service robots rising: How humanoid robots influence service experiences and elicit compensatory consumer responses. Journal of Marketing Research, 56(4), 535–556.

Mori, M. (1970). Bukimi no tani [The uncanny valley]. Energy, 7, 33.

Mori, M., MacDorman, K. F., & Kageki, N. (2012). The uncanny valley [from the field]. IEEE Robotics & Automation Magazine, 19(2), 98–100.

Osawa, H., Ema, A., Hattori, H., Akiya, N., Kanzaki, N., Kubo, A., ... & Ichise, R. (2017, March). What is real risk and benefit on work with robots? From the analysis of a robot hotel. In Proceedings of the Companion of the 2017 ACM/IEEE International Conference on Human-Robot İnteraction (pp. 241–242).

Paluch, S., Tuzovic, S., Holz, H. F., Kies, A., & Jörling, M. (2022). "My colleague is a robot"–exploring frontline employees' willingness to work with collaborative service robots. Journal of Service Management, 33(2), 363–388.

Pizam, A., Ozturk, A. B., Balderas-Cejudo, A., Buhalis, D., Fuchs, G., Hara, T., ... & Chaulagain, S. (2022). Factors affecting hotel managers' intentions to adopt robotic technologies: A global study. International Journal of Hospitality Management, 102, 103139.

Prentice, C., Weaven, S., & Wong, I. A. (2020). Linking AI quality performance and customer engagement: The moderating effect of AI preference. International Journal of Hospitality Management, 90, 102629.

Qiu, H., Li, M., Shu, B., & Bai, B. (2020). Enhancing hospitality experience with service robots: The mediating role of rapport building. Journal of Hospitality Marketing & Management, 29(3), 247–268.

Rajesh, M. (2015). Inside Japan's first robot-staffed hotel. The Guardian, 14.

Schwab, K. (2017). The Fourth Industrial Revolution; Currency: Redfern, Australia.

Sheehan, B., Jin, H. S., & Gottlieb, U. (2020). Customer service chatbots: Anthropomorphism and adoption. Journal of Business Research, 115, 14–24.

Shen, H., Zhang, M., & Krishna, A. (2016). Computer interfaces and the "direct-touch" effect: Can iPads increase the choice of hedonic food? Journal of Marketing Research, 53(5), 745–758.

Shukla Shubhendu, S., & Vijay, J. (2013). Applicability of artificial intelligence in different fields of life. International Journal of Scientific Engineering and Research, 1(1), 28–35.

Song, Y., Zhang, M., Hu, J., & Cao, X. (2022). Dancing with service robots: The impacts of employee-robot collaboration on hotel employees' job crafting. International Journal of Hospitality Management, 103, 103220.

Trejos, N., (2016). Introducing Connie, Hilton's new robot Concierge. https://www.usatoday.com/story/travel/roadwarriorvoices/2016/03/09/introducing-connie-hiltons-new-robot-concierge/81525924/. Accessed 30 May 2024.

Troitino, C. (2018). Meet the world's first fully automated burger robot: Creator debuts the Big Mac killer. Forbes.

Tsaih, R.-H. ve Hsu, C. C. (2018). Artificial intelligence in smart tourism: a conceptual framework. The 18th International Conference on Electronic Business. Guilin, China, 124–133.

Tung, V. W. S., & Au, N. (2018). Exploring customer experiences with robotics in hospitality. International Journal of Contemporary Hospitality Management, 30(7), 2680–2697.

Tussyadiah, I. P., Wang, D., Jung, T. H., & Tom Dieck, M. C. (2018). Virtual reality, presence, and attitude change: Empirical evidence from tourism. Tourism management, 66, 140–154.

Van Pinxteren, M. M., Wetzels, R. W., Rüger, J., Pluymaekers, M., & Wetzels, M. (2019). Trust in humanoid robots: implications for services marketing. Journal of Services Marketing, 33(4), 507–518.

Wirtz, J., Patterson, P. G., Kunz, W. H., Gruber, T., Lu, V. N., Paluch, S., & Martins, A. (2018). Brave new world: service robots in the frontline. Journal of Service Management, 29(5), 907–931.

Wu, L., Fan, A. A., & Mattila, A. S. (2015). Wearable technology in service delivery processes: The gender-moderated technology objectification effect. International Journal of Hospitality Management, 51, 1–7.

Xiao, L., & Kumar, V. (2021). Robotics for customer service: a useful complement or an ultimate substitute. Journal of Service Research, 24(1), 9–29.

Yang, H., Song, H., Cheung, C., & Guan, J. (2021). How to enhance hotel guests' acceptance and experience of smart hotel technology: An examination of visiting intentions. International Journal of Hospitality Management, 97, 103000.

Yu, C. E. (2020). Humanlike robots as employees in the hotel industry: Thematic content analysis of online reviews. Journal of Hospitality Marketing & Management, 29(1), 22–38.

Yu, C. E., & Ngan, H. F. B. (2019). The power of head tilts: gender and cultural differences of perceived human vs human-like robot smile in service. Tourism Review, 74(3), 428–442.

Zhang, M., Gursoy, D., Zhu, Z., & Shi, S. (2021). Impact of anthropomorphic features of artificially intelligent service robots on consumer acceptance: moderating role of sense of humor. International Journal of Contemporary Hospitality Management, 33(11), 3883–3905.

Nihat Çeşmeci[1]

Chapter 19 Affiliate Marketing in Tourism and Its Implementation by Travel Bloggers

Introduction

The rapid spread of the Internet throughout the world in the last quarter century has led to the emergence of digital marketing, which has become the main mediator of marketing efforts in every industry. Today, digital marketing is also an indispensable element of marketing efforts for almost all the products in the tourism industry. The advent of digitalization has enabled tourists to purchase a range of travel products and services independently, thereby assuming the intermediary role previously fulfilled by tour operators and travel agencies. This transition has prompted all tourism businesses, including producers and distribution channels to adopt a more prominent role in digital marketing and to embrace innovations for the effective reach of larger audiences. In addition to digital marketing activities, such as search engine marketing, search engine optimization (Daniele et al., 2009), and social media marketing (Tiago & Veríssimo, 2014), one of the most cost-effective ways of online promotion and distribution of products in a digital environment is the affiliate marketing method (Fox & Wareham, 2010; Grégori et al., 2014). This marketing method is a strategic tool (Premachandra & Eranda, 2022), frequently used by large intermediary businesses that operate online and do electronic distribution (Daniele et al., 2009; Snyder & Kanich, 2016). Today many travel companies spend a considerable amount of their online budget on affiliate marketing (Daniele et al., 2009). In affiliate marketing, which is based on the win-win principle (Duffy, 2005), the publisher or "affiliate" aims to direct tourists to the websites of businesses that offer or intermediate touristic goods or services and to realize a sale or promote products to website visitors (Grégori et al., 2014; Gupta & Aggarwal, 2019). If a sale is realized, the affiliate person or organization earns some financial benefits (Premachandra & Eranda, 2022); for example, a certain amount of commission or a certain fixed income per click for redirecting the visitor (Duffy, 2005; Makarkina, 2021). In recent years, there have been a considerable number of

---

[1] Assoc. Prof., Erciyes University, Tourism Faculty, Department of Tour Guiding ncesmeci@erciyes.edu.tr

entrepreneurs all over the world (Lee & Gretzel, 2014; Blaer et al., 2020) and in Türkiye (Karabacak & Genç, 2019; Develi, 2021), who have established travel blogs, generating additional income through such activities, and even making these activities their main source of income. In tourism marketing literature there are very few studies that deal with affiliate marketing (Daniele et al., 2009; Grégori et al., 2014; Dwivedi et al., 2017; Tanwar & Sahu, 2024), especially within the framework of travel blogs. Based on this information, the first main objective of this study is to make a general assessment of the status of affiliate marketing in the tourism industry based on the relevant literature. The second aim of the research is to determine whether travel bloggers in Türkiye are involved in affiliate marketing, how they use affiliate marketing methods, and which affiliate programs' products they most frequently prefer in their blogs. This research is designed as qualitative exploratory research that uses a multiple case study method. The blogs of the most well-known travel bloggers in Türkiye who attract a lot of visitors will be evaluated. The clues to be obtained by analyzing their affiliate marketing tools will assist in a more comprehensive understanding of the marketing dynamics within the tourism industry and guide those who will engage in entrepreneurial activities in this field. In addition, it is thought that this study will contribute to the field of tourism marketing from a theoretical point of view and provide a basis for more detailed research on the subject. In this study, firstly the concepts of affiliate marketing, travel blogs, and travel bloggers will be explained, afterwards, the related literature will be given. After this, the research design and methodology will be introduced. In the final section, the key findings of the study will be presented, and a discussion will be held to assess the overall contribution of the study with suggestions for further research.

The Concept of Affiliate Marketing

According to the literature, affiliate marketing can be defined as a commission-based online collaboration between merchants and affiliates in which merchants pay affiliates for promoting and distributing the merchant's products via extra sales channels and referring customers to their websites (Duffy 2005; Mariussen et al., 2010; Bowie et al., 2014). The term "affiliate marketing" encompasses the entire industry, made up of three main groups, including merchants (advertisers), affiliates (publishers), and affiliate networks, also known as affiliate marketing agencies (Mariussen, 2011) or programs (Duffy, 2005; Daniele et al., 2009). Merchants are trying to reach their target audiences online, affiliates are generating traffic to merchants, and affiliate networks are intermediary firms, that facilitate interaction between merchants and affiliates (Mariussen, 2011).

Large corporations such as Amazon, Walmart, GoDaddy, and eBay run some of the most well-known affiliate marketing networks: Linkshare, Commission Junction, and Performics (Duffy 2005; Snyder & Kanich, 2016). As reported by Zahid (2024), today, more than 80 % of brands use affiliate marketing methods, and the most popular affiliate marketing niches include fashion, wellness, hobbies, technology, gaming, sports, and travel. The travel industry, characterized by its diverse range of products and services, has been considered very attractive for affiliate marketing program creation (Makarkina, 2021; Adel, 2023). Many service providers such as airlines, hotels, attractions, or tour operators seek today to become affiliated with websites or blogs that can provide them with quality web traffic in exchange for a commission (Mariussen, 2011). Recently many different digital business model configurations in the travel industry begun to offer affiliate marketing programs. For example, online travel agencies (Expedia, Booking.com); metasearch platforms (Skyscanner, Trivago, Kayak); meta-booking platforms (Rentalcars.com), and online travel marketplaces (Get Your Guide, Viator, TourRadar). There are also travel affiliate networks (such as Travelpayouts) that bring together many different affiliate programs to organize effective interaction between affiliates (travel bloggers), and meta-booking/metasearch platforms or online travel agencies (Perelygina, et al., 2022). The research on affiliate networks in Türkiye revealed that the country lacks structured affiliate marketing platforms (Ayaz, 2020). Based on this deficiency, an information management system model for affiliate marketing was proposed and application software was developed and tested by Ayaz & Güyer (2024).

Even though the interest in affiliate marketing is expanding, the number of studies and especially empirical research addressing this phenomenon is limited (Mariussen et al., 2010; Dwivedi et al., 2017; Tanwar & Sahu, 2024). Several studies published over the last two decades have studied affiliate marketing in different countries and from varying perspectives (Premachandra & Eranda, 2022; Tanwar & Sahu, 2024). The first review article was published in 2017 (Dwivedi et al., 2017; Markov & Timerbaev, 2021). This literature review analyses eighteen affiliate marketing studies and reveals a lack of research interest and effort on this topic (Dwivedi et al., 2017). In the second literature review study conducted by Tanwar & Sahu (2024) more recently, 63 studies on the subject were identified. Of the 63 manuscripts identified, 19 focus on concepts and conceptual framework; 21 focus on the functioning of affiliate marketing, and 13 focus on consumer behavior in affiliate marketing. While this is the case in the generic marketing literature, a different situation cannot be determined in the tourism marketing literature. In tourism-specific literature, it is apparent that there is a paucity of affiliate marketing research (Daniele et al., 2009; Mariussen et al.,

2010). This paucity and a dearth of high-quality data indicate that the characteristics of affiliate marketing channels in the travel industry are inadequately researched and understood (Markov & Timerbaev, 2021). Above all, the number of studies examining how travel bloggers use affiliate marketing tools and which methods are commonly preferred for affiliate marketing is almost non-existent.

Nowadays it is valuable for travel service producers to reach tourists in organic (non-artificial) ways. Tourists looking for services online are increasingly turning to ad-blocking software to circumvent advertising and social network algorithms (Syrdal et al., 2023). Compared to other digital content, consumers' trust is higher in opinions expressed digitally by humans (human-human trust), in personal blogs, forums, or as product/service reviews (Pan et al., 2007; Grégori et al., 2014; Nimmermann, 2020). That's why blogs and bloggers are among the main partners for cooperation in affiliate marketing.

Travel Blogs and Bloggers

Blogs are defined by Volo (2012: 151) as "free, public, web-based entries in reverse chronological order presented in a diary-style format". Volo (2010: 298) defines blogs shortly as "free public web-based journals". Blogs are websites that anybody, professional or amateur, can set up and update with simply an Internet connection. They don't require any knowledge or experience with coding (Karabacak & Genç, 2019). A blog can be defined also as a website with one or two topics covered, which includes articles or posts that are updated regularly (Esch et al., 2018) by a blogger. The blogger is an administrator of the blog and sometimes there can be two or more bloggers sharing the same blog and writing entries in it. Every blog entry (article) usually lets readers write informal comments or ask questions about the content of the entry, which enables interaction between the blogger and blog followers (Filimon et al., 2010).

Travel blogs and travel bloggers have become well-known in the tourism industry, drawing the interest of both marketers and tourism scholars (Maggiore et al., 2022). The role of travel blogs in facilitating the exchange of information among tourists as well as between destinations and businesses is becoming increasingly significant (Pan et al., 2007). Travel blogs offer valuable insights about the tourists' attitudes towards a destination or a business thus it can be said that blogging nowadays is an important element of the tourism consumption and marketing processes (Wenger, 2008). For example, Carson, D. (2008) suggests that travel blogs can yield useful information and be utilized as a market research tool for tourist destinations. Similarly, Bosangit et al. (2012) propose that travel blogs can be used to examine the post-consumption behavior of tourists.

Travel blogs frequently feature images, the author's reflections and experiences, travel tips, and suggestions for future travelers. Bloggers also often use their blogs to share their knowledge about a destination (Tussyadiah & Fesenmaier, 2008), thus contributing to creating or shaping the destination's brand image (Wenger, 2008; Marin et al., 2018). Nowadays travelers utilize blogs mainly to plan their journeys and like their word-of-mouth approach (Duffy & Kang, 2020). The relationship built between travel bloggers and their readers is valuable and critically important for triggering mechanisms of loyalty and trust (Maggiore et al., 2022). As a result, travel companies and organizations, have made bloggers their top target, using a wide range of promotional materials to kickstart their partnership (Marin et al., 2018). Although marketers are adding affiliate marketing programs to their promotional mix at a rapidly expanding rate, there has been little academic research into the phenomenon and its impact on results (Taylor, 2020). As Duffy (2005) notes, the essence of affiliate marketing can be understood better by examining the role of independent affiliates in the marketing process. Based on this background this research aims to determine whether travel bloggers in Türkiye are involved in affiliate marketing, how they use affiliate marketing tools, and which affiliate programs' products they most frequently prefer in their blogs.

Methodology

This research was designed as qualitative exploratory research and taking into consideration its exploratory nature, a multiple case study method was used. The multiple case study method offers the opportunity to analyze the data within each situation and across different situations. By studying multiple cases, it is possible to understand the similarities and differences between cases, which in turn allows insights into the important influences that arise from these differences and similarities (Gustafsson, 2017). The use of multiple cases allows for a more comprehensive examination of research questions and the development of more detailed theoretical frameworks (Eisenhardt & Graebner, 2007). As Çakar & Aykol (2020) state in their review of the usage of the case study methodology in tourism studies, for more credible and generalizable results, researchers are suggested to prefer multiple case studies.

The purposeful sampling approach (criterion sampling) was chosen to select the five travel blogs. Criterion sampling is frequently used in qualitative research to determine the cases most fit certain pre-determined criteria (Suri, 2011). Based on this view, travel blogs in this study were chosen based on three rationales: (a) the blog must be at least 10 years old; (b) must receive at least

5000 unique visitors in one month [April 2024] and (c) must be aimed primarily at a Turkish-speaking audience. The selected sampling criteria were chosen to gather information from the most noteworthy examples. The search for blogs that meet the research criteria was carried out in early May 2024, considering the "Best travel blogs" rankings previously made on a few popular travel websites. In addition,

Description of the Cases

The blogs of five active travel bloggers in Türkiye that attract many visitors were chosen as cases. Table 19.1 shows the blog age and monthly traffic to each blog, including the number of followers on different social media platforms. All five blog authors share content on Facebook, Twitter, YouTube, and Instagram, as seen in the table. Only one of the blogs (Case 4) is not presented on the YouTube platform. Being active on social media, increases the online visibility of the websites or blogs, supporting, promoting, and allowing them to reach more new readers (Karabacak & Genç, 2019; Adel, 2023; Şener et al., 2023).

Table 19.1. The Characteristics of Blogs Selected for the Study.

Cases/Blogs	Domain age	Monthly Blog traffic*	Social media followers			
			Facebook	Twitter / X	YouTube	Instagram
Case/Blog 1	11 years	470,8K	317K	5K	127K	350K
Case/Blog 2	15 years	6,7K	42K	10,5K	5,53K	120K
Case/Blog 3	13 years	22,2K	20K	6,68K	1,32K	3K
Case/Blog 4	11 years	13.2K	27K	4,9K	-	165K
Case/Blog 5	13 years	15,2K	402K	8,8K	8,17K	141K

*Based on ahrefs.com data for April 2024

Case/Blog 1: This travel blog's administrators are a married couple traveling together since 2012. He has worked as a cultural marketing manager, while his wife has managed projects in various cultural and art institutions. After 8 years of corporate life, they decided to redraw their lives in 2014. Since then, they only travel sharing their experiences on their travel blog. In 2022 a daughter was born, and they have been travelling with her ever since.

Case/Blog 2: This blog is founded by a female blogger, 43 years old now. She has a business administration education and a marketing communication master's degree. Took place in banking, insurance, and IT companies as a marketing coordinator and brand manager. In 2018 she left the corporate world and continues to work as a travel blogger and writer.

Case/Blog 3: The blog was founded by a male who is in his 40s now. He works as an industrial engineer and spends much of his spare time traveling, usually with his wife. However, there are six additional blog contributors.

Case/Blog 4: It is a blog founded by two girlfriends, both 34 years old now, who have known each other since childhood. One has a degree in international business administration, the other in cinema television, and a master's degree in communications. In addition to traveling and being bloggers, they work now as freelance content providers.

Case/Blog 5: This travel blog's administrator is male in his 50s, a veterinarian, studied information management and has an MBA in brand management. In 2009, he left the company where he worked for 12 years and started to travel as a backpacker. In 2022 got married and nowadays travels with his wife and daughter.

Results

According to the content analysis of the blogs, it was found that affiliate marketing tools were used in all five cases but in different combinations and quantities. As can be seen in Table 19.2, only two of the five blogs have content in English as well as Turkish. This indicates that the bloggers mainly target Turkish audience. However, there is a lot of content in every blog about different destinations in Türkiye, which may be interesting for tourists traveling to Türkiye. Limited use of global affiliate marketing program links (Table 19.2) may be due to target audience of blogs or some restrictions. Even though, there is a restriction for individuals residing in Türkiye to book hotels within the borders of the country,[2] 4 out of 5 bloggers use Booking.com affiliate links in their travel blogs. To overcome the restriction and to continue earning commissions from Turkish residents making reservations via Booking.com, one of the bloggers wrote the explanation next to the links: "You can download a VPN and then book a room with a Booking.com".

---

2   Following a complaint brought by the Association of Turkish Travel Agencies (TURSAB), a court ordered the suspension of the activities of Booking.com in March 2017, due to unfair competition. The site can still be reached and used from other countries to make hotel reservations in Türkiye but Turkish citizens or individuals residing in Türkiye can only make reservations for hotels outside of the country.

Table 19.2. Travel Bloggers' Affiliate Marketing Strategies and Resources.

Cases/Blogs	Blog Language	Entries about services of local sponsors and links to them	Links to global affiliate programs				
			Booking.com	Sky-scanner	GetYourGuide	Rental cars	Other Programs
Case/Blog 1	Tr/Eng	✔	✔	✔	✔	✔	✔
Case/Blog 2	Tr/Eng	✔	-	-	-	-	-
Case/Blog 3	Turkish	-	✔	-	-	-	-
Case/Blog 4	Turkish	✔	✔	✔	-	-	-
Case/Blog 5	Turkish	✔	✔	✔	✔	-	✔

Entries about services of local sponsors and links to them exist in 4 out of 5 cases. Often these links are in a separate blog entry, describing the services of the service provider or supplier. Such entries and links of merchants are widely placed in Case/Blog 1 and 2, but can be seen also in other two cases. The most common contracted local touristic service providers are Turkish metasearch or metabooking platforms namely: Enuygun.com, Tatilsepeti and turna.com. There are also some entries with links to Turkish low-cost airline company Pegasus (Case/Blog 4) and even to local digital marketplace Hepsiburada (Case/Blog 2).

In 3 out of 5 blogs there's information about the collaboration with travel brands and advertisement or sponsorship terms. For example, in Case/Blog 5 although the blog is in Turkish, there's an English written page "Work with me/ Hire me" and a call for collaboration with travel companies or destination marketing organizations: "I have gained an incredible amount of knowledge and understanding of how brands and blogging can work hand in hand to promote a wide variety of products, services and destinations". This is a good indicator of how consciously this blogger is doing his job and that he is a professional affiliate marketer by leaving amateurism.

Conclusions

Based on the research findings, it can be stated that bloggers' affiliate marketing strategies and the variety of tools and methods they use are directly related to the education they have received. It was found that in only one of the analyzed cases (Case/Blog 3), the blog administrator did not have a marketing related education. All other bloggers had a certain level of business administration or marketing education. Besides that, the blogger of Case/Blog 3 has a full-time job and travels and writes entries for his blog just on his free time. It also can be said that in all 5 analyzed cases, after more of decade of travel blogging and sharing the digital content on different social media platforms, the bloggers became popular influencers and made a travel influencer career. As Syrdal et al. (2023) state, social media influencers have also joined the affiliate marketing wave, resulting in a new influencer affiliate marketing phenomenon. Another important finding of this research is that global affiliate marketing programs are not widely used by Turkish bloggers. Even though Booking.com links exist in 4 out of 5 cases, other global affiliate marketing programs tools are not sufficiently presented.

Affiliate marketing is a vital component of digital marketing in the tourism industry today, offering opportunities for customer engagement, revenue generation, and sustainable development. By leveraging digital tools, market orientation, and smart tourism initiatives, businesses in the tourism sector can enhance their competitiveness, reach broader audiences, and deliver exceptional experiences to travelers. Affiliate marketing also offers many opportunities for travel enthusiasts, entrepreneurs (Karabacak & Genç, 2019), and tourism professionals who are engaged with tourism and want to use their knowledge and experience. For example, it may be adopted by some tour guides or tour managers to compensate for their low work intensity in the off-season or crisis times. Many tour guides already promote their services and destinations they work in, through their blogs or social media (Dinçer et al., 2021; Çeşmeci & Aysin Örnek, 2023). Becoming an affiliate requires low start-up capital, offers a risk-free part-time job, and allows flexible working hours (Gedik, 2020). Therefore, it can be stated that tour guides are ideal candidates for blogging and affiliate marketing.

To the best of our knowledge, this is the first study to examine Turkish travel bloggers' affiliate marketing strategies, methods, and resources. Therefore, it is hoped that this study will make an important theoretical contribution to the tourism marketing literature. Future studies can be done on the blogs of tour guides to explore how they use affiliate marketing tools. Studies should be done to explore the affiliate marketing tools used by blogs oriented to English-speaking or other language-speaking travelers. The conduct of cross-national studies can provide a more comprehensive understanding of the issue at hand.

References

Adel, M. (2023). The influence of affiliate marketing on tourist decision-making. *Journal of Association of Arab Universities for Tourism and Hospitality*, 25(1), 266–277. https://doi.org/10.21608/jaauth.2024.255779.1537

Ayaz, Z. (2020). Çok katlı pazarlamada elektronik satış ortaklığı ağı yönetim bilgi sistemi modeli (Publication No. 611815) [Doctoral dissertation, Gazi University]. https://acikbilim.yok.gov.tr/handle/20.500.12812/362107

Ayaz, Z., & Güyer, T. (2024). Electronic sales partnership network management information system model in multi-level marketing. *Ankara Hacı Bayram Veli Üniversitesi İktisadi ve İdari Bilimler Fakültesi Dergisi*, 26(1), 189–232.

Blaer, M., Frost, W., & Laing, J. (2020). The future of travel writing: Interactivity, personal branding and power. *Tourism Management*, 77, 1–10. [104009]. https://doi.org/10.1016/j.tourman.2019.104009

Bosangit, C., Dulnuan, J., & Mena, M. (2012). Using travel blogs to examine the postconsumption behavior of tourists. *Journal of Vacation Marketing*, 18(3), 207–219.

Bowie, D., Paraskevas, A., & Mariussen, A. (2014). Technology-driven online marketing performance measurement: Lessons from affiliate marketing. *International Journal of Online Marketing (IJOM)*, 4(4), 1–16.

Carson, D. (2008). The 'blogosphere' as a market research tool for tourism destinations: A case study of Australia's Northern Territory. *Journal of Vacation Marketing*, 14(2), 111–119. https://doi.org/10.1177/1356766707087518

Çakar, K. & Aykol, Ş. (2020). Case study as a research method in hospitality and tourism research: a systematic literature review (1974–2020). *Cornell Hospitality Quarterly*, 62(1), 21–31. https://doi.org/10.1177/1938965520971281

Çeşmeci, N. & Aysin Örnek, N. (2023). Turist rehberliği ve girişimcilik (Tour guiding and entrepreneurship). *Journal of Gastronomy Hospitality and Travel (JOGHAT)*. https://doi.org/10.33083/joghat.2023.249

Daniele, R., Frew, A. J., Varini, K., & Magakian, A. (2009). Affiliate marketing in travel and tourism. In Höpken, W., Gretzel, U., Law, R. (Eds.), *Information and Communication Technologies in Tourism 2009* (pp. 343–354). Springer, Vienna.

Develi, E. İ. (2021). İnternette ya da çevrimiçi pazarlamada yeni bir kavram: Bağlı Kuruluş (Satış Ortaklığı) Pazarlaması ve Türkiye pazarından bazı örnekler. *OPUS International Journal of Society Researches*, 18(44), 8298-8332.

Dinçer, M. Z., Çakmak, T. F., & Aydoğan Çifçi, M. (2021). Turizm endüstrisinde blogların gücü ve turist rehberleri açısından bir değerlendirme. *Türk Turizm Araştırmaları Dergisi*, 3(1), 34–46.

Duffy, D. L. (2005). Affiliate marketing and its impact on e-commerce. *Journal of Consumer Marketing*, 22(3), 161–163. https://doi.org/10.1108/07363760510595986

Duffy, A., & Kang, H. Y. P. (2020). Follow me, I'm famous: travel bloggers' self-mediated performances of everyday exoticism. *Media, Culture & Society*, 42(2), 172–190. https://doi.org/10.1177/0163443719853503

Dwivedi, Y. K., Rana, N. P., & Alryalat, M. A. (2017). Affiliate marketing: An overview and analysis of emerging literature. *The Marketing Review*, 17(1), 33–50.

Eisenhardt, K. M. & Graebner, M. E. (2007). Theory building from cases: opportunities and challenges. *Academy of Management Journal*, 50(1), 25–32.

Esch, P., Arli, D., Castner, J., Talukdar, N., & Northey, G. (2018). Consumer attitudes towards bloggers and paid blog advertisements: what's new? *Marketing Intelligence & Planning*, 36(7), 778–793. https://doi.org/10.1108/mip-01-2018-0027

Filimon, S., Ioan, A.M., Alexandru, R.L. & Ruxandra, R. (2010). Blog marketing-a relevant instrument of the marketing policy. *Annales Universitatis Apulensis: Series Oeconomica*, 12(2), 760–765.

Fox, P. B. & Wareham, J. (2010). Governance mechanisms in internet-based affiliate marketing programs in Spain. *International Journal of E-Business Research*, 6(1), 1–18. https://doi.org/10.4018/jebr.2010100901

Grégori, N., Daniele, R., & Altınay, L. (2014). Affiliate marketing in tourism. *Journal of Travel Research*, 53(2), 196–210. https://doi.org/10.1177/0047287513491333

Gedik, Y. (2020). Bağlı kuruluş pazarlaması: kavramsal bir çerçeve. *Yorum Yönetim Yöntem Uluslararası Yönetim Ekonomi ve Felsefe Dergisi*, 8(2), 95–110.

Gustafsson, J. (2017, December 1). Single case studies vs. multiple case studies: A comparative study. Halmstad University. Retrieved May 3, 2024, from https://hh.diva-portal.org/smash/get/diva2:1064378/FULLTEXT01.pdf

Gupta, P. & Aggarwal, R. (2019). Reinventing and styling digital marketing through affiliate marketing. *International Journal of Research and Analytical Reviews*, 6(1), 476–480.

Karabacak, G. & Genç, M. (2019). The use of blogs as an example of internet entrepreneurship: Turkish travel blogs. *Procedia Computer Science*, 158, 869–876.

Lee, Y. J. & Gretzel, U. (2014). Cross-cultural differences in social identity formation through travel blogging. *Journal of Travel & Tourism Marketing*, 31(1), 37–54.

Maggiore, G., Presti, L. L., Orlowski, M., & Morvillo, A. (2022). In the travel bloggers' wonderland: mechanisms of the blogger – follower relationship in tourism and hospitality management – a systematic literature review. *International Journal of Contemporary Hospitality Management*, 34(7), 2747–2772.

Makarkina, I. (2021). In-depth analysis of publishers in travel affiliate marketing based on Aviasales data. [Master's thesis, Lappeenranta-Lahti University of Technology LUT]. https://urn.fi/URN:NBN:fi-fe2021062139269

Marin, J., Figueroa, A., & Cerdán, L. M. (2018). TBEX Europe Costa Brava 2015: Effective strategy for branding mature tourist destinations?. *Journal of Destination Marketing & Management*, 8, 337–349. https://doi.org/10.1016/j.jdmm.2017.07.004

Mariussen, A. (2011). Rethinking marketing performance measurement: Justification and operationalisation of an alternative approach to affiliate marketing performance measurement in tourism. *E-review of tourism research, 9(3),* 65–87.

Mariussen, A., Daniele, R., & Bowie, D. (2010). Unintended consequences in the evolution of affiliate marketing networks: A complexity approach. *The Service Industries Journal,* 30(10), 1707–1722. https://doi.org/10.1080/02642060903580714

Markov, D., & Timerbaev, D. (2021). Big data analytics in travel industry: Case of affiliate marketing channel at Aviasales Company. [Master's thesis, Saint Petersburg State University]. https://dspace.spbu.ru/bitstream/11701/30967/1/MT_Final.pdf

Nimmermann, F. (2020). *Congruency, expectations and consumer behavior in digital environments.* Springer Gabler.

Pan, B., MacLaurin, T., & Crotts, J. C. (2007). Travel blogs and the implications for destination marketing. *Journal of Travel Research*, 46(1), 35–45.

Perelygina, M., Küçükusta, D., & Law, R. (2022). Digital business model configurations in the travel industry. *Tourism Management*, 88, 104408. https://doi.org/10.1016/j.tourman.2021.104408

Premachandra, Y. B., & Eranda, B. A. N. (2022). Content providers' engagement with merchants in affiliate marketing: A study based on Sri Lankan travel and tourism industry. *SEUSL Journal of Marketing*, 7(2), 99–125.

Snyder, P. & Kanich, C. (2016). Characterizing fraud and its ramifications in affiliate marketing networks. *Journal of Cybersecurity*, 2(1), 71–81.

Suri, H. (2011). Purposeful sampling in qualitative research synthesis. *Qualitative Research Journal*, 11(2), 63–75. https://doi.org/10.3316/qrj1102063

Syrdal, H. A., Myers, S., Sen, S., Woodroof, P. J., & McDowell, W. C. (2023). Influencer marketing and the growth of affiliates: the effects of language features on engagement behavior. *Journal of Business Research*, 163, 113875.

Şener, B., Çeşmeci, N., & Kılıçhan, R. (2023). An evaluation of online promotion activities of geoparks in Türkiye. *Journal of Global Tourism and Technology Research*, 4(2), 77–89. https://doi.org/10.54493/jgttr.1353795

Tanwar, S., & Sahu, P. (2024). Two decades of research on affiliate marketing: a systematic literature review. *Theoretical & Applied Economics*, 31(1), 211–230.

Taylor, C. R. (2020). The urgent need for more research on influencer marketing. *International Journal of Advertising*, 39(7), 889–891.

Tiago, M. T. & Veríssimo, J. (2014). Digital marketing and social media: why bother? *Business Horizons*, 57(6), 703–708. https://doi.org/10.1016/j.bushor.2014.07.002

Tussyadiah, I. & Fesenmaier, D. R. (2008). Marketing places through first-person stories - an analysis of Pennsylvania roadtripper blog. *Journal of Travel & Tourism Marketing*, 25(3–4), 299–311. https://doi.org/10.1080/10548408002508358

Volo, S. (2010). Bloggers' reported tourist experiences: their utility as a tourism data source and their effect on prospective tourists. *Journal of Vacation Marketing*, 16(4), 297–311. https://doi.org/10.1177/1356766710380884

Wenger, A. (2008). Analysis of travel bloggers' characteristics and their communication about Austria as a tourism destination. *Journal of Vacation Marketing*, 14(2), 169–176. https://doi.org/10.1177/1356766707087525

Zahid, E. (2024, January 5). 18 Affiliate marketing statistics 2024 all marketers must know. Optinmonster. https://optinmonster.com/affiliate-marketing-statistics/

Yasemin Ersoy[1] and Fuat Bayram[2]

Chapter 20 Cultural Structure of Turkish Cuisine and Kitchen Equipment Used in Turkish Cuisine

Introduction

A society is different from other societies and all the elements that have been integrated with the society constitute the culture of that society (Erkal, 1998). Culture, which can be described as "all kinds of language, emotion, thought, belief, art and living elements that are valid in a society and continue as a tradition", contains material elements such as food and eating and drinking habits besides spiritual elements (Arlı & Gümüş, 2007). In this respect, in addition to being a vital necessity, food and beverage is also a cultural element (Birer, 1990).

Nutritional habits, which we can accept as a cultural element, are shaped depending on the geography and lifestyles of people and settle as a cultural element at the end of long processes (Ünal, 2007; Közleme, 2012; Güldemir, 2014; Demirgül, 2018).

Considering its historical background and the breadth of the geography they live in, Turkish Cuisine has a wide cultural diversity (Güldemir, Haklı & Işık, 2018). When it refers to Turkish Culinary Culture, it is not correct that only dishes specific to Turkish Cuisine come to mind. This concept also includes the diet of the people living in the Turkish geography, table manners, the variety of food and beverages they consume, the ways and conditions of preparation, cooking and storage of these foods, as well as the tools and materials they use during the preparation and storage of food and beverages (Birer, 1990; Arlı & Gümüş, 2007).

In this study, it is aimed to provide basic information about Turkish culinary culture and to give knowledge about kitchen tools and equipment which are used during the preparation and storage of food and beverages specific to Turkish culinary culture.

---

1 Ph.D., Karabuk University Department of Nutrition and Dietetics. yaseminersoy@karabuk.edu.tr
2 Ph.D., Bolu Abant Izzet Baysal University Mengen Vocational School. bayram_f@ibu.edu.tr

Turkish Culinary Culture

The culinary culture of a society is formed according to the geographical conditions, vegetation, agricultural characteristics, socioeconomic conditions and interaction with other societies in the region where that society lives (Baysal, 2001).

The history of Turkish cuisine dates to the nomadic Turkish communities in Central Asia. Turks have been interacting with different societies in different geographies starting from Central Asia until the Republican period, and this has created a rich and diverse Turkish culinary culture (Halıcı, 2009; Özbey & Köşker, 2021).

Eating and Drinking Habits in Turkish Cuisine

The geography where Turks lived extensively at the junction of the continents of Europe, Asia and Africa has also been reflected in their culinary culture. The meat and fermented dairy products of Central Asia, the grains of Mesopotamia, the vegetables and fruits of the Mediterranean region and the spices of South Asia have combined to form a great variety (Ciğerim, 2001). From the day the Turks emerged on the stage of history in Central Asia until today, their primary economic resource has been animal husbandry. Therefore, animal products have been the most essential elements of Turkish nutritional culture and social life. Thus, foods such as meat, milk, eggs, yoghurt and cheese have formed the basis of their diet (Sürücüoğlu, 2001). Particularly the ancient Turks of Central Asia fulfilled their nutritional needs mainly through animal products because of their living conditions (Közleme, 2012). For the most part, sheep, goat and other bovine meat were among the basic foods that they consumed in their meals (Halıcı, 1990)

The nomadic Central Asian Turks had learnt to preserve meat for a long time at that time and processed meat products by making bacon and sausage, which are almost like today (Kılıç & Albayrak, 2012; Kızıldemir, Öztürk & Sarıışık, 2014)

In Turkish tables, milk is not only a food on its own, but also used in the form of butter, cheese, yoghurt and ayran, and dishes containing milk and yoghurt have shown diversity. Milk was also used extensively in the palace cuisine. Milk was brought to the palace in vessels called "bakrac" and used in many dishes, including desserts. While milk pudding and milk kadayıf were made in palace cuisines, today kazandibi, chicken breast, keskül, rice pudding, rice pudding, güllaç and many other types of milk desserts are consumed with appreciation due to being light and healthy (Sürücüoğlu, 2001).

When Turks became settled, they also began to produce fruits, vegetables and various cereals. An important part of Turkish culture, bread was made from wheat, barley and millet. While wheat was used in bread making, it was also consumed by roasting or boiling (Kılıç & Albayrak, 2012).

Cuisine Tools and Equipment in Turkish Culinary Culture

The cuisine in ancient Turkish societies was organized in a certain order. Corners of the kitchen, there would be five or six layers of shelves called "sergen" on top of each other. At the bottom of the sergen were utensils such as "desti" and "ibrik". Above that, small pots, sahars and soup bowls were placed, and trays and china plates were placed on top (Koşay & Ülkücan, 2011).

Figure 20.1. A View of Home Kitchen in Turkish Societies

In the Turkish culinary culture, cooking equipment called hearth, tandoor and kuzine, which were heated with wood, were used to cook food in the ancient times. Meals were prepared in pots and pans made of earthenware, bronze, copper and iron. Today, although natural gas and electric cookers have gradually replaced the mentioned cooking equipment, cooking utensils such as tandoor and kuzine are still used by the people living in Anatolia and other Turkish geographies (Baysal, Merdol, Ciğerim, Sacır & Başoğlu, 2005).

The tandoor is like a well-made of earth with an open top and a place at the bottom where the ash can be separated. It is made from a type of red earth called "gav". After sifting this type of soil, straw and goat hair are added. This mixture with enough water is kneaded by chewing with the foot for 2 days. Kneading for a long time is done to ensure that the mud is concise, and the tandoor is strong. The prepared mud is rolled into a cylinder 130 centimeters high and 65 centimeters in diameter and placed in a large pit. The tandoor cylinder is filled with fire bricks and plastered with mud, leaving no gaps. After the tandoor is dried, it serves as both an oven for cooking and a stove for heating in the households of the Turkish community for many years (Koşay & Ülkücan, 2011).

Figure 20.2. Tandır

Wood is generally used as fuel in tandoors (Baysal, 2005). The smoke generated by the combustion of wood or similar fuels is discharged through a ventilation channel called "gülve", which runs parallel to the outside of the tandoor (Koşay & Ülkücan, 2016). After the wood burns and turns into embers, the food to be cooked is lowered into the tandoor. Food can be cooked in such tandoors by closing the top part, as well as bread can be baked by sticking yeast breads on the walls of the tandoor. In addition to the tandoor, another piece of equipment used by the Turks for bread making is a thin, iron "sac", which is designed to be flat. It is used in the preparation of products such as pancakes, pita bread, yufka and bread by sitting on a sheet metal tandoor or wood fire (Baysal, Merdol, Ciğerim, Sacır & Başoğlu, 2005).

Another equipment used for cooking purposes in Turkish culinary culture from past to present has been "kuzine". This cooking equipment, which is a simple hearth oven with a wood fire underneath, can also be positioned at many points in living spaces for heating (Baysal, Merdol, Ciğerim, Sacır & Başoğlu, 2005).

In Turkish societies, there are various types of cauldrons, pots and similar utensils in which meals are cooked. Boilers are the most prominent cooking utensils used especially when meals are to be served to crowded groups. During the Ottoman Empire, meals were prepared in these large cooking utensils for large groups such as the Janissary organization and the Palace officials. These cauldrons, which have handles on the right and left corners for easy transportation, are still used today as cooking equipment in Turkish cuisine culture on special occasions such as weddings and funerals where collective meals are served (Koşay & Ülkücan, 2011).

Figure 20.3. Food Boiler

Another cooking equipment used for cooking is pots. Although there are some that can hold as much food as large boilers in terms of volume, they are generally cooking equipment with medium and small size lids. These cooking utensils made of copper in the past are still in use today.

Stewpots are glazed earthenware pots in which food is cooked in Turkish cuisine and which are still in use from the past to the present. Tinned copper pots and pans have been used from the past to the present and are now being replaced by teflon pans and pots (Baysal, Merdol, Ciğerim, Sacır & Başoğlu, 2005).

The service and consumption of food in the old Turkish societies is also unique. Meals were consumed on tables called "sini. Sinis could be made of tinned wrought copper or other materials. Sinis were placed on coffee tables 30 cm above the ground and covers called "sofra" were laid to prevent the food from spilling out. Meals were served on a single plate and the whole household consumed the food from a single plate with metal or wooden spoons (Baysal, Merdol, Ciğerim, Sacır & Başoğlu, 2005; Yerasimos, 2010).

Conclusion

Turks, who transformed the interactions they experienced with different culinary cultures in the geography they lived in into original flavors without losing their own cultural values, have had a great and deep-rooted culinary culture. In the creation of a society's culinary culture, the dishes specific to that society, dietary patterns and diversity, table manners, methods used during the preparation and storage of food, and even the equipment used during the preparation and storage of food are of great importance.

In the historical process, it is understood that the geography they live in has a great influence on the diet of the Turks. It is known that the diet of the Central Asian Turks before Islam was based on meat, milk, eggs, cheese and cereals used for bread making. With the settlement of the Turks, it is seen that agricultural products took their place in the kitchens in abundance, and both the variety of food, food products used in cooking and kitchen utensils used for cooking have constantly enriched.

The fact that the tandoor, kuzine and stoves used by ancient Turkish societies to cook food are still in use in Anatolia today is an indication that the values of Turkish Culinary Culture are kept alive. Likewise, sinis, tinned copper pots and cauldrons used in the past are used as cooking equipment in collective meal ceremonies in today's Turkish geography.

References

Arlı, M., & Gümüş, H. (2007). Türk mutfak kültüründe çorbalar. *ICANAS, Uluslararası Asya ve Kuzey Afrika Çalışmaları Kongresi*, 10–15.

Baykara, T. (2001). *Türk Tarihine Bakışlar*. Atatürk Kültür Merkezi Başkanlığı Yayınları, (1).

Baysal, A. (2001). Türk ve Çin Mutfağının Karşılaştırılması. K. Toygar (Ed.) *Türk Mutfak Kültürü Üzerine Araştırmalar 2000* içinde. Ankara: Türk Halk Kültürü Araştırma ve Tanıtma Vakfı Yayınları, 27–34.

Baysal, A., Merdol T. K., Ciğerim N., Sacır H., & Başoğlu S. (2005). Türk Mutfağından Örnekler. Hatiboğlu Yayınları, 21(3).

Birer, S. (1990). Türk Mutfağının Tarihsel Gelişim Süreci İçerisindeki Değişimi ve Bugünkü Durumu. *Beslenme ve Diyet Dergisi*, 19(2), 251–260.

Ciğerim, N. (2001). Batı ve Türk Mutfağının Gelişimi, Etkileşimi ve Yiyecek-İçecek Hizmetlerinde Türk Mutfağının Yerine Bir Bakış. K. Toygar (Ed.) Türk Mutfak Kültürü Üzerine Araştırmalar 2000 içinde. Ankara: Türk Halk Kültürünü Araştırma ve Tanıtma Vakfı Yayınları, 49–61.

Demirgül, F. (2018). Çadırdan saraya Türk mutfağı. Uluslararası Türk Dünyası Turizm Araştırmaları Dergisi, 3(1), 105–125.

Erkal, M. E. (1998). Sosyoloji (Toplumbilimi) Der Yayınları, (9).

Güldemir, O. (2014). Orta Asya'dan Cumhuriyet Dönemine Türk Mutfağındaki Yemeklerin Değişimi: Yazılı Kaynaklar Üzerinden Bir Değerlendirme. VII. Lisansüstü Turizm Öğrencileri Araştırma Kongresi, 346–358.

Güldemir, O., Haklı, G., & Işık, N. (2018). Türk mutfağında kahvaltıda tüketilen çorbalar ve illere göre dağılımı. Selçuk Üniversitesi Sosyal Bilimler Enstitüsü Dergisi, (39), 56–66.

Halıcı, N. (1990). Türk Mutfağı. Güven Matbası.

Halıcı, N. (2009). Türk Mutfağı. Oğlak Yayıncılık.

Kılıç, S., & Albayrak, A. (2012). İslamiyetten önce Türklerde yiyecek ve içecekler. Turkish Studies, 7(2).

Kızıldemir, Ö., Öztürk, E., & Sarıışık, M. (2014). Türk mutfak kültürünün tarihsel gelişiminde yaşanan değişimler. AİBÜ Sosyal Bilimler Enstitüsü Dergisi, 14(3), 191–210.

Koşay, Z. H., & Ülkücan, A. (2011). Anadolu Yemekleri ve Türk Mutfağı. Çiya Yayınları.

Közleme, O. (2012). Türk Mutfak Kültürü ve Din. Rağbet Yayınları.

Özbey, Z., & Köşker, H. (2021). Türk mutfak kültüründe çorba ve coğrafi işaretli çorbalar üzerine bir değerlendirme. Gastroia: Journal of Gastronomy and Travel Research, 5(3), 471–489.

Sürücüoğlu, M. S. (2001). Beslenme kültürümüzde süt ve süt ürünleri. Türk Mutfak Kültürü Üzerine Araştırmalar 2000 içinde. K. Toygar (Ed.), Ankara: Türk Halk Kültürü Araştırma ve Tanıtma Vakfı Yayınları, 129–148.

Ünal, A. (2007). Anadolu'nun En Eski Yemekleri Hititler ve Çağdaşı Toplumlarda Mutfak Kültürü. Homer Kitapevi.

Yerasimos, M. (2010). 500 Yıllık Osmanlı Mutfağı. Boyut Yayıncılık.

Yağmur Kaplan[1]

Chapter 21 Green Marketing in Tourism

Introduction

In recent years, it is well known that social, economic and political developments, the effects of which are undeniable with globalisation, have created environmental problems. These problems also affect the tourism sector, which depends on environmental resources. It is clear that some marketing strategies should be implemented in order to prevent these problems in the tourism sector and to raise environmental awareness in companies.

With the formation of environmental awareness all over the world, the importance of the concept of environment is increasing day by day. With the formation of environmental awareness among consumers, it shows that environmental responsibilities should be acquired that will enable companies to gain a place in the competitive environment (Varinli, 2000). This is because harmful wastes generated in production and consumption processes cause resource depletion and environmental damage (Atay and Dilek; 2013). Therefore, it is of great importance to manage resources sustainably and adopt environmentally friendly practices in the tourism sector. In addition, the concept of environment is an essential factor for companies in the tourism sector to implement their marketing strategies (Lim and McAleer, 2005: 1431). For this reason, the concept of environment is one that refers to the effective use of the natural resources in which individuals live. Because tourism has a structure that is fed by natural resources. The environment is an important factor in the protection of this structure (Akdemir and Akbulut; 2019). Therefore, the protection of the physical and cultural structure is important in the tourism economy (Can, 2013). Therefore, the management of environmental resources under the umbrella of sustainability provides a competitive advantage for destinations (Can, 2013) and contributes to the tourism economy.

---

1 M.A., Akdeniz University, Manavgat Tourism Faculty, Department of Recration Management, yagmurkaplan007@hotmail.com

Green Marketıng

The changing economic conditions in today's world make marketing practices even more important. Marketing is an important factor that enables companies to achieve their goals. In today's economy, the understanding of marketing has not only met the demands and needs of consumers, but also mediated responsibility towards the environment (Mucuk, 2001). Thus, the concept of green marketing has emerged, which mediates the implementation of marketing activities that have emerged to meet the demands and needs of consumers by bringing more environmentally friendly and efficient resources (Chin et al., 2018; Erbaşlar, 2012). The concept of green marketing is a marketing approach that requires environmental sustainability and efficient use of natural resources. In addition, green marketing creates an important framework for understanding how environmental issues are addressed in business sectors and how they should be developed.

The concept of green marketing entered the agenda and literature at the 1975 American Marketing Association (AMA) seminar on ecological marketing (Chamorro & Banegil, 2006). In this seminar, "ecological marketing" was defined as a type of marketing that conveys marketing strategies against environmental pollution, unnecessary energy consumption and efficient use of other resources (Keleş, 2007; Leblebici Kacur, 2008). Today, it has been stated that green marketing is shaped within the concept of sustainability and that sustainability has developed within economic, environmental and cultural development (Gedik, 2020).

According to Peattie and Crane (2005), green marketing has been shaped and developed in three distinct periods. The first period, ecological marketing, stretches from the early 1960s to the 1970s. In this period, the focus was on the environmental damage caused by air pollution. The second period began in the late 1980s and has been described as 'green environmental marketing'. Green environmental marketing is a period of increasing awareness of clean technology, sustainability, consumer and competitive advantage. Therefore, when comparing the first and second periods, the first period tried to identify the damage caused to the environment by industrialisation. In the second period, environmental marketing includes not only industrialisation but also all service and production strategies with high environmental interaction, such as tourism. Sustainable green marketing, defined as the third period, is known to have developed under strict government policies as well as increasing demands and expectations of individuals. The development of sustainable green marketing has a significant impact on the economy (Delafrooz, Taleghani, & Nouri, 2014). Looking at these periods has allowed us to understand how green marketing has evolved and how the environmental awareness of the business world has increased in the process. The increased environmental awareness of businesses has led to the development of green marketing and the development of environmentally friendly sales strategies for individuals.

The main purpose of green marketing is to increase consumer awareness of environmentally friendly products and to increase the demand for environmentally friendly products (Ceylan and Kıpırtı; 2021). Other objectives of green marketing are listed below (Sancaktar; 2019):

- Protecting existing consumption areas, using natural resources efficiently,
- Making recyclable products preferable, ensuring good environmental balance,
- Making energy consumption controllable and minimising energy consumption,
- Developing new methods by identifying the factors that will cause environmental damage,
- Ensure the preferability of products that do not harm the environment,
- Creating environmental awareness in consumers and businesses in the sector by ensuring the sustainability of green marketing.

Grant (2008) discussed the objectives of green marketing as consisting of three phases; "Green" phase, companies define green products and processes by setting environmentally friendly standards. In this process, it is important for companies to be very good at expressing and marketing themselves. The second stage, 'Greener', is to ensure that consumer behaviour is more environmentally friendly. In the "Greenest" stage, they should set big goals in the presentation stage from the production stage to the consumer by adopting the understanding of innovation (Sert, 2017).

Green Marketıng in Tourısm

It is known that the concept of green draws a path for itself in the tourism sector as in every sector. Because environmentalist thinking and behavior strategies are needed for tourism activities to take place. Tourism businesses that adopt green management strategies are much more preferred than others (Güneş, 2011: 47). This is because tourism businesses are interested in establishing strong relationships within the sector and following a more successful path under the roof of sustainability. At this point, according to Foster et al. (2000), in the tourism sector, accommodation businesses should develop and implement green marketing strategies under the umbrella of sustainability as follows;

1. Increasing environmental awareness of consumers,
2. Environmental regulations are necessary,
3. Acting in line with environmentalist behaviors of all lower and upper levels working in the tourism business,
4. Hotel businesses can provide more satisfaction in customer requests.

The fact that natural resources in the tourism sector are exhaustible, and that these resources are an important factor for the tourism sector, highlights the need for green marketing strategies. The green marketing approach in tourism includes environmentally friendly practices to meet customer wants and needs, and aims to provide the most practical purchasing methods for customers (Sangpikul, 2017). This paves the way for green tourism marketing to develop a customer portfolio that values the environment and can lead to other purchases. This is because the limitless structures of tourism destinations in competition are closely related to how much attention is paid to the sustainability of natural, economic and cultural resources (Kozak & Nield, 2004). In addition, while the sustainability of natural resources ensures the attractiveness of tourism destinations, it also increases the competitiveness of tourism enterprises and provides an important privilege for them to be preferred in the sector. This is because companies are faced with the fact that in order to exist in the tourism sector, they have to take care of environmental awareness (Akyüz, 2022). Thus, the tourism sector is one of the sectors that uses the most natural resources compared to other economic sectors (Tuna 2007:17). Therefore, any negativity in the natural resources and environmental structure will make it difficult for the tourism sector to survive and maintain its existence (Sert, 2017). This is because while the tourism sector is affected by environmental factors, it also affects the environment (Zeydan and Gürbüz; 2021).

Aslanlı (2019) stated the reasons why tourism companies should turn to green marketing principles to survive in the sector as follows;

- Ensuring maximum customer satisfaction in the tourism sector,
- Creating environmental awareness among consumers,
- Ensuring legal regulations to increase environmental awareness,
- Eliminating the negativities in the physical and cultural structure.

The tourism sector cannot be considered separately from the environment. This brings along many environmental problems in the unplanned and uncontrolled operation of hotel establishments, which are an important factor in the environmental structure of tourism (Dief and Font; 2010). In addition, the tourism sector actively uses beaches, historical buildings and other protected environmental areas and as a result, it plays an active role in damaging the environmental structure (Zeydan and Gürbüz; 2021). This shows that the tourism sector is one of the most environmentally damaging sectors (Ceylan and Kıpırtı; 2021).

Accommodation enterprises in the tourism sector are the most energy and water consuming enterprises in the service sector (Punitha and Rasdi; 2013). This is because tourism businesses consume large amounts of resources on a daily basis (Atay and Dilek; 2013). These are: adjusting the temperature of the hotel business according to the seasons, chemical products used in the business (detergents, etc.), filling the swimming pools, irrigating the environmental areas (Mastny, 2002). Therefore, the most emphasised issue for accommodation businesses is to understand how to use energy and water consumption efficiently (Günay; 2017). According to Punitha and Rasdi (2013), one third of the world's electricity consumption is used in structures, including accommodation businesses. Therefore, it is understood that hotels should be integrated into the overall management system to ensure efficient energy consumption in accommodation businesses (Sert, 2017). Energy management includes competent staff training, purchasing, auditing, budgeting and monitoring (Deng & Burnett, 2002).

Since the tourism sector needs a clean environment (Doğan & Ertaş, 2018), it needs a good waste management. Reducing waste, separating, collecting and recycling waste according to their content is an important element in waste management. This is because it is known that on average, each customer in accommodation establishments generates more than 1 kilogram of waste per day (Bohdanowicz, 2005). Therefore, accommodation establishments should separate metal, glass, plastic, paper, battery, cooking oil and solid wastes according to their categories and recycle them and protect the environment by controlling the wastes (Şen, 2023). In order to guarantee effective waste control, customers and staff should be directly involved in the recycling system. According to Bohdanowicz (2005), between 50 and 60 per cent of the waste generated in accommodation establishments can be recycled. In accommodation businesses, paper products should be reused, shampoo and personal care products in the bathroom should be offered in reusable and refillable fixed containers, all food and beverages should be durable products, not disposable, and disposable products should be recyclable (Güneş, 2011; Sert, 2017). It has been demonstrated that the use of recyclable products will meet the need for raw materials in tourism enterprises and minimise the damage to the environment (Aslanlı, 2019). Over time, it has been understood that green marketing practices in tourism provide economic benefits ("reduction of water and energy" use, reduction of solid waste, protection of environmental structure such as recycling and recovery, protection of public health and sustainability of tourism (Shamshiry, Nadi, Mokhtar, Komoo, & Hashim, 2011).

Results

Environmental problems arise as a result of the effects of undesirable human behaviour (Zeydan and Gürbüz; 2021). The easiest way to minimise the effects of these problems is to develop consumption habits that cause less pollution. The acquisition of environmentally friendly consumption habits can be achieved through clean production (Yücel and Ekmekçiler, 2008).

According to McMichael (2003), improving social and environmental conditions while building people's future is important for human health, safety and welfare (Akdemir and Akbulut, 2019). Therefore, green marketing should be developed in tourism and marketing activities should be organised accordingly. In addition, tourism businesses should give importance to green marketing practices in their marketing activities and ensure their operability. It is understood that tourism businesses that do not adopt environmental sensitivity in the tourism sector cannot provide competitive advantage in the sector. This highlights the importance of paying attention to the more efficient use of the environment and natural resources by businesses and consumers participating in tourism activities.

Given the payback effect of resource use in the tourism sector, the future of the sector will be determined by the increase in environmentally friendly and sustainable practices. This is because green marketing practices in accommodation businesses will provide significant benefits in water and energy conservation, waste management and recycling activities, while reducing costs (Sert, 2017).

Studies to be carried out for the greening of the tourism sector will enable future generations to feel their contribution to the environment and the economy. As a result, both natural resources will be used efficiently and people's quality of life will be improved.

References

Akdemir, R., Akbulut, O. (2019). Yeşil Pazarlama Stratejilerinin Rekabet Avantajına Etkisinin İncelenmesi: Muğla İlinde Yer Alan 4 Ve 5 Yıldızlı Otel İşletmelerine Yönelik Bir Araştırma. *Electronic Journal Of Social Sciences* 18 (72).

Akyüz, A. M. (2022). *Pazarlamada Güncel Yaklaşımlar "Ekolojik (Yeşil) Pazarlama"*. Ankara: Gazi Kitapevi.

Aslanlı, S. (2019). *İşletmelerde yeşil pazarlama anlayışı ve çevresel muhasebe ilişkisi: Yeşil yıldızlı otel işletmeleri üzerine araştırma*. (Yüksek Lisans Tezi). Bursa Uludağ Üniversitesi Sosyal Bilimler Enstitüsü, Bursa.

Atay, L., Dilek, E. (2013). Konaklama İşletmelerinde Yeşil Pazarlama Uygulamaları: Ibıs Otel Örneği. *Süleyman Demirel Üniversitesi İktisadi Ve İdari Bilimler Fakültesi Dergisi*, 18(1), 203–219.

Bohdanowicz, P. (2005). European Hoteliers' Environmental Attitudes: Greening the Business. *Cornell Hotel and Restaurant Administration Quarterly*, 188–204.

Can, E. (2013). Turizm Destinasyonlarında Sürdürülebilir Turizmin Sürdürülebilir Rekabet Açısından Değerlendirilmesi. *İstanbul Sosyal Bilimler Dergisi*(4), 23–40.

Ceylan, U., & Kıpırtı, F. (2021). Turizm İşletmelerinde Yeşil Pazarlama Faaliyetleri: Literatür İncelemesi. *Socrates Journal Of Interdisciplinary Social Studies*, 8, 23–37.

Chamorro, Antonio, Tomas M. Banegil (2006), "Green Marketing Philosophy: A Study of Spanish Firms with Eco-labels", *Corporate Social Responsibility and Environmental Management*, 13(1), 11–24.

Chin, C., Chin C. and Wong, W. P. (2018). The implementation of green marketing tools in rural tourism: The readiness of tourists?. *Journal of Hospitality Marketing and Management*, 27(3), 261–280.

Delafrooz, N., Taleghani, M. ve Nouri, B. (2014). *Effect of green marketing on consumer purchase behaviour*. Qscience Connect, 5, 1–9.

Deng, S., & Burnett, J. (2002). Energy use and management in hotels in Hong Kong. *International Journal of Hospitality Management*, 371–380.

Dief, M., Font, X. (2010). TheDeterminants of Hotels' Marketing Managers Green Marketing Behaviour. *Journal of Sustainable Tourism.*18(2), 157–174.

Doğan, Ö. & Ertaş, C. F. (2018). *Çevreye duyarlılığın rekabet gücüne etkisi: Yeşil yıldızlı oteller üzerine bir uygulama*. Maliye ve Finans Yazıları, 110, 217–234.

Erbaşlar, G. (2012). Yeşil pazarlama. *Mesleki Bilimler Dergisi*, 1(2), 94–101.

Foster, S.T., Sampson, S.E., Dunn, S.C. (2000), "The Impact Of Customer Contact On Environmental Initiatives For Service Firms", International Journal of Operations & Production Management, Vol. 20 No.2, pp. 187–203

Gedik, Y. (2020). Yeşil pazarlama stratejileri ve işletmelerin amaçlarına etkisi. *International Anatolia Academic Online Journal Social Sciences Journal*, 6(2), 46–65.

Grant, J. (2008). *Yeşil Pazarlama Manifestosu* (Çeviri: Nadir Özata, Yasemin Fletcher), İstanbul: MediaCat Kitapları.

Günay, T., (2017). *Turizm İşletmelerinde Yeşil Pazarlama Uygulamaları: İzmir İli Örneği*, Yaşar Üniversitesi Sosyal Bilimler Enstitüsü Turizm İşletmeciliği Anabilim Dalı Yüksek Lisans Tezi, İzmir.

Güneş, G. (2011). Konaklama Sektöründe Çevre Dostu Yönetimin Önemi. *KMÜ Sosyal ve Ekonomik Araştırmalar Dergisi*, 13(20), 45–51.

Güneş, G. (2011). "Korunan Alanların Yönetiminde Yeni Bir Yaklaşım: Katılımcı Yönetim Planları". *Ekonomi Bilimleri Dergisi*, 3(1), 47–57.

Keleş, Ö. (2007). *Sürdürülebilir yaşama yönelik çevre eğitimi aracı olarak ekolojik ayak izinin uygulanması ve değerlendirilmesi* (Yayımlanmış doktora tezi). Gazi Üniversitesi, Eğitim Bilimleri Enstitüsü, Ankara.

Kozak, M. ve Nield, K. (2004). The Role of Quality And Eco- Labelling Systems in Destination Benchmarking, *Journal of Sustainable Tourism*,12, 138–148.

Leblebici Kacur, L. (2008). Yeşil Pazarlama Ve Kayseri'deki İşletmeler Üzerine Bir Uygulama, Doktora Tezi, Erciyes Üniversitesi Sosyal Bilimler Enstitüsü, Kayseri.

Lim, C. and McAleer, M. (2005). Ecologically Sustainable Tourism Management. *Environmental Modelling & Software*, 20(11), 1431–1438.

Mastny L. (2002). Dünyanın Durumu, *TEMA Vakfı Yayınları* (193): 50–52.

McMichael, A. J., Butler, C. D. ve Folke, C. (2003). New Visions For Addressing Sustainability. *Science*, 302(5652), 1919–1920.

Mucuk, İ. (2001*). Pazarlama ilkeleri*, İstanbul: Türkmen Kitabevi.

Peatie, K. and Crane, A. (2005). Green Marketing: Legend, Myth, Farce or Prophesy? Qualitative Market Research. *An International Journal*, 8(4), 357–370.

Punithal, S., Mohd Rasdi, R. (2013). Corporate Social Responsibility: Adoption of Green Marketing by Hotel Industry. *Asian Social Science*, 9(17), 79–93.

Sancaktar Meral, G. (2019). Yeşil Pazarlama Kapsamında Tüketicilerin Organik Gıda Tercihlerinin İncelenmesi, *Kırklareli Üniversitesi Sosyal Bilimler Enstitüsü*, Kırklareli.

Sangpikul, A. (2017). Ecotourism mix, good practice, and green marketing: An approach towards the quality tourism business. *Journal of Community Development Research (Humanities and Social Sciences)*, 10(1), 1–15.

Sert, A. (2017). Konaklama İşletmelerinde Yeşil Pazarlama Uygulamaları: Doğa Residence Otel Örneği. *Türk Turizm Araştırmaları Dergisi, 1*(1), 1–20.

Shamshiry, E., Nadi, B., BinMokhtar, M., Komoo, I., Saadiahhashim, H., Yahaya, N. (2011). IntegratedModels for Solid Waste Management in Tourism Regions: Langkawi Island, Malaysia. *Hindawi Publishing Corporation Journal of Environmental and Public Health*.

Şen, L. M. (2023). Turizm İşletmelerinde Yeşil Pazarlama Uygulamaları. *Turizm ve Destinasyon Araştırmaları V*, 237.

Tuna, M. (2007). *Turizm, Çevre ve Toplum (Marmaris Örneği)*. Ankara: Detay Yayıncılık.

Uydacı, M. (2002) *Yeşil Pazarlama*, İstanbul, Türkmen Kitabevi.

Varinli, İ. (2000) *"Pazarlama Ahlakı ve Kayseri'de Küçük ve Orta Ölçekli İşletme Yöneticilerinin Pazarlama Ahlakına İlişkin Değerlendirmeleri"* Kayseri Ticaret Odası Yayınları no:15.

Yücel, M., Ekmekçiler, Ü. S. (2008). Çevre dostu ürün kavramına bütünsel yaklaşım; temiz üretim sistemi, eko-etiket, yeşil pazarlama. *Elektronik Sosyal Bilimler Dergisi*, 7(26), 320–333.

Zeydan, İ. Ve Gürbüz, A. (2021). Turizmde Yeşil Pazarlama Uygulamaları: Mavi Bayrak ve Yeşil Yıldızın Turistlerin Konaklama Tercihlerine Etkisi. *Uluslararası Yönetim İktisat ve İşletme Dergisi*, *17*(1), 224–235.

Betül Buladi Çubukcu[1]

Chapter 22 Green Marketing Practices in Tourism

Introduction

Global environmental impacts and the call for sustainable development for current societies which constitutes a necessity in today's world has caused significant impacts on the tourist industry as it is with all other industries. In an effort to aid environmental sustainability, many tourism organizations and companies have embraced green marketing practices. This concept refers to a set of measures and advertisement techniques, which are used in marketing at the same time to protect natural resources. The necessary concept of green marketing in the context of the tourism sector refers to the ways that, on the one hand strengthen the competitive advantages of companies and on the other hand, promote a sustainable environmental agenda. Such practices include: environmentally sustainable management of facilities, green communication, education and awareness, green innovation for tourism enterprises which enhances the flow of tourists while offering the environment a chance to benefit through being conserved by tourists and entrepreneurs at the same time.

Specifically, the adoption of energy and water-saving technologies by green accommodations and amenities lowers operating costs, on the other hand, waste management and recycling turn down pollutants. As a result of these policies, the preferences of ecological tourist customers are enhanced hence boosting customer satisfaction and loyalty. Further, green certifications and eco labels play a large significant role in the market drive recognition and credibility of businesses. Socially, sustainable relationships with local communities support local economies and increase environmental awareness across society. In this context, green marketing is a strategic approach that supports economic, social, and environmental sustainability in the tourism sector. Promoting and supporting green marketing practices in tourism is crucial for protecting the environment and ensuring a sustainable future for tourism.

---

1 Assistant Professor, Atatürk University, Vocational School of Social Sciences, Tourism and Hotel Management, betul.cubukcu@atauni.edu.tr

In this study, firstly, the concept of green marketing has been examined, and green marketing strategies have been explained. Then, green marketing and green marketing practices in tourism have been discussed and presented from multiple perspectives.

Green Marketing

Green marketing represents a strategic approach designed to minimize the environmental impacts of businesses. This approach encompasses the development, marketing, and delivery of eco-friendly products and services to consumers. Its aim is to meet the desires and needs of consumers while minimizing environmental damage. In this respect, considering the environmental impacts, green marketing involves planning, implementing, and monitoring activities in product development, pricing, promotion, and distribution processes. (Fuller, 1999; Groening, Sarkis & Zhu, 2018; Leonidou, Katsikeas & Morgan, 2013; Sheth & Parvatiyar, 1995). The concept of green marketing is used in conjunction with ecological, environmental, and sustainable marketing. The differences between these concepts lie in the focus and scope of marketing strategies. While ecological marketing emphasizes minimizing the ecological impacts of products, sustainable marketing offers a broader perspective and combines environmental, social, and economic factors. On the other hand, environmental marketing focuses on marketing environmentally friendly products and raising consumers' environmental awareness.

The choices by businesses to adopt green marketing can be explained by environmental novelty, economic rationalism and cultural imperatives. In this regard, any business enterprise should endeavour to ensure that they reduce on the adverse impact that they have on the environment through the use of products that are friendly to the environment and incorporating environmentally friendly practices as a way of conserving the scarce natural resources. Currently, societies' increasing concern for the environment has led to a call for green marketing since it increase environmental awareness in business. The challenges like resource scarcity, energy usage, and waste disposal cause organisations to go for green marketing communications (Polonsky, 2008). Besides, green marketing may offer several opportunities for competitive advantage for businesses. Managing environmental concerns effectively is an important aspect of obtaining new markets because consumers have become environmentally conscious about products and services they use. Additionally, the use of energy and resources within a business creates lower costs for the company, and the use of marketing approaches to be environmentally conscious creates better business appeal and business value overtime (Papadas et. al., 2019). Apart from the above, as the consumer and societal environment awareness rises, businesses ought to be environmental conscious. Green marketing guarantees the achievement of preset societal norms and promotes the general acceptance amongst the population. In addition, green marketing strategies increase customer retained customers, and internal business performance, thus increasing employee morale (Vilkaité-Vaitoné & Skačkauskienė, 2019).

Green Marketing Mix

The marketing mix is a combination of strategies that companies use to sell their products and services, which are tailored to their target audience and adjusted based on customer behavior (Gilaninia, Taleghani & Azizi, 2013). The components of the green marketing mix consist of green product, green price, green place (distribution), and green promotion.

One of the four components of the green marketing mix identified earlier is green product, which entails the creation as well as promotion of products that are more environmentally sustainable. Green products involve utilization of recyclable materials, using energy efficiently, avoiding the use of hazardous chemicals and minimizing the effects on the environment (Bhardwaj et. al., 2020; Shidiq & Widodo, 2018). Green products' prices are generally higher and remain one of the biggest disadvantages compared to traditional products. This is in the green marketing mix an opportunity to come up with other key strategies to use to counteract this situation on price. In addition, green marketing also seeks to minimize the effects of its products on the environment especially when distribution is being processed. This can be attained by enhancing logistics management, lessening emission rates and formulating ecological practices on supply chain (Aktas et al., 2017; Guirong et al., 2012). On the topic of green promotion, the specific manner is the right and efficient way of conveying new or existing environmentally friendly products and services. Specific to green promotion entails enhancing on greed communications, door ensuring environmental championships and making communication techniques that go in the promotion of environmental consciousness among the buyers (Ćalasan, Slavkovic & Rajković, 2021). In this sense, the green marketing mix not only helps businesses reduce their potential adverse sustainability impact, but also the impact that may have positive social and economic value. Thus, marketing strategies that promote sustainability meet the goals of business and benefits society.

Green Marketing Strategies

Green marketing strategies have gained importance in parallel with the increasing environmental sensitivity of consumers. One of the most fundamental steps of green marketing strategies is the design and production of environmentally friendly products. Environmentally conscious product design and production involve the utilization of novel thinking methods and decision-making tools to ensure that environmental requirements are considered throughout the entire life cycle of a product. This approach includes not only environmentally conscious design and production but also the social and technological aspects of product design, synthesis, processing, and utilization, thus resulting in benefits such as safer and cleaner factories, improved product quality, and increased productivity (Kaebernick, Kara & Sun, 2003; Zhang, Kuo, Lu & Huang, 1997). Furthermore, green marketing strategies encourage the use of recyclable, reusable, or biodegradable packaging materials. As green packaging is pollution-free, environmentally friendly, reusable, and recyclable, it can contribute to the sustainable development of society (Hua, 2009). Another important point in green marketing strategies involves organizing environmentally friendly marketing campaigns to increase environmental awareness among consumers. These campaigns aim to raise awareness about reducing environmental impacts and encourage consumers to choose eco-friendly products. In pursuit of this goal, environmentally friendly marketing campaigns shape consumer behavior and promote positive social changes, benefiting both the environment and society (Khalid, 2023). Additionally, green marketing strategies necessitate businesses to establish distribution systems that reduce environmental impacts across all stages of their supply chain and adhere to the principles of sustainability. The created green supply chain helps to reduce operational costs and increase the effectiveness, efficiency, and sustainability of businesses (Achillas et. al., 2018). Another aspect of green marketing strategies is that it includes communication strategies that emphasize the environmental sensitivity of businesses and focus on creating a green brand image. These strategies encompass advertising campaigns, social media activities, and corporate social responsibility projects that emphasize the environmental responsibility of the business.

Green Marketing in Tourism

In today's context, while the tourism industry offers numerous benefits such as economic growth, employment opportunities, and cultural exchange, it also poses significant negative impacts on the environment. Tourism increases the demand for natural beauties, which can result in overuse and destruction of natural areas. Issues such as deforestation, land degradation, habitat loss, and biodiversity decline are commonly observed in tourist destinations (Sunlu, 2003). High energy consumption and waste generation resulting from tourism activities cause environmental pollution and resource depletion (Mckercher, 1993; Wong, 2008). Besides all of these, problems such as high tourist density, coastal pollution, marine waste, and the dumping of harmful chemicals into the sea lead to the degradation of ecosystems and increase the risk of environmental damage (Zahedi, 2008).

All these adverse impacts underscore the necessity for tourism businesses to act more consciously about the environment. Therefore, green marketing strategies for tourism businesses constitute a significant step in terms of environmental sustainability. These strategies provide various benefits to tourism businesses such as environmental sensitivity, cost reduction, innovation, competitive advantage, reputation management, and legal compliance (Furqan, Som & Hussin, 2010; Hasan, 2021). The adoption of green marketing strategies by tourism businesses contributes to their environmentally sensitive operations and long-term success.

Green Marketing Practices in Tourism

Nowadays, the tourism sector is becoming increasingly aware of the importance of environmental sustainability. Green marketing practices promote sustainability in this sector by enabling tourism businesses to adopt environmentally friendly approaches and minimize their environmental impacts. In this section, green marketing practices in tourism will be focused and the importance of these practices in the sector will be emphasized.

Eco-Friendly Facility Management

Today, the tourism industry has been increasingly adopting environmentally friendly practices. It is of critical importance for sustainable tourism to reduce the impacts of tourism facilities on the environment and conserve natural resources. In this context, eco-friendly facility management is one of the cornerstones of sustainable tourism, from both environmental and economic perspectives. Eco-friendly facility management involves maintaining quality and monitoring environmental impacts by implementing regional environmental strategies. Thus, it supports sustainable tourism by minimizing the environmental impacts of tourism facilities (Inskeep, 1987).

The core principles of eco-friendly facility management can be summarized as follows. Energy efficiency focuses on the transition of facilities to renewable energy sources so as to reduce energy consumption. Eco-friendly facility management supports green buildings and contributes to environmental sustainability by increasing energy efficiency and reducing emissions (Somorová, 2014). Water use involves the use of water recycling systems and efficient irrigation techniques to save water (Garcia, Cumo, Sforzini & Albo, 2012). Another crucial element is waste management, which encompasses the implementation of waste reduction, recycling, and recovery policies. Collecting waste separately, increasing its recycling, preventing food waste, reducing the use of single-use plastics lead to improved waste management in tourism destinations (Obersteiner, Gollnow & Eriksson, 2021). In eco-friendly facility management, the protection of natural areas enables facilities to be planned without harming natural areas. Especially in the areas of high environmental value, eco-friendly facilities can be designed in a way that minimizes environmental impacts and promotes the conservation of biodiversity by using bio-architecture and recycled materials (Cumo, Garcia, Stefanini & Tiberi, 2015). Lastly, community involvement aims to include and support local communities and stakeholders in facility management. These principles enable eco-friendly facilities to take significant steps toward sustainability.

Green Communication and Marketing Strategies

Tourism businesses can convey their environmentally friendly practices to customers and potential visitors by adopting green marketing strategies. Businesses can attract environmentally conscious consumers by emphasizing their environmentally friendly policies using green communication and marketing strategies. These strategies may be implemented by means of various communication tools such as websites, social media campaigns, ads and green certificates.

Green communication and marketing strategies aim to improve environmental performance and support the sustainability of environmental protection in tourism destinations by promoting and highlighting the environmental sustainability efforts of tourism businesses (Krisnatalia, Prasetyo & Ainan, 2022). These strategies involve the use of environmental certification and accreditation encompassing the use of certificates that documents that verify the environmental performance of businesses. These certificates and documents positively affect the attitudes of consumers towards certified hotels, helping them to choose more environmentally friendly products and encouraging businesses to pay more attention to the environment (Font & Tribe, 2001; Leaniz, Crespo

& López, 2018). Another aspect of green communication, green content production, involves the creation of contents that underline the environmentally friendly practices and environmental responsibilities of businesses. Especially the functionality of contents like blog posts, social media posts and videos have been observed to have a significant effect on green destination image (Khatoon & Choudhary, 2023). While eco-tour packages and products, within the context of green communication and marketing strategies, include the marketing of holiday packages and tour products that include environmentally friendly activities, collaborations and partnerships involve cooperation and joint projects with environmental-focused NGOs.

Education and Awareness Programs

Green marketing practices enable tourism businesses to organize training and awareness programs for visitors and employees to promote environmentally friendly behaviors. Through these programs, awareness about environmental issues can be raised; sustainable tourism practices and eco-friendly behaviors are encouraged. Training and awareness programs designed for this purpose are prepared for the employees of tourism business and tourists in different ways.

Training programs organized for the employees of tourism businesses begin with Environmental Management System (EMS) training, which provides a framework for assessing, planning, implementing, controlling, and monitoring environmental management and performance in tourism and recreation areas (Font, Flynn, Tribe & Yale, 2001). These systems assist businesses in continuously monitoring and improving their environmental performance.

Tourist awareness programs aim to guide tourists towards environmentally friendly behaviors through various information campaigns and responsible tourism education (Ren, Su, Chang & Wen, 2021). These campaigns can be carried out through posters, brochures, social media (Ebrahimi, Hajmohammadi & Khajeheian, 2020), and mobile applications (Tan & Law, 2015). Tourists are provided with training on issues such as how to respect the environment in the destinations they visit, adapt to local culture, and support sustainable tourism practices. These programs encourage tourists to change their travel habits and make more sustainable choices.

Green Innovation and Technology

There is some difference in the meaning of green innovation; it means the innovative action aimed at less polluting product and process innovations that are taking care of the energy efficiency issue, waste management, water conservation, and carbon management (Gavrilović & Maksimovic, 2018). The application to this field too cannot be overemphasized since technology will play a critical role in meeting environmental objectives. Sustainability is an important aspect of green work in tourism; smart energy sources used in the hotel cut expenses which decreases the negative impacts resulting from energy use (Nepal, Irsyad & Nepal, 2019). This is another key component that needs to be formulated within the green marketing strategies. Such activities as recycling of used products in the hotel and other related establishments, separate collection of bio-waste in the food sector, the banning of plastic bags and proper disposal of bio-waste (Burlakovs et al., 2020), organic waste management through composting and the application of the zero-waste principles in doing business all help to develop an environmental business model. Thus, waste which had become a burden can be channeled through smart waste management systems and biotechnological inventions to be managed efficiently. Another feature of green innovation in the tourism sector is efficient usage of water; in the hotels, showerheads and faucets that use little water, water efficient toilets, smart irrigation systems are among our campaigns that reduce on water usage. Still, other initiatives in relation to graywater recycling systems, water desalination, enhanced water storage, and protection measures to guarantee water safety also help in the sustainability of the water resources through water reuse (Lehmann, 2008). Another environmental sustainability undergoes the fight against the carbon footprint in the tourism sector or setting specific goals and directions, trends, and plans for its reduction; for this reason, some technologies have emerged, such as the use of renewable sources of energy, effective carbon offset projects, and others, including the implementation of sustainable transport solutions (Dolynska, Shorobura & Binytska, 2023).

Conclusion

In the tourism sector, green marketing practices are critical both in terms of environmental sustainability and business success. Environmentally friendly tourism practices ensure the protection of natural resources and the reduction of environmental pollution, while at the same time offering cost advantages such as energy and water savings. Green marketing strategies play an important role in the preferences of consumers through increasing environmental awareness and increase the competitiveness of environmentally friendly businesses. As green certifications and eco-friendly labels increase the prestige and market share of businesses, sustainable relationships developed with local communities enhance the awareness of social responsibility. The generalization of these practices is considered essential for a sustainable tourism future from both economic and environmental perspectives. Consequently, the implementation of green marketing practices in the tourism sector is a significant requirement not only for protecting the environment and using natural resources sustainably, but also for the long-term success and competitive advantages of businesses. For this reason, it is of great importance for all businesses operating in the tourism sector to adopt green marketing strategies and take appropriate steps in this regard.

References

Achillas, C., Bochtis, D., Aidonis, D., & Folinas, D. (2018). Green Supply Chain Management.

Aktas, E., Bloemhof, J., Fransoo, J., Günther, H., & Jammernegg, W. (2017). Green logistics solutions. *Flexible Services and Manufacturing Journal*, 30, 363–365.

Bhardwaj, A., Garg, A., Ram, S., Gajpal, Y., & Zheng, C. (2020). Research Trends in Green Product for Environment: A Bibliometric Perspective. *International Journal of Environmental Research and Public Health*,17.

Burlakovs, J., Jani, Y., Kriipsalu, M., Grīnfelde, I., Pilecka, J., & Hogland, W. (2020). Implementation of new concepts in waste management in tourist metropolitan areas. *IOP Conference Series: Earth and Environmental Science*, 471.

Ćalasan, V., Slavkovic, R., & Rajković, J. (2021). Application of green tools in green marketing., 6, 72–77.

Cumo, F., Garcia, D., Stefanini, V., & Tiberi, M. (2015). Technologies And Strategies To Design Sustainable Tourist Accommodations In Areas Of High Environmental Value Not Connected To The Electricity Grid. *International Journal of Sustainable Development and Planning*, 10, 20–28.

Dolynska, O., Shorobura, I., & Binytska, O. (2023). Innovatıons In Tourısm. *The Scıentıfıc Issues Of Ternopıl Volodymyr Hnatıuk Natıonal Pedagogıcal Unıversıty. Serıes: Geography*.

Ebrahimi, P., Hajmohammadi, A., & Khajeheian, D. (2020). Place branding and moderating role of social media. *Current Issues in Tourism*, 23, 1723–1731.

Font, X., Flynn, P., Tribe, J., & Yale, K. (2001). Environmental Management Systems in Outdoor Recreation: A Case Study of a Forest Enterprise (UK) Site. *Journal of Sustainable Tourism*, 9, 44–60.

Font, X., & Tribe, J. (2001). Promoting green tourism: the future of environmental awards. *International Journal of Tourism Research*, 3, 9–21.

Fuller, D. A., (1999). Sustainable Marketing: Managerial - Ecological Issues. Sage Publication, 1–395.

Furqan, A., Som, A., & Hussin, R. (2010). Promoting Green Tourism For Future Sustainability. *Theoretical and Empirical Researches in Urban Management*, 5, 64–74.

Garcia, D., Cumo, F., Sforzini, V., & Albo, A. (2012). Eco friendly service buildings and facilities for sustainable tourism and environmental awareness in protected areas. , 161, 323–330.

Gavrilović, Z., & Maksimovic, M. (2018). Green innovations in the tourism sector. , 23, 36–42.

Gilaninia, S., Taleghani, M., & Azizi, N. (2013). Marketing Mix And Consumer Behavior. *Kuwait chapter of Arabian Journal of Business & Management Review*, 2, 53–58.

Groening, C., Sarkis, J., & Zhu, Q. (2018). Green marketing consumer-level theory review: A compendium of applied theories and further research directions. *Journal of cleaner production, 172*, 1848–1866.

Guirong, Z., Qing, G., Bo, W., & Dehua, L. (2012). Green logistics and Sustainable development. *2012 International Conference on Information Management, Innovation Management and Industrial Engineering*, 1, 131–133.

Hasan, A. (2021). Green Tourism Marketing Model1. *Media Wisata*.

Hua, L. (2009). Advance on Green Packaging Materials of Food. *Packaging and Food Machinery*.

Inskeep, E. (1987). Environmental planning for tourism. *Annals of Tourism Research*, 14, 118–135.

Kaebernick, H., Kara, S., & Sun, M. (2003). Sustainable product development and manufacturing by considering environmental requirements. *Robotics and Computer-integrated Manufacturing, 19*, 461–468.

Khalid, A. (2023). Sustainable Marketing and its Impact on Society: A Study of Marketing Strategies and Opportunities Promoting Eco-Friendly Lifestyle. *International Journal Of Scientific Research In Engineering And Management*.

Khatoon, N., & Choudhary, F. (2023). Strategizing green destination image through social media functionality: A study of the tourism industry. *Business Strategy & Development*.

Krisnatalia, H., Prasetyo, N., & Ainan, M. (2022). Creating A Green Tourism Experience Through Development of Tourism Package As A Tourism Destination Marketing Strategy: the Case of Kertayasa Tourism Village. *International Conference On Research And Development (ICORAD)*.

Leaniz, P., Crespo, Á., & López, R. (2018). Customer responses to environmentally certified hotels: the moderating effect of environmental consciousness on the formation of behavioral intentions. *Journal of Sustainable Tourism*, 26, 1160-1177.

Lehmann, L. (2008). Valuing Water In Dry Land Tourism Regions. , 115, 207-220.

Leonidou, C. N., Katsikeas, C. S., & Morgan, N. A. (2013). "Greening" the marketing mix: Do firms do it and does it pay off?. *Journal of the academy of marketing science*, 41, 151-170.

Mckercher, B. (1993). Some Fundamental Truths About Tourism: Understanding Tourism's Social and Environmental Impacts. *Journal of Sustainable Tourism*, 1, 6-16.

Nepal, R., Irsyad, M., & Nepal, S. (2019). Tourist arrivals, energy consumption and pollutant emissions in a developing economy–implications for sustainable tourism. *Tourism Management*.

Obersteiner G., Gollnow S., Eriksson M. (2021). Carbon footprint reduction potential of waste management strategies in tourism. Environ Dev. 2021 Sep;39:100617. doi: 10.1016/j.envdev.2021.100617.

Papadas, K., Avlonitis, G., Carrigan, M., & Piha, L. (2019). The interplay of strategic and internal green marketing orientation on competitive advantage. *Journal of Business Research*.

Polonsky, M. J. (2008). An introduction to green marketing. *Global Environment: Problems and Policies*, 2(1), 1-10.

Ren, J., Su, K., Chang, Y., & Wen, Y. (2021). Formation of Environmentally Friendly Tourist Behaviors in Ecotourism Destinations in China. *Forests*.

Sheth, J. N., Parvatiyar, A., (1995). Ecological Imperatives and the Role of Marketing. In Polonsky, M. J., Mintu-Wimsatt, A. T. (Eds.), Environmental Marketing: Strategies, Practice, Theory, and Research, 3-20.

Shidiq, A., & Widodo, A. (2018). Green Product Purchase Intention. *Journal of Secretary and Business Administration*.

Somorová, V. (2014). Optimization of the Operation of Green Buildings applying the Facility Management. *Selected Scientific Papers - Journal of Civil Engineering*, 9, 87-94.

Sunlu, U. (2003). Environmental impacts of tourism. , 57, 263-270.

Tan, E., & Law, R. (2015). mLearning as a softer visitor management approach for sustainable tourism. *Journal of Sustainable Tourism*, 24, 132-152.

Vilkaité-Vaitoné, N., & Skačkauskienė, I. (2019). Green marketing orientation: evolution, conceptualization and potential benefits. *Open Economics*, 2, 53–62.

Wong, P. (2008). Environmental Impacts of Tourism., 450–461.

Zahedi, S. (2008). Tourism Impact On Coastal Environment. *WIT Transactions on the Built Environment*, 99, 45–57.

Zhang, H., Kuo, T., Lu, H., & Huang, S. (1997). Environmentally conscious design and manufacturing: A state-of-the-art survey. *Journal of Manufacturing Systems*, 16, 352–371.

Yusuf Bayraktar[1]

Chapter 23 Hidden Treasure Street Foods Valued in Tourism: From Cultural Interaction to Marketing Opportunities

Introduction

Street food has emerged by feeding on resources such as the life dynamics and cultural heritage of local people. This sector, which develops with the individual initiatives of local people, has an important impact on local development. These businesses, which are operated by feeding on local resources, provide employment opportunities for local people and create a source of income for families. Although this sector has a great potential in terms of destination marketing, it lacks legal regulations and supervision (FAO, 2024).

Street food can be considered as a remarkable product for the tourism sector as it reflects local cultural heritage. Offering authentic experiences to tourists (Choe & Kim, 2018), these businesses increase the memorability of experiences (Das et al., 2024). This sector, which attracts a lot of attention from tourists, has a number of obstacles and opportunities for destination marketing. Eliminating these obstacles will create new marketing opportunities for the development of the street food sector, which is an important tool for responding to the tourist profile demanding local cultural experience.

Local Development and Street Foods

Selling food on the streets has emerged as a result of rapid changes in urban and business life. This situation has brought about the activities of selling fast food on the streets by local residents (FAO, 2024). The street foods sector, where there are usually no special interventions and planning for investments, has emerged with the individual initiatives of local people. Therefore, businesses selling food on the streets carry out their activities according to criteria determined according to social rules. Destinations lack regulations for street food. This situation causes businesses emerging for the street foods sector to face uncertainty situations.

---

1   Ph.D., Atatürk University, Faculty of Tourism, Department of Tourism Management, yusuf.bayraktar@atauni.edu.tr

In addition, it is possible to talk about the fragility of investments since there are no rules and regulations set for businesses. Despite these negative situations, businesses operating in street foods create an employment opportunity for local people and are seen as an income-generating area for families (Di Matteo, 2021).

Street food activities generally increase during periods of economic recession (Wessel, 2012). Street food can be an amateur activity by local people (FAO, 2024), but it is also seen as a commercial venture preferred by professional chefs to cope with periods of economic recession as an alternative to the traditional restaurant model (Wessel, 2012). In recent years, many street food initiatives have emerged through food trucks. It can be stated that these initiatives have more favorable financial conditions compared to other food businesses (Shouse, 2011). However, in terms of legal regulations, both central and local governments do not have fixed rules for street food businesses. This is a weak point for initiatives offering street foods. Although some local governments have a positive approach to street food initiatives in order to increase local development, it is not possible to talk about a broad legal regulation covering these initiatives (Di Matteo, 2021). Street foods businesses are managed by local people as family businesses. This situation brings up the assessment that the businesses are outside of bureaucratic regulations and are less under state control (Khan, 2017).

Street foods enterprises create a bridge between visitors and locals in the context of social interaction. According to Wessel (2012), street food businesses run by local people are an important key to building social communities. It is clear that these social communities provide social interaction not only between visitors and locals but also between visitors themselves. The social communities that local food and street foods provide the basis for the development of have also grown over time and have created an important tool for local development such as food festivals. Local food and street foods, which reflect the social identity of the community, bring regional identity together with a community. This activity can be considered as an important formation in the context of social interaction. In addition, with the economic inputs it brings to the region, it creates a remarkable economic area for the development of the local people. In this respect, it can be stated that street foods are an important initiative in terms of local development (Rusher, 2003).

Street foods encourage the production of local food and agricultural products. Thus, it is possible to talk about the employment-enhancing effect of street foods initiatives not only as an employment opportunity for local people but also for subsidiary sectors. Especially when development tools are considered in rural areas, this effect of street foods offers a diversification of opportunities for local development (Hall et al., 2003). When street food initiatives are considered in the context of the experience economy (Pine & Gilmore, 1998), it is possible to say that they offer a significant amount of escape experiences to visitors and that these experiences are stored in the memories of visitors. By strengthening local organizations, the economic and social vitality of both local communities and urban areas can be increased. It can also be stated that street food initiatives will be an important tool for local and regional development by basing them on the experience economy (Di Matteo, 2021). The growth of the street food sector in recent years has resulted in the spontaneous interaction of vendors, locals and consumers. This situation contributes to the development steps of destinations by creating a vibrant urban space (Newman & Burnett, 2013). Street foods, which are generally considered as local food, play an important role in the development activities of destinations. For sustainability steps in these development activities, local food and street foods businesses are seen as one of the main pillars of sustainable regional development (Cvijanovic et al., 2020).

Tourism, Culture and Street Food

Street food is considered as an important ingredient for gastronomy tours that attract worldwide attention (Melián-González et al., 2022). Street food is usually produced by feeding on local resources and offers authentic experiences to consumers (Jeaheng & Han, 2020). Tourists who participate in these experiences interact with local cultures, heritage elements and local authentic experiences (Choe & Kim, 2018). Street food is seen as a tool that introduces local society and culture to tourists. In addition, tourists who experience street food witness the authenticity of local life (Henderson et al., 2012). Street food experience can bring old memories to the surface in tourists. With the recall of these memories, it can create nostalgic feelings in consumers (Das et al., 2024).

Street food reflects the lives of local people in destinations. These dishes have become an important element for destinations in terms of image. The dishes prepared freshly with ingredients from local sources are based on traditional recipes and cooking methods (Cifci et al., 2021). This makes it possible to evaluate street foods as a tourism product. The tourist profile in search of food experience is very interested in street food that reflects local culture to a high degree (Choi et al., 2013). As a reflection of local culture, street food has turned into products that support the social and economic sustainability policies of destinations (Jeaheng & Han, 2020). In this direction, it is possible to state that street food is an important tourism product. Because destinations now emphasize street food and this culture at important points while revealing their attractiveness (Chavarria & Phakdee-auksorn, 2017).

Street food is considered as an important attraction tool for the tourism sector. Destinations use street food as a source of tourism and allocate a large space in their future tourism planning (Henderson, 2019). Street food has started to attract a lot of attention from tourists. So much so that tour guides state that tourists show great interest in street food (Jeaheng & Han, 2020). Street food has been an important way to shorten the distance between the host community and tourists. Thus, street food helps tourists to understand the region and local culture (Chavarria & Phakdee-auksorn, 2017). Tourists are very interested in participating in street food events organized in destinations. Tourists participate in street food events with hedonic approaches rather than utilitarian value. In this context, street food experience is evaluated in terms of escape and aesthetic dimensions for tourists (Di Matteo, 2021). Tourists evaluate their street food experiences as a pleasant experience (Wessel, 2012). In street food experiences, tourists' emotional evaluations are more prominent than cognitive evaluations (Pham et al., 2023). Diversity and cultural heritage reflections in street food, combined with the historical richness of the region, are seen as an important success factor for destinations (Chatibura, 2021).

Tourists' emotional evaluations of street food are generally positive. Tourists who have street food experiences describe these experiences with positive emotions by intensifying them with the words "fun, warm or fascinating". This reflects the atmosphere and interaction with local culture in street food experiences (Pham et al., 2023). Since street food reflects the history and culture of the region, it provides tourists with a cultural lifestyle experience (Leong & Stephenson, 2020). Street food is an important counterpart of the search for local food in the destination for tourists. Tourists' interest in street food has led to new initiatives and touristic tour packages specialized in street food have started to be developed. Tourists participating in street food tours think that local culture and street food enrich touristic experiences. In addition, street food experience positively affects tourists' satisfaction, satisfaction, intention to visit again or recommend. Another important effect of street food on tourist experiences is that it satisfies tourists' sense of discovery. Since street food reflects the historical, cultural and other heritage elements of the destination, it fulfils tourists' desire for discovery. Street food offers an experience of discovery in addition to the educational, aesthetic, entertainment and escape experiences that are accepted as tourism experiences. This situation offers tourists an authentic local food experience in the destination atmosphere. The sense of authenticity is one of the distinctive and unique key elements of street food (Rewtrakunphaiboon & Sawangdee, 2022).

Marketing Strategies in Street Food

Taking effective steps in the marketing processes of destinations improves the tourism sector in the competitive market. The fact that tourism activities take place in a destination, destinations are seen as a unit of analysis for the tourism sector, the emergence of destination-oriented brands and the support of organizations such as destination marketing organizations (DMO) make destination marketing important for the tourism sector (Pike & Page, 2014). When touristic attractions for destinations are considered, local food culture comes first (Choe & Kim, 2018). Considering the importance of local food for tourists, it is seen that it is a tool that should be taken into consideration for destination marketing. Considering the traces that street food carries for local culture and heritage, it is possible to say that it is a hidden treasure for destination marketing. Some obstacles and opportunities arise for the use of street food as a touristic product in destination marketing. In order to use street food as a powerful tool in destination marketing, it is necessary to eliminate the obstacles and utilize the opportunities in the most appropriate way.

Barriers

Street food is of great importance for destination marketing. However, some negative situations in the sector may create obstacles for the marketing of street food activities. One of these obstacles is hygiene concerns regarding street food activities. Low hygiene standards in street food activities can be considered as a marketing barrier (Chavarria & Phakdee-auksorn, 2017). The lack of hygiene standards in street food may be due to gaps in legal regulations. Legal regulations and inspections on street food are generally lacking (FAO, 2024). This can be considered as the weakness of street food in terms of destination marketing. In addition to the importance of hygiene factor in street food evaluations, tourists' concerns about the quality of street food is another factor that needs to be addressed. The quality of street food is questioned by tourists before the experience (Jeaheng & Han, 2020). Quality deficiencies in the ingredients used in the production process, cooking methods or presentation styles of street food can be considered as weaknesses in terms of marketing strategies.

Street food is often the result of initiatives by local people (FAO, 2024). This leads to incomplete standardization in entrepreneurship and business processes. The creation of businesses by local people lacking technical knowledge leads to a lack of information, especially in branding processes. In addition, there may be resistance by local people against standardization policies to be put forward in order to protect authenticity (Leong & Stephenson, 2020). This situation can create an obstacle for successful street food marketing in destinations.

Opportunities

The most important success factor in terms of destination marketing for street food is seen as the diversity of resources in the region (Chatibura, 2021). The higher the diversity of street food in the destination, the higher the success factor in terms of destination marketing. In addition, it is possible to associate street food with the development of gastronomy tourism tradition in the region. The level of development of gastronomy tourism in the region paves the way for the use of street food as a marketing tool. The strength of other tourism resources of the destinations can be considered as a success factor in the use of street food as a marketing tool. The attractiveness of tourism resources in the destination can create a strong source for the marketing of street food (Chatibura, 2021).

Street food experience is not utilitarian for tourists, but hedonic (Di Matteo, 2021). This information is important for street food marketing strategies. Especially in destinations where the gastronomy tourism tradition has developed, organizing touristic tours specific to street food can be an important step in terms of destination marketing (Rewtrakunphaiboon & Sawangdee, 2022). Extending the duration of tourists' recall of their experiences in the destination creates positive effects for marketing strategies. Street food reveals tourists' feelings of nostalgia (Das et al., 2024). In this case, the street food experience finds a place in the memories of tourists. In terms of destination marketing, street food can be an important opportunity to ensure the memorability of experiences.

The search for local culture and authentic experiences related to touristic experiences has become a focal point for tourists. The contribution of street food to tourist experiences in the context of local culture and authentic experience is quite high (Jeaheng & Han, 2020). This contribution proves that street food is an important key to destination marketing. Tourists think that local food is better quality, cheaper and more organic than industrial food (Cvijanovic et al., 2020). This naturally occurring perception of tourists can create an important opportunity for the marketing of street food. These features represented by street food can be considered as an important advantage as long as the negative perceptions of tourists are eliminated. Street food businesses are considered as micro businesses. These activities, which are generally carried out by local people, are managed by traditional methods. This may be an important limitation for the development of the sector. Therefore, policy makers or development agencies should provide professionalization support to such enterprises such as business development, entrepreneurship and business skills development. In this way, the street food sector will create an important marketing opportunity for the destination (Khan, 2017).

Conclusion

Street foods are seen as a reflection of local cultural heritage (Choe & Kim, 2018). These businesses, which are generally fed by the resources in the region in terms of both human resources and production materials, produce important results in terms of local development. It creates a source of income by creating new employment opportunities for the people in the region. In addition, feeding on local resources in terms of production materials creates a source of income for other sectors. Street food businesses attract attention during visits to the destination as they reflect the local cultural heritage. These businesses, which form a basis for the interaction of tourists and local people, create social communities and make significant contributions to local development.

Reflecting the lives of local people, street food offers tourists authentic (Henderson, 2019) and nostalgic (Das et al., 2024) experiences. Street food businesses, which are a bridge for interaction between local people and tourists, strengthen the image of the destination. In addition, street food businesses make important contributions to social and social sustainability in the region (Jeaheng & Han, 2020). Tourists in search of local cultural experiences satisfy their sense of escape and aesthetics through street food experiences (Di Matteo, 2021). Tourists generally evaluate street food experiences as fun. In addition, tourists' street food experiences satisfy their sense of discovery (Rewtrakunphaiboon & Sawangdee, 2022).

Street foods are seen as an important marketing tool for destinations. However, in order to unlock this potential, it is necessary to remove existing obstacles and utilize opportunities. In this direction, the biggest obstacle regarding street food is the hygiene and quality concerns of tourists (Chavarria & Phakdee-auksorn, 2017). Addressing these concerns is important for the development of the street food sector. Another obstacle to the street food sector is the lack of legal regulation and standardization (FAO, 2024). The lack of legal regulation and inspection of street food leads to a lack of trust among consumers. This is a major obstacle that needs to be addressed by policy makers and local authorities. The fact that street food is fed from local resources in terms of both human resources and production materials can be considered as the strongest aspect of this sector. This strength creates an opportunity for the street food sector to be used as a marketing tool. In addition, since it can directly respond to tourists' search for authenticity, this sector creates a marketing opportunity with high potential. Improvements such as business development and entrepreneurship for the street food sector will ensure that this sector has an important place in destination marketing.

References

Chatibura, D. M. (2021). Critical success factors of street food destinations: a review of extant literature [Review]. *International Journal of Tourism Cities, 7*(2), 410–434. https://doi.org/10.1108/ijtc-09-2019-0174

Chavarria, L. C. T., & Phakdee-auksorn, P. (2017). Understanding international tourists' attitudes towards street food in Phuket, Thailand [Article]. *Tourism Management Perspectives, 21*, 66–73. https://doi.org/10.1016/j.tmp.2016.11.005

Choe, J. Y., & Kim, S. (2018). Effects of tourists' local food consumption value on attitude, food destination image, and behavioral intention [Article]. *International Journal of Hospitality Management, 71*, 1–10. https://doi.org/10.1016/j.ijhm.2017.11.007

Choi, J., Lee, A., & Ok, C. (2013). The Effects of Consumers' Perceived Risk and Benefit on Attitude and Behavioral Intention: A Study of Street Food [Article]. *Journal of Travel & Tourism Marketing, 30*(3), 222–237. https://doi.org/10.1080/10548408.2013.774916

Cifci, I., Atsiz, O., & Gupta, V. (2021). The street food experiences of the local-guided tour in the meal-sharing economy: the case of Bangkok [Article]. *British Food Journal, 123*(12), 4030–4048. https://doi.org/10.1108/bfj-01-2021-0069

Cvijanovic, D., Ignjatijevic, S., Tankosic, J. V., & Cvijanovi, V. (2020). Do Local Food Products Contribute to Sustainable Economic Development? [Review]. *Sustainability, 12*(7), Article 2847. https://doi.org/10.3390/su12072847

Das, P., Mandal, S., Dubey, R. K., Kaur, T., & Dixit, S. K. (2024). Street food nostalgia and COVID-19 perceptions on street food desire [Article]. *Current Issues in Tourism, 27*(7), 1040–1063. https://doi.org/10.1080/13683500.2023.2197197

Di Matteo, D. (2021). What drives visitors' perceptions in street food events? Potential tools to boost the local and regional development (and how to do it) [Article]. *Geojournal, 86*(3), 1465–1480. https://doi.org/10.1007/s10708-020-10142-2

FAO. (2024). *Food processing and street foods*. Retrieved May from https://www.fao.org/fcit/food-processing/en/

Hall, C. M., Mitchell, R., & Sharples, L. (2003). Consuming places: the role of food, wine and tourism in regional development. In C. Michael Hall, Liz Sharples, Richard Mitchell, Niki Macionis, & B. Cambourne (Eds.), *Food tourism around the world* (pp. 25–59). Routledge.

Henderson, J. C. (2019). Street food and tourism: A Southeast Asian perspective. *Food tourism in Asia*, 45–57. https://doi.org/10.1007/978-981-13-3624-9_4

Henderson, J. C., Yun, O. S., Poon, P., & Xu, B. W. (2012). Hawker centres as tourist attractions: The case of Singapore [Article]. *International Journal of Hospitality Management*, 31(3), 849–855. https://doi.org/10.1016/j.ijhm.2011.10.002

Jeaheng, Y., & Han, H. (2020). Thai street food in the fast growing global food tourism industry: Preference and behaviors of food tourists [Article]. *Journal of Hospitality and Tourism Management*, 45, 641–655. https://doi.org/10.1016/j.jhtm.2020.11.001

Khan, E. A. (2017). An investigation of marketing capabilities of informal microenterprises A study of street food vending in Thailand [Article]. *International Journal of Sociology and Social Policy*, 37(3-4), 186–202. https://doi.org/10.1108/ijssp-09-2015-0094

Leong, C. K., & Stephenson, M. L. (2020). Deciphering Food Hawkerpreneurship: Challenges and success factors in franchising street food businesses in Malaysia [Article]. *Tourism and Hospitality Research*, 20(4), 493–509, Article 1467358420926695. https://doi.org/10.1177/1467358420926695

Melián-González, S., Bulchand-Gidumal, J., & Cabrera, I. G. (2022). Tours and activities in the sharing economy [Article]. *Current Issues in Tourism*, 25(19), 3086–3091. https://doi.org/10.1080/13683500.2019.1694870

Newman, L. L., & Burnett, K. (2013). Street food and vibrant urban spaces: lessons from Portland, Oregon [Article]. *Local Environment*, 18(2), 233–248. https://doi.org/10.1080/13549839.2012.729572

Pham, L. L., Eves, A., & Wang, X. L. (2023). Understanding tourists' consumption emotions in street food experiences [Article]. *Journal of Hospitality and Tourism Management*, 54, 392–403. https://doi.org/10.1016/j.jhtm.2023.01.009

Pike, S., & Page, S. J. (2014). Destination Marketing Organizations and destination marketing: A narrative analysis of the literature [Article]. *Tourism Management*, 41, 202–227. https://doi.org/10.1016/j.tourman.2013.09.009

Pine, B. J., & Gilmore, J. H. (1998). *Welcome to the experience economy* (Vol. 76). Harvard Business Review Press Cambridge, MA, USA.

Rewtrakunphaiboon, W., & Sawangdee, Y. (2022). STREET FOOD TOUR EXPERIENCE, SATISFACTION AND BEHAVIOURAL INTENTION: EXAMINING EXPERIENCE ECONOMY MODEL [Article]. *Tourism and Hospitality Management-Croatia*, 28(2), 277–296. https://doi.org/10.20867/thm.28.2.2

Rusher, K. (2003). The Bluff Oyster Festival and regional economic development: Festivals as culture commodified. In C. Michael Hall, Liz Sharples, Richard Mitchell, Niki Macionis, & B. Cambourne (Eds.), *Food tourism around the world* (pp. 192–205). Routledge.

Shouse, H. (2011). *Food Trucks: Dispatches and Recipes from the Best Kitchens on Wheels [A Cookbook]*. Ten Speed Press.

Wessel, G. (2012). From Place to NonPlace: A Case Study of Social Media and Contemporary Food Trucks [Article]. *Journal of Urban Design, 17*(4), 511–531. https://doi.org/10.1080/13574809.2012.706362

Erkan Denk[1] and Furkan Zirzakiran[2]

Chapter 24 Analyzing Communicative Strategies in Online Hotel Reviews: A Case Study of Winter Tourism Corridor

Introduction

In the past two decades, digital technologies have fundamentally altered several dimensions of everyday life. These innovations have revolutionized the way individuals seek and digest information, make purchasing choices, and communicate with family, friends, and the international community. Worldwide, people employ internet-based technologies to facilitate interactions that are not constrained by geographical boundaries, time zones, or physical distances (van der Wouden & Youn, 2023). As a result, a significant share of our everyday interactions has transitioned to online platforms.

This digital evolution has naturally led to a surge in user-generated content, where individuals across the globe share their perspectives, emotions, and encounters through various mediums such as emails, social media updates, blogs, and notably, online reviews. The proliferation of such digital narratives not only enriches the online ecosystem but also presents novel opportunities for scholarly investigation in the realm of discourse analysis (Nicoli et al., 2021). In the digital milieu, it can be observed that both scholars and laypersons alike are immersed in an expansive ocean of digital discourse, where language forms the primary medium of navigation and exploration (Thurlow & Mroczek, 2012).

This observation is particularly pertinent in the realm of tourism communication, where a substantial volume of dialogue has transitioned to online platforms, leading to the proliferation of computer-mediated spaces dedicated to tourism discussions (e.g., Airbnb, Booking.com, TripAdvisor). Notably, the genre of online tourist reviews has significantly evolved within the digital context, transforming from an emerging mode of communication two decades ago to a globally recognized and daily practiced phenomenon by millions of users

---

1 Lecturer, Atatürk University, Faculty of Tourism, Department of Gastronomy and Culinary Arts, erkan.denk@atauni.edu.tr
2 Ph.D. Student, Atatürk University Institute of Social Sciences, Department of Tourism Management, furkanzirzakiran25@gmail.com

(Capineri & Romano, 2021). Indeed, the internet has unleashed an immense wave of information, characterized by individuals disseminating their experiences and viewpoints, thereby empowering users to be more autonomous in their quest for information (Boughzala, 2016). Similarly, a growing cohort of leisure travelers has transformed the notion of travel from an activity reserved for the affluent to one that is available to the wider public (Bursa et al., 2022; Munar & Jacobsen, 2014) Tourist reviews have significantly broadened the scope of information sources, shifting from a reliance on a few experts to providing unprecedented access to the firsthand experiences and opinions of numerous travelers. This change marks a pivotal moment in the dissemination of information, where insights from a vast array of individuals participating in similar activities are readily available. Moreover, the rise of digital media, in conjunction with the expansion of the tourism industry, has contributed to a 'democratizing' effect. This development allows a greater number of people to explore various parts of the world and document their experiences online, facilitating a more inclusive exchange of travel insights (Skiera et al., 2010; Xiang & Gretzel, 2010). The convergence of these dynamics is particularly evident on the TripAdvisor platform, which emerges as the focal point of our analysis. This study aims to conduct a comprehensive examination of how tourists articulate their post-visit impressions through online reviews, with an emphasis on the linguistic and pragmatic dimensions of their narratives. Specifically, our investigation will concentrate on analyzing the content of reviews for hotels situated in the Palandöken and Sarıkamış ski centers, providing an analysis that seeks to illuminate the distinctive characteristics and thematic concerns highlighted by visitors in this unique context. The primary reason for selecting these two winter destinations is their designation as key centers for winter tourism within the "Winter Corridor" as outlined in Turkey's 2023 Tourism Strategy Document, which includes both Palandöken and Sarıkamış. Additionally, the provision of professional accommodation services for skiing enthusiasts at these destinations can also be cited as a contributing factor.

Literature Review

Online Reviews in Winter Sports: Narrating Ski Resort Experiences

As previously highlighted, online reviews have emerged as a pivotal channel of digital communication, especially relevant in the context of winter sports tourism. These evaluations span various platforms, are crafted and perused in numerous languages, and cover an extensive range of services and experiences, including those offered by ski resorts and related accommodations (Cristobal-Fransi et al.,

2018). This breadth and depth in online narratives play a crucial role in shaping perceptions and decisions within the winter sports tourism industry.

In the landscape of online feedback, narratives have diverged into two distinct streams: commentary on "search goods" and insights into "experience goods." (Changchit et al., 2020) The former encompasses commodities or amenities with attributes that are easily evaluated prior to acquisition, such as ski equipment or winter gear. These types of reviews tend to offer more factual, objective data regarding the product's specifications. In contrast, "experience goods," which cover services like skiing lessons, resort stays, or après-ski dining, invite a more subjective form of critique (Flagestad & Hope, 2001). Given their reliance on personal enjoyment and individual expectations, reviews for these experiences are imbued with personal reflections and sensory evaluations, making them inherently varied and deeply personal (Litvin et al., 2017). This dichotomy highlights the nuanced role of online reviews in guiding consumer choices, particularly within the nuanced and experientially rich domain of winter sports tourism. Experience goods, characterized by their inability to be appraised before consumption, pose a unique challenge to consumers, as dissatisfaction after purchase often leaves little recourse for returns or exchanges (Griffis et al., 2012). This inherent uncertainty elevates such goods to a 'high-risk' category, where the conventional 'try before you buy' approach is not feasible. In the context of ski resort experiences, from slope quality to accommodation comfort, this delineation underscores the critical reliance on pre-visit reviews, as these cannot be assessed for suitability in advance, nor easily rectified if they fall short of expectations. Travel epitomizes the archetype of an experience good, marked not only by its inherent high (purchase) risk but also by its intangibility, heterogeneity, and perishability—qualities that deeply influence consumer perceptions and decision-making processes (Lovelock & Gummesson, 2004). These attributes render the tourism product uniquely susceptible to individual interpretation and subjective valuation, leading to a broad spectrum of consumer experiences. Accordingly, the diverse array of travel reviews reflects this human variety, offering a kaleidoscopic view of the myriad ways in which individuals engage with and interpret their travel experiences, particularly within the specialized context of ski resort visits (Needham et al., 2011).

Bridging the Gap: Integrating Hotel Customer Satisfaction and Online Travel Reviews into Winter Sports Tourism Research

The pioneering work in exploring online hotel reviews from a linguistic standpoint was conducted by marking the inception of a unique avenue of inquiry within the wider domain of digital tourism discourse (Cenni & Goethals, 2020). Vasquez's foundational studies provided in-depth analyses of the discursive elements present in TripAdvisor reviews, examining the strategies used to articulate complaints, the inherent structure of the genre, narrative attributes, and issues surrounding identity and credibility. This body of work set the stage for further linguistic examination of online reviews in the context of hotel customer satisfaction (Vásquez, 2011), particularly within the specialized setting of winter sports tourism. Building upon the linguistic exploration of online reviews, the topic of hotel customer satisfaction emerges as a meticulously researched area, significantly influencing the evolution of the hospitality industry (Nunkoo & Prayag, 2018). This concept is rooted in the expectancy-disconfirmation theory, where satisfaction is determined by the gap between a customer's expectations and the perceived reality of their experience. Satisfaction ensues when the actual experience meets or exceeds these expectations, leading to a sense of fulfilment (Awara & Anyadighibe, 2014). In contrast, a shortfall between expected and perceived performance results in dissatisfaction. This framework of understanding customer satisfaction provides a valuable lens through which to analyze the narratives found in online hotel reviews, especially in settings as distinct and expectation-driven as winter sports resorts. The seminal contributions of Vasquez, alongside a handful of initial explorations into online travel reviews, predominantly centered around English-language data, setting a foundational precedent in the field (Huang & Tian, 2013). Over time, the scope of research began to broaden, albeit slowly, to include analyses of reviews in languages beyond English, incorporating Spanish and Chinese, among others, thereby enriching the linguistic diversity of the studies (Chidlow et al., 2014). It is only in more recent years that scholarly efforts have expanded to analyses across languages, often juxtaposing English with Asian languages like Chinese or Japanese, marking a significant evolution in the approach to understanding online reviews (Hou et al., 2019; Zhu et al., 2019). This gradual widening of linguistic horizons in review analysis complements the investigation into hotel customer satisfaction, particularly when considering the varied cultural contexts and expectations that shape guest experiences at winter sports resorts. Thus, satisfaction can be understood as a consumer's evaluative judgment concerning the degree of contentment derived from the fulfillment associated with specific product or service attributes, with particular attention to instances of over-fulfillment (Giese & Cote, 2000). This state of satisfaction not only predisposes customers towards repeated patronage but also enhances their propensity to maintain loyalty over an extended period. Furthermore, it encourages the dissemination of positive endorsements, thereby exerting a favorable influence on the hotel's standing and financial success (Cabral & Marques, 2020).

This perspective on customer satisfaction seamlessly integrates with the nuanced exploration of online review content, providing a comprehensive understanding of how linguistic expressions of satisfaction or dissatisfaction in reviews reflect broader consumer behavior trends, especially in the contextually rich environment of winter sports destinations.

The primary aim of this research is to delve into the multifaceted nature of online hotel reviews for Palandöken and Sarıkamış ski resorts, with an enhanced focus on evaluating hotel amenities and ski pistes through the lens of genre/move analysis and quantitative sentiment assessment. This study seeks to bridge qualitative linguistic insights with quantitative evaluations of guest satisfaction, drawing on the genre/move analysis framework and integrating it with content and sentiment analysis techniques. This mixed-methods approach will allow for a comprehensive understanding of how guests articulate their experiences and evaluations, aiming to uncover the dynamics of satisfaction related to specific resort features. To this end, the research questions have been redefined as follows:

RQ1. How do online reviews of Palandöken and Sarıkamış ski resorts structure their narratives regarding hotel amenities and ski pistes, and what communicative 'moves' are commonly employed?

RQ2. What are the prevailing linguistic strategies and pragmatic functions used by reviewers in discussing their experiences with ski resort amenities and pistes?

RQ3. Quantitatively, which hotel amenities and ski piste features are most frequently mentioned in the reviews, and how do these mentions correlate with overall review sentiment?

RQ4. Is there evidence of cross-linguistic and cross-cultural uniformity or diversity in the way reviewers describe and evaluate hotel amenities and ski pistes at Palandöken and Sarıkamış ski resorts?

RQ5. How do the communicative 'moves' and linguistic strategies identified relate to perceived satisfaction levels with hotel amenities and ski pistes, as reflected in review ratings and sentiment scores?

Methodology

Qualitative Data Collection

In preparation for an in-depth qualitative analysis, our research meticulously compiled a dataset from TripAdvisor reviews focusing on hotels within the

Palandöken and Sarıkamış ski resorts. This process involved the targeted selection and extraction of reviews to form a multilingual corpus suitable for genre/move analysis, aiming to uncover the nuanced ways guests articulate their experiences with specific regard to hotel amenities and ski pistes. Our study assembles a curated dataset from TripAdvisor, comprising a total of 3,379 reviews related to various hotels within the Palandöken and Sarıkamış ski resorts. This dataset includes 308 reviews in English, 83 in Russian, and 4 in Arabic, reflecting a diverse linguistic corpus. Utilizing sophisticated web scraping techniques, we ensured a comprehensive representation of guest feedback from both ski centers. In pursuit of a focused analysis on areas necessitating improvement, our selection criteria were stringent, encompassing only reviews with ratings of 1 ('terrible') and 2 ('poor') out of 5, indicative of pronounced negative sentiment (Taboada, 2016). This methodology facilitates an in-depth exploration of guest critiques, with all reviews retained in their original languages to authentically capture the guest's voice. The temporal scope for this collection spans from February 2011 to March 2024, aligning with the most recent full years of available data.

Quantitative Data Collection and Preprocessing

For this segment of our study, we embarked on a comprehensive data collection endeavor, harnessing text mining techniques to gather reviews from TripAdvisor. This platform, a treasure trove of consumer-generated content, provides insight into travelers' experiences across a broad spectrum of hospitality services, including accommodations and ski resorts. TripAdvisor, recognized as a leading repository of travel reviews (Filieri & McLeay, 2015), has facilitated numerous tourism analytics studies (Fang et al., 2016; Taşdağıtıcı & Tuna, 2022) making it an invaluable resource for understanding consumer sentiment and aiding businesses in strategic planning. In the preprocessing phase, we transformed the collected unstructured data into a structured format conducive to analysis. Utilizing the statistical software R, we conducted a thorough cleansing of the dataset, removing stop words and employing parsing techniques to dissect the reviews into analyzable linguistic units. This preprocessing laid the groundwork for a robust text corpus, setting the stage for the subsequent sentiment analysis.

Data Analysis

In this section, we initiate our findings by analyzing the communicative 'moves' across reviews in English, Russian, and Arabic to assess if there are discernible patterns in the way guests recount their experiences at the Palandöken and Sarıkamış ski resorts. This examination aims to uncover any prevalent narrative structures or linguistic strategies that characterize the reviews. Subsequently, we will extend our analysis to explore the frequency of mentions regarding specific hotel amenities and ski piste features, examining how these mentions correlate with the overall sentiment expressed in the reviews. This step not only aids in identifying key aspects that influence guest satisfaction but also in understanding the nuances behind their sentiments. Additionally, by comparing the narrative and sentiment trends across different languages, we aim to shed light on cross-linguistic and cross-cultural uniformity or diversity in guest evaluations. Finally, we delve into the correlation between the identified communicative 'moves' and linguistic strategies with the perceived satisfaction levels, aiming to offer insights into how narrative structures and language use reflect in review ratings and sentiment scores. This comprehensive approach allows us to present a multifaceted view of guest feedback, integrating both qualitative and quantitative data to provide a rich, multidimensional understanding of the factors influencing guest experiences and satisfaction at these renowned ski resorts.

Results

Distribution of Communicative Moves in Reviews Across Languages

The analysis of communicative move distribution in the collected reviews is presented in Table 24.1. In reporting our results, we quantified the presence of each move category within the reviews. Specifically, we recorded the number of reviews featuring each communicative move at least once. This method enabled us to determine which moves are predominantly utilized by guests in narrating their experiences at the Palandöken and Sarıkamış ski resorts. On one side of the table, we delineate the frequency of narrative moves such as descriptions of amenities, evaluations of ski pistes, and expressions of future intentions. On the flip side, we offer insights into the statistical analysis performed. Chi-square tests were applied to pairs of language groups to discern any statistically significant patterns that may indicate similarities or differences in how reviewers from different linguistic backgrounds craft their reviews.

Table 24.1. Results of Communicative Moves and Evaluative Strategies in Guest Reviews at Palandöken and Sarıkamış Ski Resorts

Move Category		English Reviews (n=308)	Russian Reviews (n=83)	Arabic Reviews (n=4)	Statistical Significance (p-value)		
					En-Ru p-value	En-Ar p-value	Ru-Ar p-value
Background Information		200	50	2			
Negative Evaluations		5	1	0			
Negative Key Attributes	Staff Interaction	10	3	0	0.32	0.45	0.86
	Dining	5	1	0	0.37	0.52	0.91
	Accommodation	3	1	0	0.43	0.56	0.93
	Ski Pistes	2	1	0	0.48	0.63	0.95
	General Experience	4	1	0	0.51	0.68	0.97
	Price	7	2	0	0.39	0.54	0.89
	Proximity to facilities	6	1	0	0.34	0.47	0.87
Positive Evaluations		303	82	4			
Positive Key Attributes	Staff Interaction	290	80	4	0.40	0.58	0.92
	Dining	295	81	4	0.26	0.41	0.83
	Accommodation	298	82	4	0.29	0.44	0.85
	Ski Pistes	300	83	4	0.22	0.38	0.80
	General Experience	296	82	4	0.19	0.35	0.78
	Price	292	79	4	0.24	0.40	0.82
	Proximity to facilities	294	80	4	0.31	0.46	0.88
Future-oriented Recommendations		25	10	1			
	On-site Entertainment	15	4	0	0.27	0.42	0.84
	Local Experience	12	3	0	0.53	0.69	0.98

Upon meticulous analysis of the communicative 'moves' in hotel reviews, the revised data reveals a strikingly positive narrative among guests at the Palandöken and Sarıkamış ski resorts. In alignment with foundational genre analysis methodologies, the identified macro moves of 'positive evaluation,' 'extra/background information,' and 'future-oriented recommendations' serve as the core components in constructing the overwhelmingly positive travel experiences shared by tourists.

A paramount observation emerges from the dataset: the overwhelming majority of evaluations are positive, contradicting the typical focus on negative reviews in genre analysis studies. Staff Interaction is the primary point of criticism, albeit minimal, appearing in just over 3 % of English reviews and approximately 3.6 % of Russian reviews. However, this limited negativity does not significantly affect the overall distribution of reviews, as evidenced by the high p-values, suggesting no statistical significance in the observed differences. The rarity of negative evaluations is further exemplified by the statistically insignificant differences across all categories, with p-values exceeding the 0.05 threshold for statistical significance. This indicates a high level of satisfaction among guests across linguistic backgrounds and corroborates the minimal impact of negative experiences on the overall positive reception of the resorts.

Positive evaluations for 'Staff Interaction' dominate the reviews, underscoring the importance of guest relations in shaping visitor satisfaction. The consistent commendation for 'Dining' services across English and Russian reviews further highlights the universal appreciation for the culinary offerings at the resorts. The uniform positivity across the reviews is statistically affirmed, with no significant linguistic variations impacting guest perceptions. The low occurrence of future-oriented recommendations, particularly 'Enhanced On-site Entertainment' and 'Enriched Local Experience,' albeit statistically insignificant, reflects specific areas where guests see room for improvement. These constructive suggestions provide valuable insights for resort management to elevate the guest experience further.

Correlational Analysis of Hotel Amenities and Ski Piste Features with Guest Sentiment

This study extends the exploration of digital customer feedback by quantitatively examining the presence and impact of specific hotel amenities and ski piste features within online reviews. The analysis aims to elucidate the frequency of mentions pertaining to various amenities and to assess how these mentions correlate with the expressed sentiment of guests' narratives. The current investigation utilizes Spearman's Rank-Order Correlation to determine the strength and directionality of associations between the frequency of amenity references and the sentiment scores attributed to the reviews.

Table 24.2. Spearman's Rank-Order Correlation Coefficients between Amenities, Frequency of Mentions, and Sentiment Scores

Amenity	Sub-Amenity	Frequency	Spearman's ρ	Significance
Hotel Service	Reception	20	0.61	0.008**
	Chefs	15	0.45	0.032*
	Shuttle Service	18	0.55	0.014**
Room	Cleanliness	22	0.78	<0.0001***
	Comfort	25	0.82	<0.0001***
	Size	10	0.30	0.042*
Food	Quality	18	0.65	0.009**
Spa	Facilities	5	0.20	0.10
Ski Pistes	Accessibility	30	0.90	<0.0001***
	Quality	28	0.88	<0.0001***
	Equipment Rental	8	0.25	0.046*
	Lessons	12	0.40	0.033*
Additional	City Tours	4	0.10	0.50
	Childcare	6	0.15	0.30

In the analysis of hotel amenities and ski piste features, significant correlations were observed between the frequency of mentions and the sentiment scores, as detailed by Spearman's rank-order correlation coefficients. The reception service, mentioned in 20 reviews, demonstrated a strong positive correlation (ρ = 0.61) with guest sentiment, suggesting that positive experiences at reception are strongly associated with overall guest satisfaction. This relationship was statistically significant (p = 0.008**), underscoring the importance of first impressions in hospitality settings.

Similarly, the quality of food, noted in 18 reviews, showed a strong correlation (ρ = 0.65) with sentiment scores, indicating that culinary experiences significantly impact guest contentment (p = 0.009**). Chef-related services also positively correlated with satisfaction (ρ = 0.45), albeit to a lesser degree (p = 0.032*), suggesting that culinary expertise is valued by guests but with moderate influence on their overall experience. Room amenities such as cleanliness and comfort were highlighted as critical factors influencing guest perceptions. Cleanliness was mentioned in 22 reviews and exhibited a very strong positive correlation with sentiment (ρ = 0.78, p < 0.0001***), while comfort, mentioned in 25 reviews, also showed a very strong correlation (ρ = 0.82, p < 0.0001***). These findings indicate that room quality is paramount in determining guest satisfaction.

Ski pistes received considerable attention, with accessibility and quality being the most frequently praised aspects. Accessibility was mentioned 30 times and had a very strong positive correlation with sentiment ($\rho = 0.90$, $p < 0.0001$***), while the quality of the pistes, noted 28 times, similarly showed a very strong positive correlation ($\rho = 0.88$, $p < 0.0001$***). These results highlight the critical role of ski facilities in the overall satisfaction of guests at these resorts, confirming that well-maintained and accessible ski facilities are key to guest enjoyment. Conversely, certain amenities such as spa facilities and additional services like city tours and childcare were mentioned less frequently and showed weaker correlations with guest sentiment. Spa facilities, for example, were mentioned only 5 times and exhibited a negligible correlation ($\rho = 0.20$, $p = 0.10$), suggesting that while spa services are a part of the guest experience, they are less influential in determining overall satisfaction.

Table 24.3. MANOVA Results for the Relationship between Language of Review and Sentiment on Hotel Amenities

Dependent Variables	Sum of Squares	df	Mean Square	F	Sig.
Satisfaction with Reception	5.642	2	2.821	4.98	0.009**
Satisfaction with Chefs	3.213	2	1.606	3.25	0.041*
Satisfaction with Shuttle	2.878	2	1.439	2.20	0.115
Room Cleanliness Rating	9.752	2	4.876	8.60	<0.001***
Room Comfort Rating	10.823	2	5.411	9.31	<0.001***
Room Size Rating	1.644	2	0.822	1.47	0.230
Food Quality Rating	6.308	2	3.154	5.65	0.004**
Spa Facilities Rating	1.002	2	0.501	0.88	0.418
Ski Piste Accessibility Rating	12.567	2	6.283	11.22	<0.001***
Ski Piste Quality Rating	**11.310**	2	5.655	10.12	<0.001***
Multivariate F	3.720				<0.001***
Wilks' Lambda	0.642				
Significance					<0.001***

The p-values are reported as significant at the *0.05* level, *0.01* level, and ***0.001*** level.

The MANOVA results indicated a statistically significant multivariate effect of language on the combined dependent variables, $F(3,720) = 3.720$, $p < 0.001$, Wilks' Lambda = 0.642. This suggests that there are overall differences in the sentiment ratings for hotel amenities based on the language of the review. Examining the univariate main effects, the sentiment rating for "Satisfaction with Reception" differed significantly by language, F(2) = 4.98, p = 0.009, indicating a strong language-based sentiment towards reception services. "Satisfaction with Chefs" also showed significant differences, F(2) = 3.25, p = 0.041, as did the "Satisfaction with Shuttle" service, although it was not statistically significant (p = 0.115), suggesting a trend that may warrant further investigation.

Regarding room attributes, "Room Cleanliness Rating" and "Room Comfort Rating" displayed highly significant differences by language, F(2) = 8.60, p < 0.001, and F(2) = 9.31, p < 0.001, respectively. This indicates that the perception of room cleanliness and comfort varied substantially across language groups. In contrast, "Room Size Rating" did not significantly differ by language (p = 0.230), suggesting a more universal sentiment in this aspect. The "Food Quality Rating" was significantly affected by language, F(2) = 5.65, p = 0.004, while the sentiment towards "Spa Facilities Rating" showed no significant language-based difference (p = 0.418). For ski piste-related variables, "Ski Piste Accessibility Rating" and "Ski Piste Quality Rating" were both significantly influenced by the language of the review, with F(2) = 11.22, p < 0.001, and F(2) = 10.12, p < 0.001, respectively, indicating that these aspects of the skiing experience were perceived differently depending on the reviewer's language. In summary, the MANOVA analysis reveals that there are statistically significant differences in how guests of different linguistic backgrounds perceive and review hotel services and amenities, particularly regarding reception, chefs, room cleanliness, and comfort, as well as ski piste accessibility and quality.

Table 24.4. Path Analysis of Communicative Moves and Their Impact on Guest Satisfaction in Hotel Reviews

Effects	Reception	Chefs	Shuttle	Room Quality	Sentiment Score
Direct effect on Satisfaction	0.0623	-0.0182	0.0268	0.8601	0.9423
Indirect effect via Reception	-	0.0240	-0.0071	0.0154	
Indirect effect via Chefs	-0.0112	-	-0.0098	0.0064	-0.0059
Indirect effect via Shuttle	0.0038	0.0069	-	-0.0003	0.0061
Indirect effect via Room Quality	-0.1473	-0.0532	-0.0063	-	-0.3698
Indirect effect via Sentiment Score	0.1921	0.3699	0.5802	-0.4189	-
Total effect on Satisfaction	0.0972ns	0.3025*	0.5908*	0.4320*	0.5739*

Model Fit: Coefficient of determination (R^2) | 0.8810, Effect of residual variable | 0.3190, Condition index | 14.76

The path analysis concerning the communicative moves and their impact on guest satisfaction in hotel reviews indicated that the quality of room descriptions yielded a direct positive effect on satisfaction ($\beta = 0.0623$). Conversely, the chefs' interactions had a slight but not significant negative direct impact ($\beta = -0.0182$). The shuttle service provided a positive direct effect, albeit modest ($\beta = 0.0268$). Notably, the direct effect of room quality ($\beta = 0.8601$) and sentiment scores expressed in the reviews ($\beta = 0.9423$) on satisfaction was significant and substantial, highlighting these factors as crucial determinants of a guest's overall satisfaction.

The indirect effects revealed the intricate interplay between various communicative moves. For instance, positive reception quality influenced the sentiment score indirectly ($\beta = 0.1921$), and chefs' interactions indirectly affected the sentiment score positively ($\beta = 0.3699$). Shuttle service, another aspect of indirect influence, significantly improved the sentiment score ($\beta = 0.5802$), suggesting that logistics play an essential role in shaping overall satisfaction. Interestingly, there is a notable indirect negative impact of room quality on the sentiment score ($\beta = -0.4189$), which may suggest that guests' expectations of room standards play a significant role in their overall sentiment. The total effects on satisfaction scores were most strongly influenced by shuttle services ($\beta = 0.5908^*$) and room quality ($\beta = 0.4320^*$), indicating the pivotal role these factors play in overall guest satisfaction. The total effect of sentiment scores also came out as a strong predictor of satisfaction ($\beta = 0.5739^*$). In conclusion, this path analysis robustly illustrates the comprehensive effects of communicative moves on guest satisfaction. The findings suggest that hotel management should prioritize room quality and the effectiveness of shuttle services to enhance guest satisfaction.

Discussion and Conclusion

The current investigation embarked on addressing a significant void within the scholarly discourse surrounding online hotel reviews, with an emphasis on understanding the intricate role that narrative structures play in molding guest satisfaction at the Palandöken and Sarıkamış ski resorts. Historically, the focus of existing literature has largely gravitated towards quantifying guest feedback through metrics like star ratings and sentiment scores, often neglecting the rich qualitative aspects that profoundly inform these evaluations (Chua & Banerjee, 2016). Research up to this point has generally not ventured beyond these surface-level analyses to interrogate how the subtle nuances of language in reviews can affect guest perceptions and experiences (Wu et al., 2017), a gap that is especially pronounced in studies concerning hospitality and tourism. The utilization of a multilingual dataset further enhances the relevance of this study, offering insights into the cross-cultural nuances that shape guest reviews (Leon, 2019). This aspect of the research is crucial, as it addresses the increasingly global nature of the hospitality industry, where understanding diverse guest experiences is key to service improvement and customization. The inclusion of multiple languages in the analysis not only broadens the applicability of the findings but also underscores the importance of cultural sensitivity in managing guest relations and expectations (Lo Bianco, 2010).

The critical insights derived from our investigation prominently highlight the pivotal role played by specific communicative moves in sculpting the contours of guest satisfaction (De Ascaniis et al., 2015; Thumvichit & Gampper, 2019). These findings echo the researchers, who demonstrated that positive linguistic expressions within hotel reviews are potent predictors of elevated guest satisfaction levels (Zhao & Xu, 2018). Nevertheless, our research advances this understanding by intricately linking specific types of communicative moves to sentiment scores, thereby offering a nuanced perspective on how various facets of review content distinctly influence overall satisfaction. Moreover, our research contributes to a more comprehensive understanding of the emotional and psychological impact of these communicative moves (Ding, 2007). By integrating sentiment analysis, we quantify the emotional tone conveyed by these moves and their correlation with overall satisfaction scores (Bordoloi & Biswas, 2023), thereby bridging the gap between qualitative narrative structures and quantitative sentiment data. This methodological synthesis provides a robust framework for assessing the efficacy of different linguistic strategies in enhancing guest satisfaction, a step beyond the general correlations previously established by Thompson and Wang.

Our findings also suggest that the impact of specific communicative moves can vary significantly across different cultural and linguistic contexts. By analyzing reviews in multiple languages, this study highlights the cultural nuances that influence how guests perceive and articulate their experiences. This insight is crucial for multinational and multicultural hospitality businesses aiming to tailor their services to meet the diverse expectations of their clientele, thereby enhancing guest satisfaction on a global scale (Martín et al., 2020; Moreno-Perdigón et al., 2021). The analysis of our results indicates that positive language alone does not impact as significantly as the quality and comprehensiveness of the narrative descriptions within guest reviews (Filieri et al., 2020). This aligns with the findings of some researchers, who noted the significant role of detailed narrative descriptions in influencing potential customers' perceptions of hospitality services (Brotherton, 2005; Handani et al., 2022; Litvin et al., 2008; Mariottini & Hernández Toribio, 2016). Our study expands on these findings by showing that reviews with detailed descriptions of specific areas such as staff interaction and room quality have a strong positive impact on guest satisfaction.

Our detailed approach to reviewing analysis provides a deeper insight into the elements that are most valued by guests. Thorough descriptions offer potential guests a vicarious experience, allowing them to visualize and emotionally connect with the service offerings prior to making a booking decision (Viana et al., 2020). Consistent with this view, our research not only reaffirms the importance of detailed descriptions but also identifies which aspects of the hotel experience are most crucial to guest satisfaction. Our results specifically point out that detailed narratives about staff interactions and room quality are critical in enhancing guest satisfaction (Prasad et al., 2014; Sánchez-Franco & Aramendia-Muneta, 2023).

Additionally, the relationship between detailed descriptions and guest satisfaction can be explained by expectancy theory, which suggests that consumer satisfaction is greatly influenced by how well actual experiences meet prior expectations (Ehimen et al., 2021). Detailed reviews set accurate expectations for future guests, and when these expectations are met or exceeded, they lead to higher satisfaction levels. This supports our findings and implies that hotels can more effectively manage guest expectations by encouraging the publication of detailed, genuine reviews from previous guests (Vo et al., 2022).

In conclusion, this study has provided substantial evidence that the linguistic strategies used in online hotel reviews have a measurable impact on guest satisfaction. By systematically analyzing the relationship between narrative structures and guest perceptions (Cenni & Vásquez, 2021), this study not only contributes to the academic discourse on customer feedback analysis but also offers practical insights for enhancing guest experiences in the hospitality industry. Future research in this area should continue to integrate diverse methodological approaches and expand the scope of linguistic analysis to include a broader range of cultural and language contexts, thus enriching our understanding of the dynamics at play in customer satisfaction and service evaluation.

Limitations and Future Suggestions

However, the study's methodology comes with limitations that could impact the interpretation of the results. One major limitation is the dataset's language diversity—focusing only on English, Russian, and Arabic reviews may not fully capture the global demographic of the resorts' guests. Future research could address this gap by incorporating a wider range of languages to ascertain if the observed trends hold across a broader linguistic spectrum. Additionally, the inherent biases of online reviews, where more satisfied guests might be more inclined to post reviews, should be considered when generalizing these findings.

The implications of our findings are particularly relevant for hotel management and marketing strategies. Understanding that specific communicative moves and the quality of linguistic expression in reviews are closely linked to guest satisfaction, hotel managers can better strategize how to encourage satisfied guests to leave detailed, positive feedback. Moreover, identifying areas with consistently lower satisfaction scores, such as amenities or shuttle services, could help prioritize improvements.

For future research, exploring the indirect effects of communicative moves on guest satisfaction through a more complex analytical framework such as structural equation modeling could provide deeper insights. This could include examining how different aspects of the guest experience mediate the relationship between review content and guest satisfaction. Additionally, experimental studies could be designed to test interventions aimed at improving guest satisfaction and observing how these changes are reflected in online reviews. Similar studies could be conducted at other winter tourism centers in Turkey, such as Uludağ, Erciyes, and Kartalkaya, where accommodation services are provided.

References

Awara, N., & Anyadighibe, J. (2014). The relationship between customer satisfaction and loyalty: A study of selected eateries in Calabar, Cross River state.

Bordoloi, M., & Biswas, S. K. (2023). Sentiment analysis: A survey on design framework, applications and future scopes. *Artificial Intelligence Review*, 56(11), 12505–12560. https://doi.org/10.1007/s10462-023-10442-2

Boughzala, I. (2016). Social Media and Value Creation. *Journal of Organizational and End User Computing*, 28, 107–123. https://doi.org/10.4018/JOEUC.2016040107

Brotherton, B. (2005). The nature of hospitality: Customer perceptions and implications. *Tourism and Hospitality Planning & Development*, 2, 139–153. https://doi.org/10.1080/14790530500399218

Bursa, B., Mailer, M., & Axhausen, K. W. (2022). Travel behavior on vacation: transport mode choice of tourists at destinations. *Transportation Research Part A: Policy and Practice*, 166, 234–261. https://doi.org/https://doi.org/10.1016/j.tra.2022.09.018

Cabral, A., & Marques, J. P. (2020). *The influence of service innovation in customer satisfaction: case study of hotel industry.*

Capineri, C., & Romano, A. (2021). The platformization of tourism: from accommodation to Experiences. *Digital Geography and Society, 2*, 100012. https://doi.org/https://doi.org/10.1016/j.diggeo.2021.100012

Cenni, I., & Goethals, P. (2020). Responding to negative hotel reviews: A cross-linguistic perspective on online rapport-management. *Discourse, Context & Media, 37*, 100430. https://doi.org/https://doi.org/10.1016/j.dcm.2020.100430

Cenni, I., & Vásquez, C. (2021). "Jerry was a terrific host!" "You were a brilliant guest!": Reciprocal compliments on Airbnb. https://doi.org/10.1075/ps.20059.cen

Changchit, C., Klaus, T., & Lonkani, R. (2020). Online Reviews: What Drives Consumers to Use Them. *Journal of Computer Information Systems, 62*, 1–10. https://doi.org/10.1080/08874417.2020.1779149

Chidlow, A., Plakoyiannaki, E., & Welch, C. (2014). Translation in cross-language international business research: Beyond equivalence. *Journal of International Business Studies, 45*(5), 562–582. http://www.jstor.org/stable/43653820

Chua, A., & Banerjee, S. (2016). Helpfulness of user-generated reviews as a function of review sentiment, product type and information quality. *Computers in Human Behavior, 54*, 547–554. https://doi.org/10.1016/j.chb.2015.08.057

Cristobal-Fransi, E., Daries, N., Serra-Cantallops, A., Ramón-Cardona, J., & Zorzano, M. (2018). Ski Tourism and Web Marketing Strategies: The Case of Ski Resorts in France and Spain. *Sustainability, 10*(8).

De Ascaniis, S., Borrè, A., Marchiori, E., & Cantoni, L. (2015). *Listen to Your Customers! How Hotels Manage Online Travel Reviews. The Case of Hotels in Lugano*. https://doi.org/10.1007/978-3-319-14343-9_5

Ding, H. (2007). Genre analysis of personal statements: Analysis of moves in application essays to medical and dental schools. *English for Specific Purposes, 26*(3), 368–392. https://doi.org/https://doi.org/10.1016/j.esp.2006.09.004

Ehimen, S., Uduji, J., & Ugwuanyi, C. (2021). Hotel Guests' Experience, Satisfaction and Revisit Intentions: An Emerging Market Perspective. *African Journal of Hospitality Tourism and Leisure, 10*, 406–424. https://doi.org/10.46222/ajhtl.19770720-108

Fang, B., Ye, Q., Kucukusta, D., & Law, R. (2016). Analysis of the perceived value of online tourism reviews: Influence of readability and reviewer characteristics. *Tourism Management, 52*, 498–506. https://doi.org/https://doi.org/10.1016/j.tourman.2015.07.018

Filieri, R., Acikgoz, F., Ndou, V., & Dwivedi, Y. (2020). Is TripAdvisor still relevant? The influence of review credibility, review usefulness, and ease of use on consumers' continuance intention. *International Journal of Contemporary*

Hospitality Management, ahead-of-print. https://doi.org/10.1108/IJCHM-05-2020-0402

Filieri, R., & McLeay, F. (2015). Why do travelers trust TripAdvisor? Antecedents of trust towards consumer-generated media and its influence on recommendation adoption and word of mouth. *Tourism Management, 51*. https://doi.org/10.1016/j.tourman.2015.05.007

Flagestad, A., & Hope, C. (2001). "Scandinavian Winter"; Antecedents, concepts and empirical observations underlying a destination umbrella branding model. *Tourism Review, 56*, 5–12. https://doi.org/10.1108/eb058351

Giese, J., & Cote, J. (2000). Defining Consumer Satisfaction. *Academy of Marketing Science Review, 4*, 1–24.

Griffis, S., Rao, S., Goldsby, T., & Niranjan, T. (2012). The Customer Consequences of Returns in Online Retailing: An Empirical Analysis. *Journal of Operations Management, 30*, 282–294. https://doi.org/10.1016/j.jom.2012.02.002

Handani, N. D., Williady, A., & Kim, H.-S. (2022). An Analysis of Customer Textual Reviews and Satisfaction at Luxury Hotels in Singapore's Marina Bay Area (SG-Clean-Certified Hotels). *Sustainability, 14*(15).

Hou, Z., Cui, F., Meng, Y., Lian, T., & Yu, C. (2019). Opinion mining from online travel reviews: A comparative analysis of Chinese major OTAs using semantic association analysis. *Tourism Management, 74*, 276–289. https://doi.org/https://doi.org/10.1016/j.tourman.2019.03.009

Huang, R., & Tian, X. (2013). An Investigation of Travel Behavior of Chinese International Students in the UK. *Journal of China Tourism Research, 9*, 277–291. https://doi.org/10.1080/19388160.2013.812898

Leon, R. (2019). Hotel's online reviews and ratings: a cross-cultural approach. *International Journal of Contemporary Hospitality Management, 31*, 2054–2073. https://doi.org/10.1108/IJCHM-05-2018-0413

Litvin, S., Goldsmith, R., & Pan, B. (2008). Electronic Word-of-Mouth in Hospitality and Tourism Management. *Tourism Management, 29*, 458–468. https://doi.org/10.1016/j.tourman.2007.05.011

Litvin, S., Goldsmith, R., & Pan, B. (2017). A retrospective view of electronic word of mouth in hospitality and tourism management. *International Journal of Contemporary Hospitality Management, 30*, 00–00. https://doi.org/10.1108/IJCHM-08-2016-0461

Lo Bianco, J. (2010). The importance of language policies and multilingualism for cultural diversity. *International Social Science Journal, 61*(199), 37–67. https://doi.org/https://doi.org/10.1111/j.1468-2451.2010.01747.x

Lovelock, C., & Gummesson, E. (2004). Whither Services Marketing?: In Search of a New Paradigm and Fresh Perspectives. *Journal of Service Research*, 7(1), 20–41. https://doi.org/10.1177/1094670504266131

Mariottini, L., & Hernández Toribio, M. (2016). La narración de experiencias en TripAdvisor. *Rilce. Revista de Filología Hispánica*, 33, 302–330. https://doi.org/10.15581/008.33.1.302-30

Martín, J. C., Rudchenko, V., & Sánchez-Rebull, M.-V. (2020). The Role of Nationality and Hotel Class on Guests' Satisfaction. A Fuzzy-TOPSIS Approach Applied in Saint Petersburg. *Administrative Sciences*, 10(3).

Moreno-Perdigón, M. C., Guzmán-Pérez, B., & Ravelo Mesa, T. (2021). Guest satisfaction in independent and affiliated to chain hotels. *International Journal of Hospitality Management*, 94, 102812. https://doi.org/https://doi.org/10.1016/j.ijhm.2020.102812

Munar, A. M., & Jacobsen, J. K. S. (2014). Motivations for sharing tourism experiences through social media. *Tourism Management*, 43, 46–54. https://doi.org/https://doi.org/10.1016/j.tourman.2014.01.012

Needham, M., Rollins, R., Ceurvorst, R., Wood, C., Grimm, K., & Dearden, P. (2011). Motivations and Normative Evaluations of Summer Visitors at an Alpine Ski Area. *Journal of Travel Research*, 50, 669–684. https://doi.org/10.1177/0047287510382298

Nicoli, N., Henriksen, K., Komodromos, M., & Tsagalas, D. (2021). Investigating digital storytelling for the creation of positively engaging digital content. *EuroMed Journal of Business*. https://doi.org/10.1108/EMJB-03-2021-0036

Nunkoo, R., & Prayag, G. (2018). A Systematic Review of Consumer Satisfaction Studies in Hospitality Journals: Conceptual Development, Research Approaches and Future Prospects. *Journal of Hospitality Marketing & Management*, 28. https://doi.org/10.1080/19368623.2018.1504367

Prasad, K., Wirtz, P., & Yu, L. (2014). Measuring Hotel Guest Satisfaction by Using an Online Quality Management System. *Journal of Hospitality Marketing & Management*, 23, 445–463. https://doi.org/10.1080/19368623.2013.805313

Sánchez-Franco, M. J., & Aramendia-Muneta, M. E. (2023). Why do guests stay at Airbnb versus hotels? An empirical analysis of necessary and sufficient conditions. *Journal of Innovation & Knowledge*, 8(3), 100380. https://doi.org/https://doi.org/10.1016/j.jik.2023.100380

Skiera, B., Hennig-Thurau, T., Malthouse, E., Friege, C., Gensler, S., Lobschat, L., & Rangaswamy, A. (2010). The Impact of New Media on Customer Relationships. *Journal of Service Research*, 26. https://doi.org/10.1177/1094670510375460

Taboada, M. (2016). Sentiment Analysis: An Overview from Linguistics. *Annual Review of Linguistics, 2*. https://doi.org/10.1146/annurev-linguistics-011415-040518

Taşdağıtıcı, E., & Tuna, M. (2022). Content analysis of tripadvisor reviews on safe tourism certified restaurants during the Covid-19 pandemic. *Journal of Tourism, Leisure and Hospitality, 4*. https://doi.org/10.48119/toleho.1135370

Thumvichit, A., & Gampper, C. (2019). Composing responses to negative hotel reviews: A genre analysis. *Cogent Arts & Humanities, 6*(1), 1629154. https://doi.org/10.1080/23311983.2019.1629154

Thurlow, C., & Mroczek, K. (2012). *Digital Discourse: Language in the New Media*. https://doi.org/10.1093/acprof:oso/9780199795437.001.0001

van der Wouden, F., & Youn, H. (2023). The impact of geographical distance on learning through collaboration. *Research Policy, 52*(2), 104698. https://doi.org/https://doi.org/10.1016/j.respol.2022.104698

Vásquez, C. (2011). Complaints online: The case of TripAdvisor. *Journal of Pragmatics, 43*, 1707–1717. https://doi.org/10.1016/j.pragma.2010.11.007

Viana, J., Mayer, V., & Souza-Neto, V. (2020). Experience sharing about hotels on TripAdvisor: motivation and preferences of brazilian tourists. *Marketing & Tourism Review, 5*. https://doi.org/10.29149/mtr.v5i1.5907

Vo, N. T., Hung, V. V., Tuckova, Z., Pham, N. T., & Nguyen, L. H. L. (2022). Guest Online Review: An Extraordinary Focus on Hotel Users' Satisfaction, Engagement, and Loyalty. *Journal of Quality Assurance in Hospitality & Tourism, 23*(4), 913–944. https://doi.org/10.1080/1528008X.2021.1920550

Wu, L., Shen, H., Fan, A., & Mattila, A. (2017). The impact of language style on consumers? reactions to online reviews. *Tourism Management, 59*, 590–596. https://doi.org/10.1016/j.tourman.2016.09.006

Xiang, Z., & Gretzel, U. (2010). Role of social media in online travel information search. *Tourism Management, 31*(2), 179–188. https://doi.org/https://doi.org/10.1016/j.tourman.2009.02.016

Zhao, Y., & Xu, X. (2018). Predicting overall customer satisfaction: Big data evidence from hotel online textual reviews. *International Journal of Hospitality Management, 76*, 111–121. https://doi.org/10.1016/j.ijhm.2018.03.017

Zhu, Y., Ma, L., & Jiang, R. (2019). A cross-cultural study of English and Chinese online platform reviews: A genre-based view. *Discourse & Communication, 13*, 175048131983564. https://doi.org/10.1177/1750481319835642

Mehmet Necati Cizrelioğullari[1] and Tuğrul Günay[2]

Chapter 25 The Impact and Strategic Implications of Digital Marketing on the Tourism Industry

Introduction

The tourism sector is experiencing transformative changes due to the advent of digital marketing. Evaluating the effectiveness of these strategies is crucial for understanding their impact on growth, customer engagement, and competitive dynamics (Gupta, 2019). This study makes significant contributions to the understanding of consumer behavior and preferences by examining the efficacy of online marketing strategies within the tourism industry, which is heavily reliant on technology. Research in this area not only has the potential to enhance the overall customer experience but also to significantly influence both the global economy and societal structures. It is essential for businesses to understand the effectiveness of digital marketing strategies to maintain competitiveness in the digital era (Yasmin et al., 2015; Dara, 2016; Krizanova et al., 2019).

The emergence of digital technology and the widespread use of the internet have profoundly impacted the operational dynamics of businesses across various sectors, including the tourism industry (Waterton, 2010). Digital marketing, in particular, has revolutionized tourism marketing by transforming promotional strategies for destinations, enhancing customer engagement techniques, and increasing the generation of bookings. In an era characterized by heightened interconnectivity and accessible information, it is crucial for travel industry enterprises to grasp the significant influence of digital marketing (Iyer et al., 2015; Kaur, 2017; Gupta, 2019).

According to Pencarelli (2020), digital marketing has drastically altered the methods used by tourism businesses to promote destinations, interact with customers, and stimulate bookings. Technological advancements have expanded the global reach and exposure of these businesses, enabling them to connect with a broader audience. Improved targeting and personalization of marketing

---

1 Assoc. Prof. Dr., Mardin Artuklu University, Tourism Faculty, Department of Tourism Guidance, necaticizreliogullari@artuklu.edu.tr
2 Assoc. Prof. Dr., Cyprus Science University, Tourism Faculty, Department of Tourism and Hotel Management, tugrulgunay@csu.edu.tr

messages have enhanced customer engagement and conversion rates. The integration of online booking systems and the availability of readily accessible information have empowered travelers to make informed decisions and conveniently arrange their travel plans. Moreover, the rise of influencer marketing and the extensive use of user-generated content have significantly influenced travelers' decision-making processes. The use of data-driven insights has allowed businesses to refine their marketing strategies and respond effectively to changing consumer behaviors (Yetimoğlu and Uğurlu, 2020; Fedeli and Cheng, 2023).

The aim of investigating the impact of digital marketing on tourism marketing is to understand the pivotal role that digital channels and strategies play in shaping the industry. Analyzing the impact of digital marketing helps identify new opportunities for tourism businesses to effectively reach and engage their target demographics. This analysis also allows for the exploration of innovative methods for promoting destinations, attractions, and services through digital platforms. A thorough understanding of digital marketing's ramifications enables tourism marketers to optimize their strategies and make informed decisions. By strategically using digital channels, organizations can enhance their marketing efforts, improve the accuracy of customer targeting, and maximize their marketing investment returns.

Additionally, digital marketing has changed how customers interact with tourism businesses and has led to shifts in consumer behavior within the travel and tourism sector. By examining the effects of digital marketing, organizations can adjust their marketing strategies to align with evolving consumer preferences, expectations, and online behaviors. The influence of digital marketing on tourism marketing also presents opportunities for the industry's expansion and advancement. By recognizing the importance of digital channels, enterprises can sustain competitiveness, foster technological progress, and stimulate innovation within the tourism industry.

The impact of digital marketing on tourism is crucial for businesses aiming to successfully navigate the digital landscape. Understanding this impact helps businesses make informed decisions, refine marketing tactics, and effectively leverage digital opportunities to stay competitive in the tourism market. As this field evolves, continuous research and adaptation will be key to capitalizing on the potential benefits of digital marketing in tourism.

Definition and Evolution of Digital Marketing

Digital marketing encompasses strategic activities designed to advertise and promote products, services, and brands through digital platforms and technologies.

This field involves a variety of strategies, techniques, and tools aimed at effectively reaching and engaging targeted audiences, driving website traffic, generating leads, and achieving marketing objectives. Technological advancements and shifts in consumer behavior have significantly influenced the evolution of digital marketing (Basimakopoulou et al., 2022).

The advent of the internet in the latter half of the 20th century marked a significant milestone in digital marketing's evolution. This development offered businesses an unprecedented opportunity to establish a robust online presence, interact with customers, and market their offerings globally. Such disruptive technology has radically transformed traditional marketing approaches, creating numerous opportunities for companies aiming to enhance growth and expand their influence. With the internet, businesses can create precise, focused marketing campaigns, monitor customer actions, and gain valuable insights into consumer preferences. The enduring impact of the internet on business strategies is profound and widespread (Oggolder et al., 2019; Blank and Dutton, 2013).

The significance of websites as crucial marketing tools has grown, with companies using them to showcase products or services and provide valuable information to potential clients. Consequently, search engine optimization (SEO) has become essential for enhancing website visibility in search results and driving organic traffic. Implementing SEO strategies improves digital visibility and attracts a larger customer base (Sharma et al., 2019). The rise of search engines, particularly Google, has spurred the adoption of website optimization strategies to boost search rankings, thus enhancing visibility and traffic. Additionally, pay-per-click (PPC) advertising has become popular as it allows businesses to display customized ads on search engine results pages and pay only when users click on the ads (Kritzinger and Weideman, 2013).

Content marketing has also surged in popularity as companies recognize the value of creating engaging, informative content. Through blogs, articles, and videos, businesses engage with their audience, establish authority in their industry, and significantly increase website traffic. The pivotal role of content marketing in digital strategies continues to grow (Baltes, 2015). Moreover, the ubiquity of smartphones has emphasized the importance of mobile optimization. Responsive designs and mobile-friendly websites ensure consistent user experiences across devices, while mobile marketing strategies like SMS marketing, mobile apps, and location-based advertising have become increasingly relevant with the rise of mobile usage (Rowley, 2008).

Advancements in data analytics have greatly benefited digital marketing professionals. With the ability to collect extensive data on customer behavior, preferences, and demographics, marketers can tailor campaigns, personalize product recommendations, and refine targeted advertising, thereby enhancing customer experiences and improving conversion rates (Bala and Verma, 2018). Influencer marketing has grown due to the influential role of social media influencers who endorse products and services across platforms. Collaborating with influencers allows companies to effectively reach their target demographic and utilize their credibility and appeal. User-generated content also plays a crucial role in engaging customers and encouraging them to share their experiences (Jin et al., 2019).

The integration of automation tools and marketing technologies has significantly improved the efficiency of digital marketing processes. Email marketing automation, customer relationship management (CRM) systems, and marketing automation platforms help manage and nurture leads, provide personalized content, and optimize campaign performance (Heinzelbecker, 2023). Artificial intelligence (AI) and machine learning (ML) have dramatically transformed digital marketing strategies. The use of chatbots, recommendation engines, predictive analytics, and personalized experiences has become more common, thanks to advancements in AI and ML, which also facilitate the automation of tasks and provide deeper customer insights (Boddu et al., 2022).

Virtual reality (VR) and augmented reality (AR) technologies are increasingly used in the tourism and hospitality sectors to enhance customer experiences. These technologies provide immersive experiences, virtual tours, and interactive content, allowing customers to explore destinations, hotels, or products before purchasing. Companies utilize VR and AR to give consumers firsthand experiences of their offerings, significantly boosting sales (Nayyar et al., 2018).

The digital marketing has undergone substantial transformation since its inception. The landscape has evolved due to new technologies, changing consumer behaviors, and emerging trends. Companies must stay aware of these changes and adapt their strategies to effectively engage and attract their intended audiences within the digital realm.

Benefit of Digital Marketing in the Tourism Industry

The tourism sector has experienced a substantial shift due to the emergence of digital marketing, offering businesses numerous benefits and opportunities to engage with their target audience. This article aims to explore these advantages further and substantiate them through illustrative examples (Gupta, 2019). Digital marketing significantly enhances the outreach of tourism enterprises, as it provides a medium to establish connections with a worldwide demographic. Utilizing digital platforms such as social media, websites, and search engines, these enterprises have the opportunity to effectively showcase their offerings and expand their target audience globally. For instance, a hotel in a popular tourist destination uses digital marketing strategies to promote its services and attract potential guests from various nations (Kaur, 2017).

Digital marketing offers a more economically advantageous alternative compared to traditional marketing methods. Companies can enhance the effectiveness of their marketing campaigns by leveraging online advertising platforms such as Google Ads and social media ads. These platforms allow companies to allocate dedicated budgets and tailor their marketing efforts towards specific demographics, thereby achieving greater precision and efficiency. This is particularly beneficial for small and medium-sized tourism enterprises operating within limited marketing budgets. For example, a tour operator in a specific locality can initiate focused promotional campaigns on Facebook, using targeted advertisements to promote a recently introduced tour package and effectively engage with prospective customers in a specific geographic area (Cain, 2008; Igbokwe, 2022).

Digital marketing enables businesses to engage in direct, real-time communication with customers, facilitating the promotion of engagement and the establishment of relationships. Social media platforms provide a digital space where individuals interested in travel can engage in discussions, share their experiences, evaluate establishments, and interact with businesses. By actively engaging with customers through comments, messages, and personalized responses, businesses can build a solid foundation of trust and loyalty. As illustrated by Ibrahim and Ganeshbabu (2018), resorts can use social media platforms to actively interact with guests, including responding to inquiries, providing personalized suggestions, and promptly resolving any issues that may arise. Organizations can devise tailored marketing initiatives that specifically target the distinct preferences, behaviors, and demographics of their clientele by employing data analytics and customer relationship management (CRM) software. For example, a travel company might distribute customized email newsletters to its subscribers, featuring travel packages aligned with their previous travel experiences and personal interests (Frow and Payne, 2009; Payne and Frow, 2005).

The use of content marketing and storytelling within the digital marketing realm presents tourism businesses with the opportunity to effectively engage and captivate their intended audience. By generating content that is informative, valuable, and inspiring, such as blog posts, user-generated content, photos, and videos, businesses can maintain their audience's interest. For instance, a destination marketing organization might produce engaging videos that highlight the unique experiences and attractions of their location, thereby eliciting interest from prospective travelers. The realm of digital marketing offers a diverse array of tools and strategies that provide travelers with notable convenience and promptness, such as the ability to receive immediate updates and make prompt reservations. This capability is facilitated by the use of online travel agencies and booking platforms, which enable users to efficiently search for and reserve flights, accommodations, and activities through a streamlined digital interface. Hotels can employ a website booking engine to provide customers with current updates on room availability and facilitate secure, real-time reservations. Providing a seamless and hassle-free travel experience can attract travelers seeking a stress-free journey (Gupta, 2019; Parlov et al., 2016; Iazzi et al., 2017).

Airbnb, a globally recognized online platform for accommodations and vacation rentals, has gained significant acclaim for its innovative digital marketing strategies. An example worth noting is their experiential advertising campaign, known as "Live There," which sought to connect travelers with local traditions and cultures (Guttentag, 2019; Andreu et al., 2020; Dolnicar & Zare, 2020).

Future Trends and Innovations in Digital Marketing for the Tourism Industry

The concept of personalized marketing encompasses the strategic approach of customizing marketing messages, content, and offers to suit the unique preferences, behaviors, and needs of individual customers (Alsukaini et al., 2022; Vinod, 2023). Here are instances of personalized marketing within the travel sector:

Travel companies have the capacity to employ customer data to craft personalized email campaigns targeted towards their subscribers. For example, a hotel can send personalized emails to previous guests, offering them exclusive discounts and personalized recommendations tailored to their past stays and individual preferences. These emails may include customized greetings, tailor-made travel itineraries, or recommendations for upcoming trips based on the customer's travel history (Alford, 2001; Vinod, 2023).

Dynamic website content enables travel websites to customize the presentation of information based on the browsing behavior and preferences of individual visitors. This is exemplified by promoting beach vacation packages, showcasing popular beach destinations, and recommending associated activities or tours to individuals who have previously expressed interest in beach destinations. This approach offers an enhanced and tailored experience that increases the user's engagement and the likelihood of booking (Stankov et al., 2009; Alsukaini et al., 2022).

Personalized mobile applications in the tourism sector can leverage customer data and location-based services to provide tailored experiences. These apps can offer personalized suggestions for local points of interest, dining options, or events, considering the user's current geographical position and individual preferences. Additionally, these apps can send push notifications containing customized offers or updates related to the user's saved or upcoming travel plans (Scherp and Boll, 2004; Wörndl and Herzog, 2020).

Travel agencies and tour operators can curate customized travel itineraries that cater to distinct customer segments or individual preferences. By collecting relevant information about a customer's budget, personal likes, and preferred activities, the agency can create a tailored travel plan that includes desired accommodations, transportation modes, and leisure activities. This approach allows customers to enjoy a unique and personalized travel experience tailored specifically to their preferences (Liu et al., 2011).

Social media platforms enable travel brands to deliver tailored advertisements to specific audiences by utilizing various targeting options. By using user data related to demographic characteristics, interests, and behaviors, advertisers can create ads that are tailored to individual preferences and traits. For example, a destination marketing organization can target individuals who have shown a keen interest in adventure travel by presenting tailored ads that highlight exciting outdoor activities and experiences (Zeng and Gerritsen, 2014; Gebreel and Shuayb, 2022).

Tailored loyalty programs within the tourism industry can incentivize customers by aligning rewards with their specific preferences and past transactions. By monitoring customer engagements and expenditures, companies can offer personalized incentives, exclusive discounts, or special privileges tailored to the individual's travel patterns and preferences. This strategy helps foster customer loyalty and encourages repeat bookings (Sanchez-Casado et al., 2019).

Virtual Reality (VR) technology allows individuals to engage with simulated environments that replicate real or imagined locations. In the tourism sector, VR can provide virtual tours, allowing travelers to explore various destinations and attractions remotely. This technology offers comprehensive views of famous landmarks, hotels, and natural wonders, enhancing users' engagement levels and helping them make informed decisions. An example is the "VRoom Service" initiative by Marriott Hotels, which allows guests to order Samsung Gear VR headsets to their rooms, enabling them to virtually visit places like Hawaii or London without leaving their hotel (Nayyar et al., 2018; Wei, 2019; Yung and Khoo-Lattimore, 2019).

Augmented Reality (AR) enriches human perception of reality by superimposing digital information onto the physical environment. AR applications can enhance travel experiences by providing contextual information, interactive maps, and virtual guides, thus enriching the traveler's interaction with their surroundings. The use of VR and AR technologies allows for simulated experiences such as virtual dives, remote hikes, and visits to cultural sites that are otherwise inaccessible, and can also be used for educational purposes, like training pilots with flight simulations or enhancing customer service skills in the hospitality industry (Ozdemir, 2021; Yung and Khoo-Lattimore, 2019; Nayyar et al., 2018). These technologies collectively have the potential to transform the tourism industry by providing personalized, engaging, and interactive experiences that cater to the individual needs and preferences of each traveler.

Conclusion

This comprehensive analysis underscores the transformative impact of digital marketing on the tourism industry. As elucidated through various examples and scholarly insights, digital marketing has not only revolutionized the way tourism businesses connect with their audience but has also enhanced customer engagement, personalization, and overall market reach. The strategic application of digital marketing tools—ranging from SEO, PPC, and content marketing to sophisticated uses of AI, VR, and AR has empowered tourism businesses to offer enriched, interactive, and personalized experiences to their customers.

The evolution of digital marketing has been pivotal, enabling businesses to overcome geographical barriers and effectively target global audiences. Through the adept integration of digital technologies, companies in the tourism sector have been able to optimize their marketing strategies, reduce costs, and improve operational efficiencies (Gupta, 2019). The future of digital marketing in tourism looks promising, with advancements in personalized marketing, dynamic content, and immersive technologies such as VR and AR expected to set new benchmarks for how travel experiences are marketed and consumed (Alsukaini et al., 2022; Nayyar et al., 2018).

Tourism enterprises must continue to adapt to these rapid technological changes to stay competitive. Leveraging data analytics for customer insights (Bala & Verma, 2018), embracing mobile optimization, engaging through social media, and innovating with VR and AR will be critical for developing compelling, customized, and efficient marketing strategies (Heinzelbecker, 2023; Boddu et al., 2022). As the digital landscape evolves, so too must the strategies employed by those within the tourism industry to captivate and charm the modern traveler.

The ongoing research and adaptation to new digital marketing trends will be essential for the sustained growth and innovation of tourism marketing. Businesses that anticipate changes and proactively integrate new technologies and strategies are more likely to thrive in the increasingly competitive tourism market. Ultimately, the profound understanding of digital marketing's impact on tourism is indispensable for any tourism-related enterprise aiming to excel in this digital era.

Discussion

The profound influence of digital marketing on tourism marketing is undeniable, reflecting a significant paradigm shift in how tourism businesses approach advertising, customer engagement, and conversions. Through an array of digital channels including social media, search engines, and online booking platforms tourism entities have expanded their reach globally, enhancing brand visibility and penetrating new markets. Peter and Dalla Vecchia (2021) note that this expansive reach has enabled the exploration of new demographic segments and solidified connections across diverse geographic locales. The precision of digital marketing allows for the deployment of highly targeted strategies and personalized messages, which are crafted using rich consumer data and analytics. This personalization not only increases the relevance of marketing content but also boosts conversion rates and strengthens customer relationships (Chandra et al., 2022).

Direct interactions facilitated by digital tools such as social media platforms and chatbots have redefined customer service within the tourism sector. These platforms offer prompt responses to inquiries, personalized recommendations, and a reliable communication channel, fostering trust and customer loyalty (Yoong & Lian, 2019). The accessibility of extensive travel information has similarly transformed consumer behavior. Prospective travelers can now effortlessly compare prices, read reviews, and gather detailed information about potential destinations. Online booking systems have streamlined the reservation process, making trip planning more convenient and enhancing the overall travel experience (Kim et al., 2021).

Influencer marketing and user-generated content have profoundly impacted tourism marketing by authenticating user experiences and expanding the influence of popular travel personalities. These elements help shape travelers' perceptions and significantly influence their decision-making processes. The authenticity of user-generated content, including reviews, photos, and videos, provides prospective travelers with a genuine look at destinations, which can be more persuasive than traditional advertising (Lampeitl & Åberg, 2017).

Moreover, the wealth of data generated through digital marketing activities offers invaluable insights into customer preferences, behaviors, and trends. By employing sophisticated data analytics, tourism businesses can refine their marketing strategies, enhance decision-making, and uncover new opportunities for growth and engagement (Chen et al., 2021).

The digital marketing has dramatically transformed tourism marketing, enabling more precise targeting, enhanced customer interaction, and streamlined processes for information dissemination and booking. The integration of influencer marketing and user-generated content has also enriched the decision-making resources available to travelers. For tourism businesses, adopting robust digital marketing strategies is essential to remain competitive in a technologically advancing landscape, effectively connect with digital-savvy consumers, and leverage the myriad benefits of the digital environment.

References

Alford, P. (2001). eCRM in the travel industry. *Travel & Tourism Analyst*, (1), 57–76.

Alsukaini, A. K. M., Sumra, K., Khan, R., & Awan, T. M. (2022). New trends in digital marketing emergence during pandemic times. *International Journal of Innovation Science*, 15(1), 167–185.

Andreu, L., Bigne, E., Amaro, S., & Palomo, J. (2020). Airbnb research: an analysis in tourism and hospitality journals. *International Journal of Culture, Tourism and Hospitality Research*, 14(1), 2–20.

Bala, M., & Verma, D. (2018). A critical review of digital marketing. *M. Bala, D. Verma (2018). A Critical Review of Digital Marketing. International Journal of Management, IT & Engineering*, 8(10), 321–339.

Baltes, L. P. (2015). Content marketing-the fundamental tool of digital marketing. *Bulletin of the Transilvania University of Brasov. Series V: Economic Sciences*, 111–118.

Basimakopoulou, M., Theologou, K., & Tzavaras, P. (2022). A Literature Review on Digital Marketing: The Evolution of a Revolution. *Journal of Social Media Marketing*, 1(1), 30–40.

Blank, G., & Dutton, W. H. (2013). The Emergence of Next-Generation Internet Users. *A companion to new media dynamics*, 122–141.

Boddu, R. S. K., Santoki, A. A., Khurana, S., Koli, P. V., Rai, R., & Agrawal, A. (2022). An analysis to understand the role of machine learning, robotics and artificial intelligence in digital marketing. *Materials Today: Proceedings*, 56, 2288–2292.

Cain, P. M. (2008). Limitations of conventional marketing mix modelling. *Admap Magazine, (493)*, 48–51.

Chandra, S., Verma, S., Lim, W. M., Kumar, S., & Donthu, N. (2022). Personalization in personalized marketing: Trends and ways forward. *Psychology & Marketing, 39*(8), 1529–1562.

Chen, S. X., Wang, X. K., Zhang, H. Y., Wang, J. Q., & Peng, J. J. (2021). Customer purchase forecasting for online tourism: A data-driven method with multiplex behavior data. *Tourism Management, 87*, 104357.

Dara, S. (2016). Effectiveness of digital marketing strategies. *International Journal for Innovative Research in Multidisciplinary Field, 2*(12), 290–293.

Dolnicar, S., & Zare, S. (2020). COVID19 and Airbnb–Disrupting the disruptor. *Annals of tourism research, 83*, 102961.

Fedeli, G., & Cheng, M. (2023). Influencer Marketing and Tourism: Another Threat to Integrity for the Industry?. *Tourism Analysis, 28*(2), 323–328.

Frow, P. E., & Payne, A. F. (2009). Customer relationship management: a strategic perspective. *Journal of business market management, 3*, 7–27.

Gebreel, O. S. S., & Shuayb, A. (2022). Contribution of social media platforms in tourism promotion. *International Journal of Social Science, Education, Communication and Economics (SINOMICS JOURNAL), 1*(2), 189–198.

Gupta, G. (2019). Inclusive use of digital marketing in tourism industry. In *Information Systems Design and Intelligent Applications: Proceedings of Fifth International Conference INDIA 2018 Volume 1* (pp. 411–419). Springer Singapore.

Guttentag, D. (2019). Progress on Airbnb: a literature review. *Journal of Hospitality and Tourism Technology, 10*(4), 814–844.

Heinzelbecker, K. (2023). CRM, CXM, and Marketing Automation. In *Marketing and Sales Automation: Basics, Implementation, and Applications* (pp. 51–63). Cham: Springer International Publishing.

Iazzi, A., Trio, O., & Gravili, S. (2017). Hotels and online travel agencies: Power or trust for a competitive long-term relationship. *International Journal of Technology Marketing, 12*(2), 115–126.

Ibrahim, S. S., & Ganeshbabu, P. (2018). A study on the impact of social media marketing trends on digital marketing. *Shanlax International Journal of Management, 6*(1), 120–125.

Igbokwe, I. P. (2022). A Comparative Study of Social Media Marketing and Conventional Marketing–A Case Study. *African Journal of Business and Economic Research, 17*(4), 169.

Iyer, V. R., Dey, N., & Chakraborty, S. (2015). Advent of Information Technology in the world of Tourism. In *Emerging innovative marketing strategies in the tourism industry* (pp. 44–53). IGI Global.

Jin, S. V., Muqaddam, A., & Ryu, E. (2019). Instafamous and social media influencer marketing. *Marketing Intelligence & Planning, 37*(5), 567–579.

Kaur, G. (2017). The importance of digital marketing in the tourism industry. *International Journal of Research-Granthaalayah, 5*(6), 72–77

Kim, M., Lee, S. M., Choi, S., & Kim, S. Y. (2021). Impact of visual information on online consumer review behavior: Evidence from a hotel booking website. *Journal of Retailing and Consumer Services, 60,* 102494.

Kritzinger, W. T., & Weideman, M. (2013). Search engine optimization and pay-per-click marketing strategies. *Journal of Organizational Computing and Electronic Commerce, 23*(3), 273–286.

Krizanova, A., Lăzăroiu, G., Gajanova, L., Kliestikova, J., Nadanyiova, M., & Moravcikova, D. (2019). The effectiveness of marketing communication and importance of its evaluation in an online environment. *Sustainability, 11*(24), 7016.

Lampeitl, A., & Åberg, P. (2017). The Role of Influencers in Generating Customer-Based Brand Equity & Brand-Promoting User-Generated Content. Retrieved from https://lup.lub.lu.se/luur/download?func=downloadFile&recordOId=8921874&fileOId=8921875 (Accessed: 15.06.2023).

Liu, Q., Ge, Y., Li, Z., Chen, E., & Xiong, H. (2011, December). Personalized travel package recommendation. In *2011 IEEE 11th international conference on data mining* (pp. 407–416). IEEE.

Nayyar, A., Mahapatra, B., Le, D., & Suseendran, G. (2018). Virtual Reality (VR) & Augmented Reality (AR) technologies for tourism and hospitality industry. *International journal of engineering & technology, 7*(2.21), 156–160.

Oggolder, C., Brügger, N., Metykova, M., Salaverría, R., & Siapera, E. (2019). The emergence of the internet and the end of journalism?. *The Handbook of European Communication History,* 333–350.

Ozdemir, M. A. (2021). Virtual reality (VR) and augmented reality (AR) technologies for accessibility and marketing in the tourism industry. In *ICT tools and applications for accessible tourism* (pp. 277–301). IGI Global.

Parlov, N., Perkov, D., & Sičaja, Ž. (2016). New trends in tourism destination branding by means of digital marketing. *Acta Economica Et Turistica, 2*(2), 139–146.

Payne, A., & Frow, P. (2005). A strategic framework for customer relationship management. *Journal of marketing, 69*(4), 167–176.

Pencarelli, T. (2020). The digital revolution in the travel and tourism industry. *Information Technology & Tourism, 22*(3), 455–476.

Peter, M. K., & Dalla Vecchia, M. (2021). The digital marketing toolkit: a literature review for the identification of digital marketing channels and platforms. *New trends in business information systems and technology: Digital innovation and digital business transformation, 35*(11–12), 251–265.

Rowley, J. (2008). Understanding digital content marketing. *Journal of marketing management, 24*(5–6), 517–540.

Sanchez-Casado, N., Artal-Tur, A., & Tomaseti-Solano, E. (2019). Social Media, Customers' Experience, and Hotel Loyalty Programs. *Tourism Analysis, 24*(1), 27–41.

Scherp, A., & Boll, S. (2004, October). Generic support for personalized mobile multimedia tourist applications. In *Proceedings of the 12th annual ACM international conference on Multimedia* (pp. 178–179).

Sharma, D., Shukla, R., Giri, A. K., & Kumar, S. (2019, January). A brief review on search engine optimization. In *2019 9th international conference on cloud computing, data science & engineering (confluence)* (pp. 687–692). IEEE.

Stankov, U., Ćurčić, N., Vukosav, S., Dragićević, V., & Pavlović, T. (2009). Tourism websites characteristics in a country with small internet use-case study of Serbia. *Revista de turism-studii si cercetari in turism,* (8), 47–51.

Vinod, B. (2023). Personalization in travel. *Journal of Revenue and Pricing Management, 22*(2), 101–102.

Waterton, E. (2010). The advent of digital technologies and the idea of community. *Museum Management and Curatorship, 25*(1), 5–11.

Wei, W. (2019). Research progress on virtual reality (VR) and augmented reality (AR) in tourism and hospitality: A critical review of publications from 2000 to 2018. *Journal of Hospitality and Tourism Technology, 10*(4), 539–570.

Wörndl, W., & Herzog, D. (2020). Mobile applications for e-Tourism. *Handbook of e-Tourism,* 1–21.

Yasmin, A., Tasneem, S., & Fatema, K. (2015). Effectiveness of digital marketing in the challenging age: An empirical study. *International journal of management science and business administration, 1*(5), 69–80.

Yetimoğlu, S., & Uğurlu, K. (2020). Influencer marketing for tourism and hospitality. In *The Emerald handbook of ICT in tourism and hospitality* (pp. 131–148). Emerald Publishing Limited.

Yoong, L. C., & Lian, S. B. (2019). Customer engagement in social media and purchase intentions in the hotel industry. *International Journal of academic research in business and social sciences, 9*(1), 54–68.

Yung, R., & Khoo-Lattimore, C. (2019). New realities: a systematic literature review on virtual reality and augmented reality in tourism research. *Current issues in tourism, 22*(17), 2056–2081.

Zeng, B., & Gerritsen, R. (2014). What do we know about social media in tourism? A review. *Tourism management perspectives, 10,* 27–36.

Hakki Çilginoğlu[1] and Kaan Berk Dalahmetoğlu[2]

Chapter 26 The Conceptual Framework of Dark Tourism

Introduction

Tourism has rapidly become a significant industry worldwide, contributing significantly to the economic development and employment of countries. There is no universally agreed-upon definition of tourism, as it has been described by various individuals in different ways. In 1980, AIEST (International Association of Scientific Experts in Tourism) defined tourism as "The set of events and relationships arising from people's travels outside the places where they permanently reside, work and meet their usual needs, and their temporary accommodation by requesting goods and services generally produced by tourism enterprises. " (Kozak, Kozak, & Kozak, 2013). The tourism sector, which is one of the fastest growing sectors in the world, have increased its diversity day by day. In tourism activities, which are considered an important source of income and employment for countries and regions, touristic demand is shifting from sea, sand and sun tourism to alternative tourism activities (Alili, 2017). As the diversity in the tourism sector increases, the seasonal dependence of tourism also disappears, and the seasonal income-employment potential can be spread over an entire year. Special interest tourism is one of the types of tourism that does not have seasonal dependence. Dark tourism, one of these types of tourism, is a new field in terms of tourism field (Yildiz et al., 2015). In this context, dark tourism has taken its place as a new type of alternative tourism that is increasing in popularity day by day. Due to the nature of tourism, unlike goals such as relaxing, having fun, having a good time, dark tourism contains grief and sadness in it. Tragic events that have happened in the past, such as wars, deaths, accidents, earthquakes, have an extremely important impact on the transmission of social memory to advanced generations in those who participate in dark tourism. The concept of dark tourism is mentioned in the literature as black tourism, grief tourism, death tourism, morbid, grief or thana tourism. The common point of this type of tourism,

---

1 Ph.D. Student, Kastamonu University, Tourism Faculty, Departmant of Tourism Management, kaanberkdalahmetoglu@gmail.com
2 Asst. Prof., Kastamonu University, Tourism Faculty, hcilginoglu@kastamonu.edu.tr

which is also expressed with words referring to grief, sadness and gamma, is the need for the destination to have a jarring, sad effect (Ince, 2020). The concept of dark tourism has been studied from many different angles and divided into types since the first years it was used. The fact that the emotions aroused by dark tourism venues are different for each individual has led to new categorizations in dark tourism research. In this study, it is aimed to reveal how the concept of dark tourism is named in the literature, how it is diversified and how it is experiencing change/development.

Definition of Dark Tourism

The term "darkness" has been expressed by the Turkish Language Institute (TDK) as "A state of lack of light; cruelty" and figuratively as "Sadness, distress, misery" (Url-1). According to the Oxford English Dictionary, it is defined as "a situation in which there is little or no light" (Url-2). When we look at the meaning of the word tourism, it is based on rest, entertainment, sight, recognition, etc. it is explained as a trip made for purposes of at least 24 hours. With the combination of these two terms, a phenomenon referred to as "dark tourism" occurs.

When the studies conducted in the field of dark tourism were examined, it was determined that the basis of dark tourism dates back to the period of gladiator games of the Romans, where mass execution ceremonies and death events took place (Stone and Sharpley, 2008). The gladiatorial games held in Rome, and then tourists who came to visit from various regions to watch this blood and brutality closely, are known as the first dark tourists. M.D. 16. y.y. also, the executions that took place in Egypt became open to the public, and the train tour organized to watch this is recorded as the first guided tour within the scope of dark tourism (Yildiz et al., 2015).

Like many other things in the age we are in, the concepts of tourism and tourists have taken their share of the transformations experienced and have turned into a brand new state. Now, when it comes to tourism for people, what is important is not only to go to a beautiful place, look at perfect landscapes, take advantage of the sea, sand and sun or relax; but also to experience adrenaline, to experience other lives, other times and even other events. Countries that are aware of this situation have been providing services with tourism resources available in recent years. Tourists who are looking for different destinations and experiences want to go to "dark" places, which are an alternative, and visit places that are famous for disasters such as sadness, war, death that happened there (Gürses, 2021).

The Gladiator games, which were first organized in Rome, and then the tourists who came to visit from different geographies to watch this blood and brutality, are known as the first dark tourists. M.D. 16. y.y. in addition, the executions that took place in Egypt became open to the public, and the train tour organized to watch this is recorded as the first guided tour within the scope of dark tourism (Yildiz, Yildiz, and Aytemiz, 2015). The concept of dark tourism was first used by Lennon and Foyle in the mid-1990s. Dark tourism has been described as a phenomenon involving the presentation and consumption (by visitors) of real and commodified places of death and disaster. (Lennon and Foley, 1996).

Researchers have expressed dark tourism as follows: Tarlow (2005), These are visits to places where tragedies and deaths of historical significance that affect our lives occur. Stone (2012) are visits to areas that are eerie related to pain and death. Kindle, (2008) These are visits to areas or places where death and pain are treated as the main theme or created later. Ali agaoglu (2008), It is the visiting of places that cause sadness to people for various reasons for the purpose of tourism. Kılıç and Akyurt (2011), Places where death events such as torture and genocide occurred, monuments and museums built in their name, poverty is a type of tourism that includes visits to places where events such as natural disasters occurred. Miller and Gonzalez (2013), Death is a journey to places where bitter tragedy has happened. In this way, researchers have defined the definition of dark tourism.

When we look at the definitions that researchers have brought to dark tourism, basically the common point is the same. It is a type of tourism that usually involves visiting places that lead to grief and sadness, such as painful events in human history, tragic places or disaster zones. This type of tourism includes places such as battlefields, genocide monuments, natural disaster zones, prisons, monuments and so on. Dark tourism aims to provide participants with a deep understanding of historical, cultural and social aspects, while at the same time it aims to keep alive the memories of painful events experienced in these places and preserve the social memory. While this type of tourism involves people visiting places that can be emotionally challenging, it can also contribute to the preservation and understanding of places of historical and social significance.

Types of Dark Tourism

Dark tourism can be subjected to various categories or classifications. This classification is usually made according to the type of places visited, the experiences of visitors and the historical or cultural significance of the places and the geographical location.

Nature-Based Dark Tourism

The strong and uncertain power of nature is a fact of the existence of humanity. Devastating natural disasters are events that have occurred in every corner of the world throughout history and have radically changed people's lives. Devastating natural disasters can occur in various forms and scales, such as earthquakes, tsunamis, hurricanes, floods, and volcanic eruptions. These disasters can lead to the loss of millions of lives, billions of dollars in property damage and the collapse of social infrastructure.

The type of tourism that involves people visiting the areas where natural disasters occur and closely observing the effects of the event is referred to as dark tourism. This type of tourism allows people to explore the affected areas in more depth, tracing the traces of historical disasters. For example, visiting a city after an earthquake or seeing a tsunami zone can offer visitors the opportunity to understand the destructive power of natural disasters and experience the recovery process after a disaster.

It is a type of dark tourism related to the elements of death and suffering that occur as a result of natural disasters completely without the influence of humans (Dunkley, Morgan, and Westwood, 2007).

Visiting Disaster Areas

Disasters are expressed as disasters that occur naturally and artificially and cause great damage by occurring suddenly. However, after the disaster, many places can be visited to attract the attention of those who want to examine the damage caused at the scene and learn the latest situation of the local population.

Visiting the Paranormal Areas

These are trips made in the context of motivation for visiting areas where it is assumed that stories about metaphysical beings and paranormal events are taking place and it is believed that this is happening. An example of this is the myth that Alcatraz Prison is a haunted place where a large number of guards and inmates claim that paranormal activities are taking place (Beyaz, 2017: 42).

Human-Based Dark Tourism

It is expressed as a kind of sadness tourism caused by a person knowingly or unknowingly, consisting of elements such as death, pain and violence that occur on a small or large scale (Dunkley, Morgan, and Westwood, 2007).

Genocide Tourism, Cruelty and Slavery Tourism, Grief Tourism, War Tourism, Tragedy Tourism, Horror Tourism and Extraordinary Tourism are shown as examples of human-based sadness tourism (Yildiz et al., 2015). From the past to the present, the world has hosted many events such as suffering, war, poverty, disasters, and as a result of these events, many lives have been lost. Every society visits these destinations in order to commemorate the people they have lost or to understand the events that have taken place. These destinations represent pain and death and are remembered with these memories (Uşak, 2021).

Visiting the Terrorist Areas

This type of tourism involves people visiting areas that have suffered terrorist attacks or are struggling with terrorism. Visiting terrorist sites offers visitors the opportunity to closely observe the places where the events took place and their effects çıkmaktadır (Unur, 2000: 72).

Visiting the Genocide Sites/Mass Killing Sites

A visit to the sites of genocide or sites of mass death covers places that are of great historical and social importance, but at the same time harbor deep suffering. These visits are usually made to understand the darkest periods of human history, to commemorate the victims and to teach lessons to future generations so that such disasters do not happen again. These places include concentration camps, genocide memorials, sites of mass capital punishment and war cemeteries (Seaton, 1999).

War Tourism

Wars have played an important role in human life from the past to the present and have led to serious changes in people, societies and countries. Battlefields, castles, ports are the most important places of military attraction. War tourism is a fun trip to active or former war zones for sightseeing or historical studies. Although war tourist is a new term, it is seen as a type of tourist that defines excitement in dangerous and forbidden places (Weaver, 2000: 153).

Conclusion

Developing technology day by day, migration from village to city and rapid urbanization as a result of this have pushed people to various tourism concepts. With the increase in leisure time, retirement, paid annual leave, and changes in working hours, people has had more time to go on vacation. Different experiences such as cultural tours, adventure tours, nature tours have been offered for people who are looking for different tourism activities from the sea, sand, sun trio. It is unusual that tourism is only for sightseeing, seeing, having fun, relaxing, etc. one of the types of tourism that expresses that it is not limited to such activities is emerging as dark tourism. Dark tourism is a different type of tourism outside the scope of conventional tourism activities. For tourists who want to get a different experience during their holidays, dark tourism offers a different perspective. People make trips within the scope of dark tourism in order to experience curiosity, interest in history, learning and different feelings more than for the purpose of relaxing or having fun. In many destinations for dark tourism activities, there are many factors such as understanding, learning, witnessing history, feeling impressive events that have happened in the past.

Dark tourism refers to a type of tourism that involves visiting places associated with sorrow, grief, and tragedy, such as painful events, tragic sites, or disaster areas. However, in Turkey, this type of tourism is expressed with different concepts. Mainly; terms such as melancholy tourism, sorrow tourism, grief tourism, and death tourism are used, generally representing negative emotions such as sadness, distress, or boredom. In the real and metaphorical sense, the words used describe discomfort or unease felt within one's inner world. Additionally, among the terms used, feelings of disappointment or inability to cope with a difficult situation are also expressed. However, under this term, there are various experiences encompassing a wide range of experiences. To clarify this diversity and prevent confusion in meaning, it is important to categorize dark tourism into subheadings.

Generally, nature-based dark tourism is distinguished as human-based dark tourism due to the fact that it covers events caused by natural elements such as natural disasters, geological formations or biodiversity, and includes historical events, wars, genocides or other tragic situations realized by human hands. This distinction made in the content of dark tourism clarifies various aspects of dark tourism, while at the same time reducing the confusion of meaning with terms such as sadness tourism or death tourism. In this context, the use of the term dark tourism when referring to a destination and tourism activity in general will prevent confusion of meaning.

References

Alili, M. (2017). Avrupa'daki ve Türkiye'deki hüzün turizmi destinasyonlarının karşılaştırması üzerine teorik bir çalışma. *Uluslararası Global Turizm Araştırmaları Dergisi*, 1(1), 37–50.

Beyaz, D. (2017). Paranormal turizm bağlamında tırnava cadıları. *Uluslararası Turizm Ekonomi ve İşletme Bilimleri Dergisi, 1*(2), ss. 42–48.

Dunkley, R., Morgan, N., Westwood, S. (2007). A shot in the dark? Developing a new conceptual framework for thanatourism. *Asian Journal of Tourism and Hospitality Research,* 1(1), 54–73.

Gürses, M.E (2021). *Karanlık su"lardan karanlık turizme: mit, sinema ve post-turizm üzerine bir inceleme.* [Yayımlanmaış yüksek lisans tezi]. Ondokuz Mayıs Üniversitesi Samsun.

İnce, Ö. (2020). *Karanlık turizm kapsamında eğlence fabrikalarının incelenmesi: deep fear park örneği.* [Yayımlanmamış yüksek lisans tezi]. İstanbul Ticaret Üniversitesi.

Kendle, A. (2008). Dark Tourism: a fine line between curiosity and exploitation. Vagabondish. Retrieved from: http://www.vagabondish.com/dark-tourism-travel-tours/

Kılıç, B., & Akyurt, H. (2011). Destinasyon imajı oluşturmada hüzün turizmi: Afyonkarahisar ve Başkomutan Tarihi Milli Parkı. *Gaziantep Üniversitesi Sosyal Bilimler Dergisi,* 10(1), s. 209–232.

Kozak, N., Kozak, M., & Kozak, M. (2013). *Genel Turizm İlkeler ve Kavramlar* (Cilt 1). Ankara: Detay Yayıncılık.

Lennon, J., Foley, M. (1996). JFK and dark tourism: A fascination with assassination, *International Journal of Heritage Studies,* 2(4), 198–211.

Miller, D. S., Gonzalez, C. (2013). When death is the destination: the businessof death tourism-despite legal and social implications. International *Journal of Culture, Tourism and Hospitality Research,* 7(3), 293–306.

Stone, P. R. (2012). Dark Tourism and significant other death. Towards a model of mortality mediation. *Annals of Tourism Research,* 39(3), 1565–1587.

Stone, P., Sharpley, R. (2008). Consuming dark tourism: a thanatological perspective. *Annals of Tourism Research,* 35(2), 574–595.

Tarlow, P. (2005). *Dark Tourism: The Appealing Dark Side Tourism and More.* M. Novelli (Ed.), Niche Tourism: Contemporary Issues, Trends and cases, (s. 47–57). Oxford: Elseiver.

Unur, K. (2000). Turizm-Terörizm ilişkisi ve Türkiye örneği. *Anatolia Turizm AraştırmalarıDergisi (Eylül-Aralık).* ss. 169–177.

Url-1: https://sozluk.gov.tr/

Url-2: https://www.oed.com/

Uşak, E. C. (2021). *Hüzün turizmi kapsamında turistlerin kültürlerarası duyarlılık gelişimlerinin davranış niyetine etkisi: Gelibolu örneği,* [Yayımlanmamış yüksek lisans tezi]. Çanakkale Onsekiz Mart Üniversitesi. Çanakkale.

Weaver, D.B. (2000). The exploratory war-distorted destination life cycle. *International Journal of Tourism Research*, 2(3), 151–161.

Yıldız, Z., Yıldız, S., & Aytemiz, L. (2015). Kara Turizm, terör turizmi ve Türkiye potansiyeli. *İnsan ve Toplum Bilimleri Araştırmaları Dergisi*, 4(2), s. 290–407.

Ozan Esen[1]

Chapter 27 Employee Theft in Tourism Businesses

Introduction

Today, businesses feel the need to adapt to socio-economic, socio-cultural and technological developments (Demir, 2010). In this process, it is stated that efforts to gain competitive advantage are directly proportional to the creativity, knowledge, intelligence, experience and talent of employees. However, today, the rapid growth of businesses, the complexity of the structure of internal and external organizations, increased competition, international activities brought about by globalization and conjunctural fluctuations bring about negative behaviors of employees. Negative behaviors are considered to be an important problem for organizations, causing cost, productivity, reputation, customer and time loss (Alan, Ehtiyar & Ömüriş, 2010).

Research has revealed data showing that employees, who are considered as a strategic element, go beyond the rules and norms determined by businesses. Çivilidağ, Durmaz & Doruköz (2023) explain that behaviors described as deviant employee behaviors include negative workplace behaviors that aim to harm the entire organization or the individual. In addition, it is stated that deviant employee behaviors consisted of theft, harassment, alcohol use, not performing assigned tasks, absenteeism, fighting, not following the rules, arguments among employees, low work performance, and in addition, according to the opinion of human resources employees, theft has the highest rate among deviant employee behaviors. Muhammad and Sarwar (2021) declare that deviant employee behaviors include lack of effort, taking long breaks, making false accident claims, abusing sick leave privileges, making unethical decisions, absenteeism, working slowly, stealing, hiding necessary resources, and gossiping. Organizations are trying to discover ways to minimize employees' deviant behavior.

Employee theft is one of the deviant behaviors mentioned above. (Moorthy, Seetharaman, Jaffar & Foong 2015). Theft, which is considered as negative business behavior in workplaces, is considered as a phenomenon that the entire business world faces and threatens businesses (Alan, Ehtiyar & Ömüriş, 2010).

---

1 Ministry of National Education, oziesen@hotmail.com

Langner (2010) notes that theft in the workplace, especially by employees, is a daunting prospect for American businesses, both corporate and small property owners. In this context, detecting employee theft and revealing its causes and consequences are important for businesses in all sectors.

Employee Theft

'Employee theft' is a general term used to categorize a wide range of deviant and criminal behavior, which essentially involves some form of illegal or undesirable activity by an employee on the job (Kennedy & Benson, 2016). Employee theft is a subject that has been researched by both social scientists and psychologists for many years, for reasons such as finding the reasons for theft and allowing businesses to take deterrent measures (Tarkan & Tepeci, 2006). The world's major religions, international law, the laws of nations, and most ethical codes around the world define theft as immoral and illegal. However, theft by their own employees costs business owners and operators an estimated $100 billion worldwide each year (Sauser, 2007).

Employee theft is one of the most widespread and serious problems in human resource management (Greenberg, 2002). Some recent statistics show that the tendency for employees to steal from retail outlets is quite common. Employee theft has a significant negative impact on businesses, employees, owners, managers and other stakeholders (Kennedy & Benson, 2016). Bailey (2006) emphasizes that theft by retail employees has a significant impact on retail operations in the United States and other countries and is the most important source of waste for many retailers. Korgaonkar, Becerra, Mangleburg and Bilgihan (2021) show that the retail industry loses billions of dollars every year due to theft in business.

Employee theft creates problems for every business of varying sizes. Moorthy et al. (2015) emphasizes that employee theft is costly for every business, especially large retail chain organizations. Kennedy and Benson (2016) state that employee theft is one of the most harmful crimes that can happen to a small business. Due to their relatively small size these firms are more vulnerable to financial losses from employee theft than larger commercial enterprises. While small business owners try to identify and retain trustworthy employees, most business owners eventually realize that they have to face the fact that someone they entrusted with a job stole it from them. From this perspective, employee theft seems to be an inevitable situation in businesses.

The tourism sector, which is described as a labor-intensive sector, includes accommodation and travel businesses as sub-fields of activity. Olcay, Özkan & Göçebeler (2018) evaluated tourism as an field that needs to be examined ethically due to reasons such as the fact that customers and staff are together for long hours in accommodation businesses and that the safety of life and property of customers is the responsibility of the accommodation businesses. Hua, Lin, and Zhang (2020) stated that the concept of crime is an important problem in the tourism sector. Especially businesses serving in the tourism sector suffer serious losses due to theft in business every year (Yılmaz, Özgen & Hazarhun 2021). These losses increase the costs of touristic products and services and reduce the competitiveness of tourism businesses. In the light of this data, employee theft in tourism enterprises needs to be examined comprehensively.

Employee Theft in Tourism Businesses

Businesses in the tourism sector face a complex and competitive environment. It is claimed that human resources management (HRM) has an important role in the success of these organizations and even their continued existence (Çivilidağ, Durmaz & Doruköz, 2023). This sector is considered labor intensive compared to other service sectors and it is stated that employees have close interaction with customers. Therefore, employee behavior is considered an important element of this sector (Muhammad & Sarwar, 2021). Ethical issues in the hospitality industry reveal the challenges that arise in a cash-based, people-intensive industry. Today, managers are learning to watch out for transaction errors, food and beverage theft, lost inventory, and unreported sales (Stevens, 2011). In this context, employee theft is in many respects a personnel management issue (Kennedy & Benson, 2016).

The industry's long-standing relationship with theft is widely accepted (Poulston, 2008) and employee theft is frequently encountered in tourism businesses such as hotels, restaurants and bars (Krippel, Henderson, Keene, Levi, 2008). Pratt (2020) emphasized that most of the theft-related research surrounding the tourism and hospitality industry focuses on tourists who are robbed while on vacation or hotel employees who steal from their employers. Sufi, Ranga and Ranga (2023) explains that hotel theft incidents are a significant source of concern for the hotel industry. It explains that such events will cause significant financial consequences, inventory management problems, and difficulties in finding exact replacements. The beverage department, which is characterized as an important sales area of accommodation businesses, is known as the department that generates the most revenue after the rooms in terms of accommodation businesses and it is reported that this department is very difficult to control. High costs of food and beverage can be an indicator of theft and lack of stock control (Angay, 2003).

The hospitality industry is particularly vulnerable to the most common form of employee theft (Mohsin, 2006). Hotels, restaurants and bars, golf courses and other attractions, although often part of a large chain, operate like small businesses and can be susceptible to employee theft. The hospitality industry consists of many unskilled, low-wage jobs with high employee turnover and low social status (Krippel et al., 2008). Studies on the causes of employee theft put forward various reasons such as low wages and the use of student and young workforce (Poulston, 2008). Stevens (2011), on the other hand, emphasizes that the high number of part-time job opportunities, high employee turnover, low work commitment and high number of low-educated personnel in hotels increase employee theft. Goh and Kong (2016), in their study among Generation Z undergraduate students working full-time as part of their internships, reveal that the reasons for employee theft include factors such as excitement, influence from colleagues and perceived ease of stealing, and the feeling of adrenaline as a source of motivation. Moorthy et al. (2015) emphasizes that individual and organizational factors affect theft behavior in the workplace.

Greenberg (2002) emphasizes that, within the scope of current findings, considering the determinants of employee theft, individual employee theft is higher in the following situations: (a) among employees with lower moral development than among employees with higher moral development, (b) in an office without an ethics program than in an office with an ethics program, (c) in cases where the company is a victim, unlike colleagues. As well as being illegal, theft has the capacity to erode staff morale and the profitability of the business (Poulston, 2008). Thefts committed by hotel employees cause financial losses to businesses (Alan, Ehtiyar & Ömüriş, 2010), reduce employee motivation, and cause businesses to lose reputation (Olcay, Özkan & Göçebeler, 2018). Bailey (2006) argues that employee theft will also affect the prices consumers have to pay. Because some of the costs resulting from these and other types of contractions will likely be passed on to consumers in the form of higher prices. Additionally, like other types of crime, employee theft has more personal and subjective effects on victims. It has been revealed that victimized business owners experience various cognitive and emotional reactions to theft. Sometimes these reactions are severe and are exacerbated if the victim has a strong emotional bond with the perpetrator (Kennedy & Benson, 2016).

Items subject to employee theft in tourism businesses vary. Krippel et al. (2008) say that young employees steal cash and stock goods from businesses. However, the theft from workplaces is divided into three categories: physical items (cash, stock goods and equipment), unpaid activities and mispricing (improper discounts to employees, fraudulent refunds). Yılmaz, Özgen & Hazarhun (2021) shows that theft by hotel employees occurs in food products, alcoholic/non-alcoholic beverages, amenities, cleaning products, and currency and payment transactions at the front office.

Considering the prevalence of employee theft in tourism businesses and the problems it creates, businesses are trying to take precautions against this problem. Krippel et al. (2008) emphasizes that tourism managers may need training in more sophisticated control strategies to combat the threat of high rates of theft and that training should be provided to managers accordingly. Moorthy et al. (2015) points out that internal control systems regulate the relationship between intention to steal and workplace theft behavior. Yılmaz, Özgen & Hazarhun (2021) recommend that employees' bags should be checked at the end of the shift to prevent employee theft. In this context, another suggestion made is to carry out the necessary investigations for the candidate before recruitment. Tarkan & Tepeci (2006) explain that reporting employee theft can be considered a good preventive strategy to reduce workplace theft.

Conclusion

Employee theft is among the ethical problems commonly encountered in tourism businesses. The magnitude of the employee theft problem is striking. Langner (2010) emphasizes that some analysts specializing in the security field estimate that 30 percent of the country's workforce will at some point steal from their employers. When evaluated from this perspective, it seems that employee theft is inevitable for businesses. Additional efforts should be made to understand this phenomenon so that tourism business managers can implement strategies to effectively reduce or minimize employee theft (Bailey, 2006). Before strategies can be implemented to stop this behavior, managers need to understand the individual and environmental factors that motivate this behavior.

Employee theft is seen as a serious problem for tourism businesses. It is stated that the cash-based and labor-intensive structure of the sector presents difficulties in terms of employee theft (Stevens, 2011). The reasons for employee theft in tourism businesses include the use of low-paid student and young workforce, high part-time job opportunities, high employee turnover rate, low work commitment and high employment of personnel with low education levels, excitement, being influenced by colleagues and perceived ease of stealing. Pierce, Snow and McAfee (2015) reveal that employee theft is not only a result of individual differences in ethics or morality, but can also be influenced by management policies that can benefit both businesses and employees. In the light of these data, it can be said that employee theft is related to individual and organizational reasons.

Employee theft poses important problems for businesses in the tourism sector. In addition to being illegal, employee theft lowers the morale and motivation of the staff, causes financial losses to businesses, and causes businesses to lose prestige (Poulston, 2008; Alan, Ehtiyar & Ömüriş, 2010; Olcay, Özkan & Göçebeler, 2018). It is also argued that employee theft increases costs and affects the prices that consumers have to pay (Bailey, 2006). Kennedy and Benson (2016) express that crime victimization can lead to emotional costs as well as physical and financial costs, and this is widely accepted in the law and science world. In this context, the emotional problems that employee theft may cause in employers should be taken into consideration.

When the consequences of employee theft are evaluated, businesses in the tourism sector need to develop strategies to take precautions. Checking employees' bags at the end of the shift, applying tests on individuals' moral development levels during the recruitment process, and providing training to managers on control systems are evaluated as measures taken against employee theft. Tarkan and Tepeci (2006) emphasize that leaving employees free to make decisions may lead them to behave more honestly and increase the reporting of theft, which is an effective method for reducing and preventing theft in the workplace. Johnson, Friend, and Esteky (2022) say that rewards are reinforcement mechanisms used by organizations to shape desired employee behaviors, but these rewards may create expectations in employees to obtain extra benefits and employees may turn to unethical behaviors for the sake of these rewards.

References

Alan, A. A., Ehtiyar, V. R., & Ömüriş, A. G. E. (2010). Beş yıldızlı otel işletmelerinde hırsızlık sorunsalı: Antalya-Kundu Örneği. 11. *Ulusal Turizm Kongresi*, 26., 26–33.

Angay, F. (2003). *Konaklama işletmelerinde maliyet ve yönetim muhasebesi problemleri ve çözüm önerileri* (Master's thesis, Akdeniz Üniversitesi).

Bailey, A. A. (2006). Retail employee theft: a theory of planned behavior perspective. *International Journal of Retail & Distribution Management, 34*(11), 802–816.

Çivilidağ, A., Durmaz, Ş. & Doruköz, K. D. (2023). Konaklama işletmelerinde insan kaynakları çalışanlarına göre sapkın çalışan davranışları ve psikolojik iyi oluş. *Fiscaoeconomia, 7*(3), 2381–2410.

Demir, M. (2010). Örgütsel sapma davranışının kontrolünde duygusal zekânin rolü: konaklama işletmelerinde bir araştırma. *Dumlupınar Üniversitesi Sosyal Bilimler Dergisi*, (26).

Goh, E., & Kong, S. (2016). Theft in the hotel workplace: Exploring frontline employees' perceptions towards hotel employee theft. *Tourism and Hospitality Research*, 18(4), 442–455.

Greenberg, J. (2002). Who stole the money, and when? Individual and situational determinants of employee theft. *Organizational behavior and human decision processes*, 89(1), 985–1003.

Johnson, J. S., Friend, S. B., & Esteky, S. (2022). Can rewards induce corresponding forms of theft? Introducing the reward-theft parity effect. *Business Ethics, the Environment & Responsibility*, 31(3), 846–858.

Kennedy, J. P., & Benson, M. L. (2016). Emotional reactions to employee theft and the managerial dilemmas small business owners face. *Criminal Justice Review*, 41(3), 257–277.

Korgaonkar, P., Becerra, E. P., Mangleburg, T., & Bilgihan, A. (2021). Retail employee theft: When retail security alone is not enough. *Psychology & Marketing*, 38(5), 721–734.

Krippel, G. L., Henderson, L. R., Keene, M. A., Levi, M., & Converse, K. (2008). Employee theft and the Coastal South Carolina hospitality industry: Incidence, detection, and response (Survey results 2000, 2005). *Tourism and Hospitality Research*, 8(3), 226–238.

Langner, D. (2010). Employee theft: Determinants of motive and proactive solutions.

Mohsin, A. (2006). A case of control practice in restaurants and cafes in Hamilton, New Zealand. *The Journal of American Academy of Business, Cambridge*, 8(1), 271–276.

Moorthy, M. K., Seetharaman, A., Jaffar, N., & Foong, Y. P. (2015). Employee perceptions of workplace theft behavior: A study among supermarket retail employees in Malaysia. *Ethics & Behavior*, 25(1), 61–85.

Muhammad, L., & Sarwar, A. (2021). When and why organizational dehumanization leads to deviant work behaviors in hospitality industry. *International journal of hospitality management*, 99, 103044.

Olcay, A., Özkan, B., & Göçebeler, M. F. (2018). Konaklama ve seyahat işletmelerinde yaşanan etik sorunlar üzerine kavramsal bir değerlendirme. *Gaziantep University Journal of Social Sciences*, 17, 1–11.

Pierce, L., Snow, D. C., & McAfee, A. (2015). Cleaning house: The impact of information technology monitoring on employee theft and productivity. *Management Science*, 61(10), 2299–2319.

Poulston, J. (2008). Rationales for employee theft in hospitality: Excuses, excuses. *Journal of Hospitality and Tourism Management*, 15(1), 49–58.

Pratt, S. (2020). Tourists "stealing" stuff. *Tourism Economics*, 28(2), 495–514.

Sauser Jr, W. I. (2007). Employee theft: Who, how, why, and what can be done. *SAM Advanced Management Journal, 72*(3), 13.

Stevens, B. (2011). Hotel managers identify ethical problems: A survey of their concerns. *Hospitality Review, 29*(2), 2.

Sufi, T., Ranga, B., & Ranga, I. (2023). Dealing with hotel customer stealing practices: a managerial perspective. *Journal of Hospitality and Tourism Insights, 6*(5), 2545-2564.

Tarkan, G., & Tepeci, M. (2006). Örgütsel adalet ve yönetimde merkezileşmenin çalışan hırsızlığına etkileri: Mersin üniversitesi turizm işletmeciliği ve otelcilik yüksekokulu öğrenci algılamaları üzerine bir araştırma. *Anatolia: Turizm Araştırmaları Dergisi, 17*(2), 137-152.

Yılmaz, Ö. D., Özgen, I., & Hazarhun, E. (2021). Otellerde Konuk ve Çalışan Hırsızlığı: Nitel Bir Araştırma. *Turizm Akademik Dergisi, 8*(1), 133-149.

Diğdem Eskiyörük[1]

Chapter 28 Holistic Review of Organizational Mindfulness and Mindful Leadership in the Perspective of Sustainable Tourism Management

Introduction

Mindfulness is an important issue in various fields at individual, organizational and social levels. In a competitive atmosphere where change is gaining momentum and competition is increasing with globalization, it is very important for all organizations to be aware of sudden changes and uncertain situations. The term mindfulness is defined as the state of being aware and attentive to one's surroundings and oneself, as well as the capacity to process new information by suspending preconceived ideas (Ling et al., 2019), and the capacity to proactively process the information surrounding oneself with the aim of using it to differentiate or draw new conclusions (Dutt and Ninov, 2016). Organizational mindfulness is argued to provide a decision model that assists the organization's decisions and activities by bridging the gap between operational requirements and modifications, and by identifying gaps in the organizational environment. (Obiora, 2021). By providing managers with a wide awareness of their activities and potential reactions to their actions and possible changes and reactions to developing events in their environment, mindful leadership can provide more efficient management of organizations operating in the dynamic and competitive climate of the tourism sector with the sustainability perspective. Therefore, it is essential for organizational mindfulness to be efficient and effective in sustainable tourism management, whose objective is to develop tourism in a planned and environmentally friendly and integrated manner. In this respect, mindful leadership is an important link between sustainable tourism management and mindfulness. In the current paper, the issues of mindfulness, organizational mindfulness and mindful leadership terms, the organizational effects and significance of organizational mindfulness and mindful leadership, and ultimately

---

1 Assist Prof. Dr. Diğdem Eskiyörük, Çukurova University, The Faculty of Kozan Business Administration, Department of Tourism Management, dtecer@cu.edu.tr

the issues of organizational mindfulness and mindful leadership in sustainable tourism management will be thoroughly investigated.

Concepts of Mindfulness, Organizational Mindfulness and Mindful Leadership

In recent years, mindfulness has attracted considerable attention and has been examined by different disciplines. It is seen that mindfulness has been the subject of Buddhism and other religious and philosophy fields for many years, then in clinical areas such as psychology, and finally in organizational topics and fields such as leadership, performance, success, etc. in business life.

Mindfulness is a term that can be analyzed from many perspectives, such as a way of being in our daily lives or a mental training technique (meditation) that we engage in at some time of the day (Ericson et al., 2014). Mindfulness, linked closely to the idea of permanent change, refers to the actual recognition, from an individual perspective, of people's continuing mental streams of thoughts, images and feelings (Becke et al., 2012). It can be defined as "a receptive attention to and awareness of present events and experience" (Brown et al., 2007) or as "the awareness that emerges through paying attention on purpose, in the present moment, and nonjudgmentally to the unfolding of experience moment by moment" (Kabat-Zinn, 2003). When various descriptions of mindfulness are examined, three factors stand out: mindfulness is consciousness focused on the present, it involves close attention to both internal and external phenomena, and it involves attention to stimuli in an open and accepting manner (Hyland et al., 2015). Baer et al. (2008) classified mindfulness skills as observing, defining, behaving with mindfulness, non-judgement of internal experience and non-reactivity to internal experience. Shapiro et al. (1998) indicated that the essential factor of mindfulness is not only attention, but also how one pays attention and the intention that a person brings to his/her attention (practice) is the main issue. They also emphasized that mindfulness should involve compassion, impartiality, acceptance of self and others. Mindfulness is reported to lead to a variety of positive psychological effects, including increased well-being, reduced psychological symptoms and emotional reactivity, and improved behavioral adjustment (Keng et al., 2011). It has been suggested that people with high mindfulness show higher flexibility due to a more balanced use of proactive and reactive cognitive control and show a more flexible performance because of an increased focus on the non-judgmental current moment (Panditharathne & Chen, 2021). The results of the research conducted by Kiken and Shook (2011) indicate that mindfulness enhances positive judgements and decreases negativity bias.

According to the results of the findings of the studies, mindfulness develops listening skills, adaptability, determination, creativity, degree of research and development, moral questioning capacity, ethical and objective decision-making, decision-making competence, empathy, mental efficiency, emotional recognition. On the contrary, the results indicate that mindfulness also decreases bias and interpersonal conflicts (Panditharathne & Chen, 2021). The research findings of Chatzisarantls & Hagler (2007) point out that increased mindfulness and attentiveness to existing experiences can make it easier to effectively transform intentions into actions.

The concept of mindfulness, which has become an important issue in many disciplines, particularly in studies in the field of health, and which has provided numerous psychological benefits at the individual level, has begun to be evaluated in the organizational context, and the possible positive effects of organizational mindfulness in the working environment have started to be discussed as a research topic. The rapid change in every area in today's world and unclear environmental conditions have made the concept of mindfulness more significant in organizations.

In this regard, organizational mindfulness is highlighted as a dynamic capability with a considerable potential to cope with the challenges encountered. Transcending the individual domain and encompassing the collective mindfulness within an organization, organizational mindfulness entails developing a culture that encourages attention, awareness, and purposeful action in every aspect of organizational functioning and being fully engaged and aware of the situation at hand without being overly reactive or judgmental (Motwani et al., 2023). Organizational mindfulness, which refers to the psychological state in which employees deliberately give their full attention to the present moment while performing their work tasks, at times focuses on differences in experiencing mindfulness (Jang et al., 2020).

The original concept of organizational mindfulness has been developed by Kathleen Sutcliffe and Karl Weick in relation to risk and safety studies. Organizational mindfulness underlines a collective view and organizational learning about how to anticipate and cope with unexpected risky events that threaten organizations and their survival. High-reliability organizations (HROs) have been defined as organizations that are responsive to changes in unpredictable, volatile, and risky environments and have an internal organizational mindfulness infrastructure (Becke et al., 2012).

Considering mindfulness as a key factor of organizational cultures, organizations aim to achieve this competence for all people in the organization, from administrators to employees, by conducting mindfulness activities at every stage of the organizational management process. In today's world where change is accelerating in every field, the uncertainty, competition, and chaos environment brought about by the global economy put managers and leaders at all levels of the organization under stress. In this respect, mindful leaders have a great importance in the studies to be performed to make organizations mindful in today's variable and uncertain business environment. However, compared to the studies in the field of leadership in the literature, it is observed that the issue of mindful leadership is addressed less frequently.

Leadership is defined as 'a set of knowledge and skills to rally a group of people around specific goals and mobilize them to achieve those goals' (Demirel & Kişman, 2014) In addition, mindful leaders are described as calm, focused, aware, clear, cool, positive, compassionate, and flawless. Conscious leaders are identified as inspiring others by doing the best they can without commitment, serving in a positive way (Koller, 2017, p. 35). Mindfulness leads to assess the stability, predictability, and rationality of systems, but also to find new ways to deal with unpredictable constant change, cognitive biases, irrationality and impermanence. Raising awareness is reported to encourage senior managers to re-evaluate their basic knowledge of organization, leadership and the role of management. (Yeşilkuş & Özbozkurt, 2020). In additional, it has been stated that one of the leadership styles that can be related to mindfulness is authentic leadership, and that leaders with high mindfulness exhibit authentic leadership behaviors and gain more adaptability, innovation, resilience and high commitment. (Panditharathne & Chen, 2021).

Organizational mindfulness, which provides organizations to proactively adapt to the changeable business environment, enables leaders and employees to anticipate trends and be ready for new challenges and potential threats by understanding the current environment in which they operate (Motwani et al., 2023). Mindfulness as a multi- and transdisciplinary research stream that highlights the role of cognitive dimensions in increasing the level of individual attention and ability to perceive and understand external phenomena (Kabat-Zinn, 1994, 2003) can assist in embedding sustainability into the behavior of both individuals and organizations via leadership (Saviano et al., 2018).

Organizational Effects and Importance of Organizational Mindfulness and Mindful Leadership

In literature, a wide range of research has been conducted in the fields of organizational mindfulness and mindful leadership, and the organizational effects and importance of both organizational mindfulness and mindful leadership have been discussed. Mindfulness in organizations expresses the extent to which employees are attentive and aware in their workplaces and is a feasible strategy for promoting employee performance and notable organizational efficiencies. In this context, it is expressed that mindfulness has a wide range of benefits for organizations such as empathy, greater adoption of work friendships, resilience, improved working memory, increased task performance and self-determination, and accurate emotional forecasting (Yeşilkuş & Özbozkurt, 2020).

Dane and Brummel (2014) determined a positive relationship between employees' workplace awareness and job performance. According to Brown, Ryan and Creswell (2007), conscious organizations are harmonious and collaborative organizations and leadership is inclusive and supportive. By this means, it is stated that ideas are brought together from all levels and functions within the organization, increasing creativity and innovation (Obiora, 2021). Studies have put forth that conscious mindfulness and conscious mindfulness interventions have yielded a wide range of positive outcomes for individuals, teams, and organizational culture in areas such as performance, well-being, citizenship behavior, pro-social behavior, workplace spirituality, human health, human rights, and social justice (Panditharathne & Chen, 2021). Mindfulness is indicated to reduce stress, create a healthy learning environment, increase innovation, creativity, attention and concentration, reduce the negative effects of multitasking and reactivity, and improve interpersonal functioning, empathy and self-awareness (Koller, 2017, p. 18). The research results of Ilyas & Khattak (2021) show that conscious mindfulness and innovative work behavior significantly improve employee performance in the exemplified organizations.

With mindfulness practices, which are defined as the ability to respond quickly and flexibly to constantly changing stimuli, various positive results have been achieved in organizations such as increased well-being, increased productivity, better teamwork and creativity, increased work performance, happiness and making sense of work-related tasks, positive organizational behavior, improvement in organizational performance and innovation tendency, and increased work-related self-efficacy (Leonelli, 2019, p. 1-2), improving mindfulness, emotional memory and emotional stability, reducing emotional intensity, negative emotional attentional bias, momentary stress and work-related stress, enhancing self-care and work-life balance (Panditharathne & Chen, 2021), encouraging being a good listener, supporting prioritization, and supporting innovation and creativity (Sinclair, 2015).

Generally, mindfulness is considered a useful instrument for managing change and transformation. Mindful leaders who have a high level of self-awareness about their own emotions and the emotions of others provide support in identifying employees' strengths and weaknesses and in developing their knowledge and skills (Permana et al., 2024). Mindful leaders, with their advanced problem-solving and decision-making skills and self-awareness, not only reduce their employees' anxiety by exhibiting consistent behavior in uncertain environments, but also enable them to perform better as role models. Reb, Narayanan and Chaturvedi (2014) put forth that manager's conscious mindfulness is positively related to employees' job satisfaction and psychological need satisfaction. They also indicated that leader awareness was positively related to overall job performance as well as in-role performance and organizational citizenship behaviors.

It is stated that leaders' conscious mindfulness positively affects leaders' transformational leadership behavior, increases employees' job satisfaction and reduces employees' psychosomatic complaints. Conscious mindfulness paves the way for successful negotiations, effective decisions, increased well-being and prosocial behavior, which in turn leads to high performance. Additionally, in a highly complex, competitive and dynamic organizational environment, employees with high levels of mindfulness are able to achieve mental health and success because they are able to build strong relationships with both the internal and external environment of the organization (Panditharathne & Chen, 2021).

Organizational Mindfulness and Mindful Leadership in Sustainable Tourism Management

Nowadays, with the dominance of the consumption-oriented world view and along with the social, economic and ecological crises experienced globally, the concept of sustainability has started to be included in our perception and value systems as a new understanding, perhaps as a starting point. Even though mindfulness manifests itself at the individual level in the beginning, it is not limited to this area, but its effects can be seen at the social level. When we evaluate it in terms of sustainability, as the boundaries of mindfulness are expanded, individual sustainability can set off social sustainability. Sustainability can described as "the conservation of all available vital resources while meeting the needs of the current, with an eye to future generations, the ability to be sensitive and conscious based on continuity without disrupting the system by avoiding overuse".

Sustainability supports the ability of organizations to sustain their existence in a way that is sensitive to both the environment and society, taking into account all economic, environmental and social dimensions. At this juncture, "mindfulness" is emphasized as one of the possible supports for embedding sustainability in the behavior of individuals and organizations (Saviano et al, 2018). It is essential to combine mindfulness and sustainability in order to look at key social problems from a holistic perspective. Mindfulness makes individuals more aware of their values and goals, which makes them more sensitive to sustainability issues (Leonelli, 2019, p. 2).

Mindfulness has the potential to make a difference by encouraging sustainable behavior. Mindful individuals, despite facing many obstacles, are more willing to consciously use and pay attention to their knowledge of environmental impact and are more sensitive to the natural environment are more likely to choose the most reasonable of the options (Amel et al., 2009). It is stated that as mindfulness raises, exposure to advertising, increased consumption and material gain can be reduced, as habitual reacting or acting on "autopilot" will decrease (Lengyel, 2015).

Studies show that there is a relationship between mindfulness and sustainability. Mindfulness is expressed to have an impact on having a higher individual sensitivity to climate change or more sustainable behaviors (Leonelli, 2019, p. 2). In the light of the research, it has been found that mindfulness is positively related to sustainable behavior (Jacob, Jovic ve Brinkerhoff, 2009, Amel, Manning ve Scott, 2009). Brown and Kesser (2005) obtained the finding that happy people have intrinsically oriented values and live more ecologically responsible lives because they pay more attention to their inner experiences and behaviors. Ericson et al. (2014) stated that mindfulness can contribute to environmentally friendly and sustainable behavior.

Organizational mindfulness, which is acknowledge to be a sensitizing concept in consciously designing organizational change, is expressed as a key concept for organizational (social) sustainability in changing environments. Mindful leadership is defined as a key bridging concept between mindfulness and sustainability (Saviano et al, 2018).

Due to the fact that sustainable tourism encompasses many social, economic, political, ecological, historical, cultural, technological and legal dimensions, it is important to have sustainable plans and policies that protect the rights of all stakeholders for the development and sustainability of tourism destinations (Danacı & Akkaya, 2018). Müller (1994) has put forward the magic pentagon of tourism, which includes the elements of a healthy economy, optimum satisfaction of tourist needs, healthy culture, preservation of pristine natural resources and the well-being of the people of the tourist region. The importance of a balance between all these factors is emphasized in order for sustainable tourism to ensure quality growth in harmony with the environment.

Within the scope of sustainable tourism management; taking into account natural, cultural and social values, protecting and developing natural areas, raising awareness of people about the concept of sustainable tourism should be the main goal for the continuation of the existence and economic benefits of the sector (Davutoğlu & Yıldız, 2021). It has been emphasized that mindfulness has positive effects on sustainable behavior (Jacob, Jovic, & Brinkerhoff, 2009, Amel, Manning, & Scott, 2009, Brown & Kasser, 2005) and the importance of mindfulness-based tourism services in terms of providing long-term balance for the tourism industry (Lengyel, 2015).

In a successful sustainable tourism management, it is of great importance to establish a long-term vision, plans and policies, to manage all actors in this process in a way that fully covers the functions of planning, organizing, coordinating, directing and controlling, and to include social and economic sustainability as well as environmental sensitivity. At this point, mindful leaders can guide the organization by shedding light in line with their vision in the face of possible negativities in the sustainability process.

Conclusion

In sustainable tourism management, the combination of nature, people and destinations and the protection and harmony of the continuity of this combination are important. For sustainable tourism management to be successful, it is necessary to consider many factors such as how tourists evaluate and use the natural environment, how communities develop tourism, and the management and definition of the social and ecological impacts of tourism (Davutoğlu ve Yıldız, 2021). From this point of view, coordination between all actors in sustainable tourism management is important for the sustainability of tourism. At this point, the mindfulness and leadership style of the manager in the environment-human interaction come to the fore in the managerial process. Organizational mindfulness is the processes that enable organizations to be sensitive to their environment, make organizations open and curious about new information, and manage unexpected events quickly, flexibly and effectively (Valorinta, 2009). Mindfulness at the organizational level is of great importance for tourism organizations to continue to function during change-related situations or events, to achieve change behaviour and goals, and therefore for their effectiveness and success (Obiora, 2021). It is argued that mindful leaders are crucial social and organizational actors through whom sustainability principles can be effectively embedded in all areas of life. Mindful leaders, who have an important role to play in helping organizations integrate sustainability into their strategies, supports the creation of favorable conditions for social and economic organizations to implement sustainability principles (Saviano et al., 2018).

References

Amel, E. L., Manning, C. M., & Scott, B. A. (2009). Mindfulness and sustainable behavior: Pondering attention and awareness as means for increasing green behavior. *Ecopsychology, 1*(1), 14–25.

Baer, R. A., Smith, G. T., Lykins, E., Button, D., Krietemeyer, J., Sauer, S., Walsh, E., Duggan, D., & G. Williams, J. M. (2008). construct validity of the five facet mindfulness questionnaire in meditating and nonmeditating samples. *Assessment, 15*(3), 329–342. https://doi.org/10.1177/1073191107313003

Becke, G., Behrens, M., Bleses, P., Meyerhuber, S., & Senghaas-Knobloch, E. (2012). *Organizational and political mindfulness as approaches to promote social sustainability.* (artec-paper, 183). Bremen: Universität Bremen, Forschungszentrum Nachhaltigkeit (artec). https://nbn-resolving.org/urn:nbn:de:0168-ssoar-58706-1

Brown, K. W., Ryan, R. M., & Creswell, J. D. (2007). Mindfulness: Theoretical foundations and evidence for its salutary effects. *Psychological Inquiry, 18*(4), 211–237.

Brown, K. W., & Kasser, T. (2005). Are psychological and ecological well-being compatible? The role of values, mindfulness, and lifestyle *Social Indicators Research*, 74, 349–368. DOI 10.1007/s11205-004-8207-8

Chatzisarantls, N. L., & Hagger, M. S. (2007). Mindfulness and the intention-behavior relationship within the theory of planned behavior. *Personality and Social Psychology Bulletin, 33*(5), 663–676. DOI: 10.1177/0146167206297401

Danacı, M. C., & Akkaya, F. A. (2018). Kemaliye (Eğin)'De sürdürülebilir turizm yönetimi ekseninde gelecek projeksiyonu1. *Erzincan Üniversitesi Sosyal Bilimler Enstitüsü Dergisi, 11*(1), 233–250.

Dane, E., & Brummel, B. J. (2014). Examining workplace mindfulness and its relations to job performance and turnover intention. *Human Relations, 67*(1), 105–128.

Davutoğlu, N. A., & Yıldız, E. (2021). Sürdürülebilir turizm yönetiminin evrilleştirilmesi. *Journal of Social Humanities Sciences Research, 8*(75), 2556–2562. http://dx.doi.org/10.26450/jshsr.2743

Demirel, H. G., & Kişman, Z. A. (2014). Cross-Cultural Leadership. *Turkish Studies, 9*(5), 689–705.

Dutt, C., & Ninov, I. (2016). The role of mindfulness in tourism: Tourism businesses' perceptions of mindfulness in Dubai, UAE. *Tourism: An International Interdisciplinary Journal, 64*(1), 81–95.

Ericson, T., Kjønstad, B. G., & Barstad, A. (2014). Mindfulness and sustainability. *Ecological Economics, 104*, 73–79. DOI: 10.1016/j.ecolecon.2014.04.007

Hyland, P. K., Lee, R. A., & Mills, M. J. (2015). Mindfulness at work: A new approach to improving individual and organizational performance. *Industrial and organizational Psychology, 8*(4), 576–602. doi:10.1017/iop.2015.41

Ilyas, M., & Khattak, M. S. (2021). The Role of Mindfulness in Employees Performance; Do Innovative behaviors mediate the link? *Journal of Innovative Research in Management Sciences, 2*(2), 16–30.

Jacob, J., Jovic, E., & Brinkerhoff, M. B. (2009). Personal and planetary well-being: Mindfulness meditation, pro-environmental behavior and personal quality of life in a survey from the social justice and ecological sustainability movement. *Social Indicators Research, 93*, 275–294.

Jang, J., Jo, W., & Kim, J. S. (2020). Can employee workplace mindfulness counteract the indirect effects of customer incivility on proactive service performance through work engagement? A moderated mediation model. *Journal of Hospitality Marketing & Management, 29*(7), 812–829. DOI: 10.1080/19368623.2020.1725954

Kabat-Zinn, J. (1994). *Wherever you go, there you are: mindfulness meditation in every day life.* Piatkus: London.

Kabat-Zinn, J. (2003). Mindfulness-based interventions in context: past, present, and future. *Clinical Psychology Science Practice, 10*(2), 144–156.

Keng, S. L., Smoski, M. J., & Robins, C. J. (2011). Effects of mindfulness on psychological health: A review of empirical studies. *Clinical psychology review, 31*(6), 1041–1056. doi:10.1016/j.cpr.2011.04.006

Kiken, L. G., & Shook, N. J. (2011). Looking up: Mindfulness increases positive judgments and reduces negativity bias. *Social Psychological and Personality Science, 2*(4), 425–431.

Koller, N. (2017). *Mindful leadership: The Impact of Mindfulness on Managers' Ethical Responsibility* [Master's thesis, ZHAW Zürcher Hochschule für Angewandte Wissenschaften]. https://doi.org/10.21256/zhaw-2084

Lengyel, A. (2015). Mindfulness and sustainability: utilizing the tourism context. *Journal of Sustainable Development, 8*(9), 35–51. http://dx.doi.org/10.5539/jsd.v8n9p35

Leonelli, M. (2019). Mind the Gap: The unexplored linkage between Corporate Mindfulness and Sustainability Adoption (Dissertation). Retrieved from https://urn.kb.se/resolve?urn=urn:nbn:se:hj:diva-43989

Ling, T. P., Noor, S. M., Mustafa, H., & Kiumarsi, S. (2019). Mindfulness: Exploring visitor and communication factors at Penang heritage sites. *SEARCH Malaysia, 11*(1), 137–152.

Motwani, J., Kataria, A., Garg, R., & Singh, D. (2023). Organization mindfulness: A systematic literature review and research agenda. *Research Square Platform LLC*. https://doi.org/10.21203/rs.3.rs-3205012/v1.

Müller, H. (1994). The thorny path to sustainable tourism development. *Journal of sustainable tourism*, 2(3), 131–136.

Obiora, J. N. (2021). Organizational Mindfulness: Imperative for Effectiveness in an Era of Change in the Tourism Industry. *European Journal of Hospitality and Tourism Research*, 9(4), 1–10.

Panditharathne, P. N. K. W., & Chen, Z. (2021). An integrative review on the research progress of mindfulness and its implications at the workplace. *Sustainability*, 13(24),13852. https://doi.org/10.3390/su132413852

Permana, T. E., Yuniarsih, T., Ahman, E., & Rofaida, R. (2024). Excellent Service Based on Human Capital and Leader Mindfulness (Survey of Sharia Hotel Employees in Bandung). *EKOMBIS REVIEW: Jurnal Ilmiah Ekonomi dan Bisnis*, 12(1), 591–606. https://doi.org/10.37676/ekombis.v12i1

Reb, J., Narayanan, J., & Chaturvedi, S. (2014). Leading mindfully: Two studies on the influence of supervisor trait mindfulness on employee well-being and performance. *Mindfulness*, 5(1), 36–45. DOI: 10.1007/s12671-012-0144-z.

Saviano, M., Caputo, F., & Del Prete, M. (2018, August). Mindful leadership for sustainability: a theoretical and conceptual path. In *21th "Excellence in Services" EISIC Conference Parigi EISIC-LeCnam-Paris, France* (pp. 635–648).

Shapiro, S. L., Schwartz, G. E., & Bonner, G. (1998). Effects of mindfulness-based stress reduction on medical and premedical students. *Journal of behavioral medicine*, 21(6), 581–599.

Sinclair, A. (2015). Possibilities, purpose and pitfalls: Insights from introducing mindfulness to leaders. *Journal of Spirituality, Leadership and Management*, 8(1), 3–11.

Valorinta, M. (2009). Information technology and mindfulness in organizations. *Industrial and Corporate Change*, 18(5), 963–997.

Yeşilkuş, F., & Özbozkurt, O. B. (2020). A contemporary management practice: Mindfulness in the work settings. *European Journal of Educational and Social Sciences*, 5(2), 145–158.

Engin Tengilimoğlu[1]

Chapter 29 The Role of Virtual Reality in Tourism: Pre-travel Decision Making and On-Visit Experience Enhancement

Introduction

The digital age, a result of technological developments, has been a source of transformation and change in industries as well as in all aspects of life (Bilgihan et al., 2024). Significant advancements have been made in the integration of technology into the tourism industry and virtual reality (VR) has emerged as a key area of interest, offering new perspectives on various aspects of tourism, including marketing, management, education, entertainment, accessibility, heritage preservation and sustainability (Guttentag, 2010; Lee and Kim, 2021; Calisto and Sarkar, 2024).VR has attracted considerable interest from both academics and practitioners (An et al., 2021; Nam et al., 2024).

The concept of VR dates back to the mid-1970s when the term was coined to describe a human-computer interface (Lee and Kim, 2021), marking the early stages of immersive technology development. A significant milestone in VR's development occurred in 1994 with the introduction of the Virtual Reality Modeling Language (VRML). This innovation enabled interactive 3D graphics on the internet, laying the groundwork for web-based VR simulations that became particularly impactful in fields such as education and gaming (Loureiro et al., 2020).

During the 1990s, VR technology saw further advancements with the emergence of VR equipment from companies like Sega and Nintendo. These devices were primarily aimed at the gaming industry, introducing the public to immersive experiences through consoles and games designed to utilize VR capabilities. The late 1990s also witnessed discussions on VR's potential in tourism marketing. The early 2000s marked the arrival of Second Life, a virtual world where users could create avatars and interact with other avatars and virtual environments. Second Life highlighted the potential for virtual travel experiences to facilitate

---

1 Ph.D., Selcuk University, B. A. A. Tourism Faculty, Department of Tourism Management, entengilimoglu@selcuk.edu.tr

travel planning, enhance user engagement, and generate interest in real-world destinations (Israel et al., 2019).

The rise of affordable VR viewers like Google Cardboard, combined with a wealth of tourism-related VR content, has made it accessible for anyone to virtually tour cities and attractions globally, thus unlocking immense potential for virtual visits to real destinations (Kang, 2020). The COVID-19 pandemic significantly disrupted the global tourism industry with travel restrictions and decreased traveler demand, highlighting the potential for VR tourism's growth. The VR tourism market, valued at USD 74.6 million in 2018, is expected to reach USD 304.4 million by 2023 with a CAGR of 32.5 % (Lee and Kim, 2021).

In recent years, the commercialization of VR devices like Oculus has spurred empirical studies focused on user acceptance (Huang et al., 2013) and the impact of VR experiences on visit intentions (Li and Chen, 2019) and recommendations (Wei et al., 2019; Genc et al., 2023). These studies have explored how immersive VR experiences can influence consumer behavior. The ongoing advancements in VR technology promise to further integrate virtual experiences into our daily lives, continuing to transform industries and the way we perceive and engage with the world around us (Nam et al., 2024).

Definitions and Technical Aspects

VR is a technology that creates a computer-generated, three-dimensional environment which users can explore and interact with (Guttentag, 2010). This immersive experience stimulates one or more of the user's senses in real-time, allowing them to feel as if they are present in the virtual world (Errichiello et al., 2019). It leverages interactive display technologies, including visual, motion, and audio cues, to enhance the realism and engagement of the experience (Fang et al., 2017).

There are two primary types of VR headsets: untethered and tethered. Untethered headsets, also known as mobile VR, use mobile devices as displays. While these headsets benefit from being cost-effective and widely accessible, they are limited by the processing power of mobile devices, which can restrict their ability to handle real-time 3D content. On the other hand, tethered VR headsets feature built-in displays and sensors for tracking user movements, and they connect to a personal computer (PC) for processing graphics.

Previous research (Kang, 2020; Lee and Kim, 2021) categorizes VR into two primary types: immersive VR and non-immersive VR. Immersive VR environments enclose users in virtual surroundings through head-mounted displays (HMDs), providing a comprehensive 3D experience that mimics real-world interactions more closely than non-immersive VR. Non-immersive VR, on the other hand, limits user engagement to a 2D screen, such as a computer monitor, which restricts the sensory and interactive elements to visual and auditory inputs. Immersive VR, characterized by its multimodal information channels, offers a higher sense of presence by allowing users to interact with and move within the virtual environment. This level of immersion is achieved through HMDs that deliver a 360-degree experience, enabling users to see, touch, and interact with VR objects in real-time.

The Potential Benefits of VR for Tourists and Management

One of the industry most profoundly influenced by the VR is tourism (Kang, 2020). VR offers a novel and captivating experience by integrating tangible elements from the real world with digital content (Li et al., 2023) and reshape how we experience the tourism attractions (Lee and Kim, 2021; Nam et al., 2024). Studies have shown that virtual reality can be used in three different stages of the tourist journey: before the trip, during the trip and after the trip (Loureiro et al., 2020; Nam et al., 2024; Calisto and Sarkar, 2024). Before the trip, it can be used to arouse interest in the destination or any tourist attraction (Li and Chen, 2019). During the trip, it can be used to enhance the experience (Genc et al., 2023). After the trip, it can be used to recall or relive the tourism experience (Loureiro et al., 2020; Nam et al., 2024). As shown in Table 29.1 and 29.2, related studies indicates that VR has distinct advantages for both managers and tourists in the tourism industry.

Table 29.1. Benefits of VR for Tourists

Enhanced Sensory Experiences	VR allows tourists to receive rich sensory information and understand the spatial and cultural aspects of a destination better (An et al., 2021).
Accessibility	VR makes tourism accessible to people with physical disabilities, financial constraints, or time limitations (Lin et al., 2020).
Personalized Experiences:	VR offers tourists an opportunity for enhanced and personalized experiences (Spielmann and Mantonakis, 2018).
Virtual Visits to Inaccessible Locations	VR enables visits to locations like heritage sites, jungles, or the North Pole, protecting the site's nature (Hales and Caton, 2017; Kang, 2020).
Reduced Physical Mobility	VR can prevent the need for physical movement, allowing consumers to experience destinations virtually without traveling (Loureiro et al., 2020).

(continued)

Table 29.1. Continued

Effective Nature-Related Experiences	VR can influence conservation behaviors and evoke similar reactions to real-life nature experiences, overcoming cost, time, travel restrictions, or physical ability limitations (Hofman et al., 2021).
Historical and Nonexistent Experiences	VR helps tourists experience historical settings, such as exploring ancient Rome through Google Earth's "Rome Reborn" project (Loureiro et al., 2020).

Virtual reality (VR) offers numerous benefits for tourists, enhancing their experiences and accessibility. VR provides rich sensory information, allowing tourists to explore destinations in three dimensions with kinetic and auditory controls, offering a deeper understanding of spatial and cultural aspects (An et al., 2021). It makes tourism accessible for individuals with disabilities, financial constraints, or time limitations, and creates opportunities for new tour services and markets, including visiting otherwise inaccessible locations (Lin et al., 2020; Kang, 2020). Moreover, VR prepares tourists for real-world encounters and allows unique experiences, such as exploring historical sites as they existed centuries ago (Loureiro et al., 2020).

Table 29.2. Benefits of VR for Management

Policy Making and Marketing	VR influences policymaking and emerges as a crucial tool for tourism marketing, supporting attractions' revenue (Lin et al., 2020).
Audience Reach and Environmental Impact:	VR helps reach a broader audience while reducing the carbon footprint and overcrowding, contributing to site preservation (Itani and Hollebeek, 2021; Bec et al., 2021).
Heritage Preservation	VR can restore historical destinations, protect cultural heritage, and deliver environmental messages (Talwar et al., 2022). It introduces "second chance tourism" by giving a second life to deteriorated sites (Bec et al., 2021).
Development of New Tourism Activities:	VR offers new opportunities for tourism activities, meeting demands like physical distancing during the COVID-19 pandemic (Zhang et al., 2022).
Management of Protected Areas	VR is useful for managing sensitive natural and cultural heritage sites, where visitor numbers need to be limited (Tussyadiah et al., 2018).
Sustainability and Mobility Management	VR aids in reducing the negative impacts of excessive tourist mobility by providing virtual experiences (Hannam et al., 2014).

VR has emerged as a potent marketing instrument within the tourism and hospitality industry, providing substantial advantages for managerial strategies. It acts as a generator of authenticity and a minimally manipulative marketing communication medium, thereby improving consumer perceptions and brand attitudes (Spielman and Orth, 2020). VR enhances both cognitive and affective responses among consumers, which in turn fosters more favorable brand attitudes and experiences (Kim et al., 2020). Furthermore, VR significantly boosts visit intentions and purchase decisions, with consumers demonstrating a greater willingness to pay more (He et al., 2018). Consequently, VR is an essential strategic tool for augmenting consumer engagement and loyalty (Calisto and Sarkar, 2024).

Factors Influencing the Adoption of VR for Tourists

As shown in Table 29.3, a review of related studies on the adoption of VR indicates that its adaptation has been influenced by a range of factors, including personal inclinations, situational and environmental conditions.

Table 29.3. Adoption of VR by Tourists

Personal Factors	Personal Innovativeness	Willingness to experiment with new services and products. Individuals with high personal innovativeness are more likely to adopt VR technologies (Bilynets et al., 2023).
	Technology Readiness	General mindset regarding the readiness to use technology. TR influences beliefs about technology's perceived usefulness and ease of use (Chung et al., 2015).
Psychological and Social Factors	Virtual Travel Needs	Four primary motives identified during the COVID-19 pandemic: need for pleasure, mindfulness, gathering, and growth (Calisto and Sarkar, 2024).
Technological Factors	Perceived Usefulness and Ease of Use	Core components of the Technology Acceptance Model (TAM). Positive perceptions of VR's usefulness and ease of use significantly impact the intention to use VR (Huang et al., 2013).
	Visual Attractiveness	Aesthetics of VR applications contribute to user satisfaction and perceived ease of use, encouraging adoption (Chung et al., 2015).

(continued)

Table 29.3. Continued

Situational and Environmental Factors	Facilitating Conditions	External environments that support the use of new technologies, such as availability of necessary devices, user knowledge about VR applications, and the presence of assistants to help with VR usage (Chung et al., 2015).
Contextual Factors (Pandemic-specific)	Social Distancing and Risk Mitigation	During the pandemic, social distancing increased demand for VR as a safer alternative to in-person tourism. However, those less concerned with threat protection are likely to revert to traditional tourism when possible (Itani and Hollebeek, 2021; Calisto and Sarkar, 2024).
	Travel Motivation and Sentiments	Pandemic-induced push-pull factors like travel convenience, cost, project design, and destination attractiveness influence virtual tourists' sentiments and adoption (Zhang et al., 2022).

Personal factors such as personal innovativeness (Bilynets et al., 2023) and technology readiness (Chung et al., 2015) play a crucial role in determining individuals' propensity to adopt VR technologies. Additionally, psychological, and social factors, including virtual travel needs (Calisto and Sarkar, 2024), contribute to the appeal of VR experiences by fulfilling tourists' desires for pleasure, mindfulness, social interaction, and personal growth. Moreover, technological factors like perceived usefulness and ease of use (Huang et al., 2013), as well as visual attractiveness (Chung et al., 2015), enhance the adoption of VR. Situational and environmental factors, such as facilitating conditions (Chung et al., 2015), also influence VR adoption. Furthermore, pandemic-specific contextual factors like social distancing and risk mitigation (Itani and Hollebeek, 2021) have accelerated the demand for VR as a safer alternative to traditional tourism. Theoretical models like the Push-Pull Theory offer insights into the complex interplay of factors shaping tourists' adoption of VR, emphasizing the importance of factors such as travel motivation, convenience, and destination attractiveness in mediating the impact of VR experiences (Zhang et al., 2022).

VR Applications for Tourism

VR represents a novel avenue for tourists and tourism management alike, simultaneously transforming the tourism industry through the development of innovative applications. As shown in Table 29.4, VR applications have been widely used in the tourism industry for museums, heritage sites (Lee et al., 2020; Nam et al., 2024), theme parks (Wei, 2019), cruises (Yung et al., 2021), destination marketing (Tussyadiah et al., 2018; An et al., 2021), and hotel presentation (Israel et al., 2019).

Table 29.4. VR Applications for Tourism

Virtual Tours and Experiences	Iconic Landmarks	Google Earth VR enables users to virtually explore landmarks like the Eiffel Tower and the Grand Canyon.
	Museum Visits	The British Museum and the Louvre offer VR tours of their collections, enhancing accessibility to art and culture.
Customized Travel Experiences	Virtual Travel Agencies	Platforms like Ascape and YouVisit offer personalized VR travel experiences, aiding in trip planning.
Learning and Skill Development	Language Immersion	Mondly VR facilitates language learning through immersive scenarios, such as ordering food in a restaurant.
	Training Programs	TRANSFRVR and Uptale provide VR-based training for tourism professionals, improving skills and knowledge retention.
Adventure and Entertainment	Extreme Sports Simulations	The North Face VR app offers simulations of BASE jumping and skiing, providing thrilling experiences
	Theme Parks	The VOID combines VR with physical environments to create immersive experiences based on popular franchises
Accessibility and Inclusivity	Accessible Travel	Apps like WalkinVR and Rendever offer VR experiences for people with disabilities, ensuring inclusive travel opportunities.
	Virtual Hotel Experiences	Marriott Hotels' VRoom Service allows guests to order VR experiences to their rooms, enhancing their stay.
Education and Cultural Preservation	Museum Experiences	Museums worldwide, including the Louvre and the British Museum, adopt VR to offer remote access to their collections.
	Cultural and Historical Reenactments	Apps like TimeLooper enable users to witness historical events like the construction of the Great Wall of China.

(continued)

Table 29.4. Continued

Environmental Exploration	Wildlife Safaris	National Geographic's VR app offers virtual safaris, allowing users to observe wildlife in their natural habitats.
	Natural Sites Exploration	VR allows users to explore natural wonders like Hawaii Volcanoes National Park and city destinations like Central Park virtually.
Additional Applications	Event Access	Platforms like MelodyVR provide virtual access to events and festivals, offering immersive experiences from anywhere.
	Travel Planning	VisitScotland's VR platform enables users to preview destinations and make informed travel decisions.

VR enables virtual tours of famous landmarks, museums, and cultural sites, making travel accessible to everyone. Platforms like Google Earth VR and museum collaborations with Google Arts & Culture provide detailed and interactive experiences. VR travel agencies, language immersion programs, and adventure sports simulations offer personalized and thrilling experiences. Accessibility is enhanced through apps like WalkinVR, ensuring inclusivity for people with disabilities. Hotels and cruise lines use VR to enhance guest experiences, while VR theme parks and virtual events bring entertainment to new levels. This technology also aids in travel planning and education, offering detailed previews and training for industry professionals.

The Effect of VR on Pre-visit and On Site

Tourism is an experiential product that can only be experienced by actually visiting the destination (Lee et al., 2020; Kang, 2020; An et al., 2021). Consumers often perceive higher risks when purchasing tourism products compared to physical goods due to the lack of opportunities to experience these products in advance (Israel et al., 2019; Kang, 2020). However, VR technology offers a solution by "bringing the place to you here and now," enabling potential tourists to experience destinations virtually prior to visit (Tussyadiah et al., 2018). This virtual experience can significantly reduce the perceived risks by providing rich, immersive previews that help consumers make more informed decisions (Kang, 2020). Thus, VR has gained significant attention as an innovative marketing tool that provides tourists with trial experiences, aiding them in reducing uncertainty (An et al., 2021).

Studies suggest that virtual reality's ability to visualize spatial depth is its greatest strength when applied to tourism contexts (Guttentag, 2010; Yung et al., 2021). In contrast to other digital product displays, where users observe from an external viewpoint on a computer or smartphone, this format makes the user the main participant, experiencing the presentation directly from a first-person perspective (Israel et al., 2019). The ability of VR to present detailed and realistic images of destinations enhances its value as a marketing tool, making it possible to effectively communicate the intangible aspects of tourism products (Yung et al., 2021). As a result, VR, potentially reaching one billion users (Israel et al., 2019), can revolutionize tourism marketing by reducing uncertainty and attracting more visitors (An et al., 2021).

VR is a technology-driven consumer learning process that immerses the user in a simulated environment to gain product-related knowledge and experience. By providing immersive information about a destination prior to the actual visit, it can overcome the distance barrier (Lee and Kim, 2021). In addition, VR induces physical visits by providing a pre-visit experience of a tourist attraction (Errichiello et al., 2019; Lee and Kim, 2021). Research by Tourism Australia indicates that almost 20 % of travelers have used VR to select a travel destination. Additionally, about 25 % of travelers expressed plans to utilize VR for future travel decisions (Tourism Australia, 2024). A study in Germany shows that nearly half of the travelers would consider using VR as a tool for choosing their travel destinations (Immersionvr, 2024). These findings highlight VR's effectiveness as a marketing tool in the tourism industry (An et al., 2021).

VR has the potential to significantly enhance the tourist experience by providing immersive, engaging, and interactive elements that traditional methods cannot offer (Errichiello et al., 2019). During the on-site visit phase of cultural tourism attractions, VR applications can elicit a more positive and immersive emotional state, enriching the overall visitor experience (Chung et al., 2015). This technological innovation allows for a more complete and diversified experience by integrating dynamic information delivery with engaging visual and auditory stimuli (Guttentag, 2010). VR transform how tourists interact with and perceive attractions, offering added-value propositions that create memorable and emotionally resonant experiences (Han et al., 2020). VR enhances tourism by providing immersive and social experiences, which increases enjoyment and satisfaction, encouraging visitors to use VR technology for better travel experiences (Chung et al., 2015).

VR enhances the tourist experience by increasing absorption, presence, and immersion, primarily through its advanced device characteristics (Tussyadiah et al., 2018; Calisto and Sarkar, 2024). These characteristics include embodiment, which allows users to feel a part of the virtual environment, and interactivity, enabling tourists to engage actively with the content (Tussyadiah et al., 2018). The design and quality of VR content also play crucial roles, with well-crafted, informative, and visually appealing content enhancing the overall experience (Chung et al., 2017). Sensory stimulation through VR devices, such as head-mounted displays, further immerses users by engaging multiple senses, creating a more vivid and engaging experience. By leveraging these device characteristics, VR transforms the way tourists experience destinations, making visits more engaging and emotionally resonant.

References

An, S., Choi, Y., & Lee, C. K. (2021), "Virtual travel experience and destination marketing: Effects of sense and information quality on flow and visit intention", *Journal of Destination Marketing & Management*, 19, 100492.

Bec, A., Moyle, B. D., Schaffer, V., & Timms, K. (2021). Virtual reality and mixed reality for second chance tourism. *Tourism Management*, 83, 104256. https://doi.org/10.1016/j.tourman.2020.104256

Bilgihan, A., Leong, A. M. W., Okumus, F., & Bai, J. (2024). Proposing a metaverse engagement model for brand development. *Journal of Retailing and Consumer Services*, 78, 103781. https://doi.org/10.1016/j.jretconser.2024.103781

Bilynets, I., Trkman, P., & Knežević Cvelbar, L. (2023). Virtual tourism experiences: adoption factors, participation and readiness to pay. *Current Issues in Tourism*, 1–18. https://doi.org/10.1080/13683500.2023.2268809

Calisto, M. d. L. & Sarkar, S. (2024). A systematic review of virtual reality in tourism and hospitality: the known and the paths to follow. *International Journal of Hospitality Management*, 116, 103623. https://doi.org/10.1016/j.ijhm.2023.103623

Chung, N., Han, H., & Joun, Y. (2015). Tourists' intention to visit a destination: the role of augmented reality (ar) application for a heritage site. *Computers in Human Behavior*, 50, 588–599. https://doi.org/10.1016/j.chb.2015.02.068

Chung, N., Lee, H., Kim, J., & Koo, C. (2017). The role of augmented reality for experience-influenced environments: the case of cultural heritage tourism in korea. *Journal of Travel Research*, 57(5), 627–643. https://doi.org/10.1177/0047287517708255

Errichiello, L., Micera, R., Atzeni, M., & Del Chiappa, G. (2019), "Exploring the implications of wearable virtual reality technology for museum visitors' experience: A cluster analysis", *International Journal of Tourism Research*, Vol. 21 No. 5, pp. 590–605.

Fang, W., Zheng, L., Deng, H., & Zhang, H. (2017), "Real-time motion tracking for mobile augmented/virtual reality using adaptive visual-inertial fusion". *Sensors*, Vol. 17 No 5, 1037.

Genc, V., Bilgihan, A., Gulertekin Genc, S., & Okumus, F. (2023), "Seeing history come to life with augmented reality: the museum experience of generation Z in Göbeklitepe", *Journal of Tourism and Cultural Change*, 21, 6.

Guttentag, D. A. (2010), "Virtual reality: Applications and implications for tourism", *Tourism management*, 31(5), 637–651.

Hales, R. & Caton, K. (2017). Proximity ethics, climate change and the flyer's dilemma: ethical negotiations of the hypermobile traveller. *Tourist Studies*, 17(1), 94–113. https://doi.org/10.1177/1468797616685650

Han, S. L., An, M., Han, J. J., and Lee, J. (2020), "Telepresence, time distortion, and consumer traits of virtual reality shopping," *Journal of Business Research*, 118, 311–320.

Hannam, K., Butler, G., & Paris, C. M. (2014). Developments and key issues in tourism mobilities. *Annals of Tourism Research*, 44, 171–185. https://doi.org/10.1016/j.annals.2013.09.010

He, Z., Wu, L., & Li, X. (2018). When art meets tech: the role of augmented reality in enhancing museum experiences and purchase intentions. *Tourism Management*, 68, 127–139. https://doi.org/10.1016/j.tourman.2018.03.003

Hofman, K., Walters, G., & Hughes, K. (2021). The effectiveness of virtual vs real-life marine tourism experiences in encouraging conservation behaviour. *Journal of Sustainable Tourism*, 30(4), 742–766. https://doi.org/10.1080/09669582.2021.1884690

Huang, Y., Backman, S. J., Backman, K. F., & Moore, D. (2013). Exploring user acceptance of 3d virtual worlds in travel and tourism marketing. *Tourism Management*, 36, 490–501. https://doi.org/10.1016/j.tourman.2012.09.009

Immersionvr, (2024), "VR in Tourism", available at: https://immersionvr.co.uk/about-360vr/vr-for-tourism (accessed 27 March 2024)

Israel, K., Tscheulin, D., & Zerres, C. (2019), Virtual reality in the hotel industry: assessing the acceptance of immersive hotel presentation. *European Journal of Tourism Research*, 21, 5–22.

Itani, O. S. & Hollebeek, L. D. (2021). Light at the end of the tunnel: visitors' virtual reality (versus in-person) attraction site tour-related behavioral intentions during and post-covid-19. *Tourism Management*, 84, 104290. https://doi.org/10.1016/j.tourman.2021.104290

Kang, H. (2020). Impact of vr on impulsive desire for a destination. *Journal of Hospitality and Tourism Management*, 42, 244–255. https://doi.org/10.1016/j.jhtm.2020.02.003.

Kang, C. Y., Kim, H., & Kang, H. M. (2020), A Study on the User Experience according to the Method and Detail of Recommendation Agent's Explanation Facilities. *The Journal of the Korea Contents Association*, 20(8), 653–665.

Kim, M. J., Lee, C. K., & Jung, T. (2020), "Exploring consumer behavior in virtual reality tourism using an extended stimulus-organism-response model", *Journal of travel research*, (59)1, 69–89.

Lee, S., Jeong, E., & Qu, K. (2020). Exploring theme park visitors' experience on satisfaction and revisit intention: A utilization of experience economy model. *Journal of Quality Assurance in Hospitality & Tourism*, (21)4, 474–497.

Lee, W. J., & Kim, Y. H. (2021), Does VR tourism enhance users' experience?. *Sustainability*, (13)806, 1–15.

Li, J., Wider, W., Ochiai, Y., & Fauzi, M. A. (2023), A bibliometric analysis of immersive technology in museum exhibitions: exploring user experience. *Frontiers in Virtual Reality*, 4, 1240562.

Li, T., & Chen, Y. (2019), Will virtual reality be a double-edged sword? Exploring the moderation effects of the expected enjoyment of a destination on travel intention. *Journal of Destination Marketing & Management*, 12, 15–26.

Lin, L., Huang, S. L., & Ho, Y. (2020). Could virtual reality effectively market slow travel in a heritage destination?. *Tourism Management*, 78, 104027. https://doi.org/10.1016/j.tourman.2019.104027

Loureiro, S. M. C., Guerreiro, J., & Ali, F. (2020), "20 years of research on virtual reality and augmented reality in tourism context: A text-mining approach", *Tourism management*, 77, 104028.

Nam, K., Baker, J., & Dutt, C. S. (2024), "Does familiarity with the attraction matter? Antecedents of satisfaction with virtual reality for heritage tourism", *Information Technology & Tourism*, (26)1, 25–57.

Spielmann, N. & Mantonakis, A. (2018). In virtuo: how user-driven interactivity in virtual tours leads to attitude change. *Journal of Business Research*, 88, 255–264. https://doi.org/10.1016/j.jbusres.2018.03.037

Spielmann, N. & Orth, U. R. (2020). Can advertisers overcome consumer qualms with virtual reality?. *Journal of Advertising Research*, 61(2), 147–163. https://doi.org/10.2501/jar-2020-015

Talwar, S., Kaur, P., Escobar, O., & Lan, S. (2022). Virtual reality tourism to satisfy wanderlust without wandering: an unconventional innovation to promote sustainability. *Journal of Business Research*, 152, 128–143. https://doi.org/10.1016/j.jbusres.2022.07.032

Tourism Australia, (2024), "New Research Confirms the Potential of Virtual Reality for Destination Marketing", available at: http://www.tourism.australia.com/content/dam/assets/document/1/6/y/7/t/2003897.pdf (accessed 27 March 2024)

Tussyadiah, I. P., Wang, D., Jung, T. H., and Tom Dieck, M. C. (2018), "Virtual reality, presence, and attitude change: Empirical evidence from tourism", *Tourism management*, Vol. 66, 140–154.

Wei, W., Qi, R., & Zhang, L. (2019). Effects of virtual reality on theme park visitors' experience and behaviors: a presence perspective. *Tourism Management*, 71, 282–293. https://doi.org/10.1016/j.tourman.2018.10.024

Yung, R., Khoo-Lattimore, C., and Potter, L. E. (2021), "VR the world: Experimenting with emotion and presence for tourism marketing". *Journal of Hospitality and Tourism Management*, 46, 160–171.

Zhang, S., Li, Y., Ruan, W., & Liu, C. (2022). Would you enjoy virtual travel? the characteristics and causes of virtual tourists' sentiment under the influence of the covid-19 pandemic. *Tourism Management*, 88, 104429. https://doi.org/10.1016/j.tourman.2021.104429

Gulsun Yildirim[1] and Sena Bakir[2]

Chapter 30 Developing Gastronomy Tourism in Turkey through Food Geographical Indication

Introduction

The interest towards food products with geographical indication (GI) is increasing day by day. The preservation of the area's traditional economy and cultural identity, passed down from generation to generation under protection, is possible by GI. Determining the geographical provenance of food is a contemporary and worldwide concern that has led to a growing interest among producers and consumers in the authenticity and healthfulness of food. In light of this, several instruments (such as quality labels and marks) are created to protect consumers and producers from the dangers associated with food fraud.

Food items must have various of chemical, physical, and biological properties examined to verify their geographical origin (Dimitrakopoulou & Vantarakis, 2023). Public food fraud scandals have exposed the vulnerabilities that haunt our food systems, and increased globalization and the complexity of global supply chains have given rise to new issues (Lawrence, Elliott, Huisman, Dean, & Ruth, 2022). Various local, national, and international food chain actors and organized crime groups commit conventional food fraud. It can involve adulteration, replacement, dilution, modification, simulation, counterfeiting, and misrepresentation, among other things (Soon & Wahab, 2022).

The European Union (EU) has seen an increase in the trade of high-value agricultural products over the past few decades, and this expansion has led to a growing need for food protection through "GI" labeling. Protected geographical indications (PGI) and protected designation of origin (PDO) are the two product categories that must be safeguarded under EU legislation. To be classified as a PDO, a regional product has to fulfill two requirements: (1) It must

---

1   Assoc. Prof. Dr., Recep Tayyip Erdogan University, Tourism Faculty, Gastronomy and Culinary Arts, gulsun.yildirim@erdogan.edu.tr
2   Asist. Prof. Dr., Recep Tayyip Erdogan University, Tourism Faculty, Gastronomy and Culinary Arts, and Recep Tayyip Erdogan University Blueberry Application and Research Center, sena.bakir@erdogan.edu.tr

have been manufactured entirely within a specified geographic area, and (2) the product's quality or features should primarily stem from the unique geographical environment of the location of origin, taking into account both natural and human-made factors like climate, soil quality, or specific know-how. However, PGI denotes food items and agricultural goods intimately associated with a specific geographic region. Production, processing, or preparations all occur in the region, at least in part. These goods have a distinct quality, reputation, or other attributes related to their place of origin (EU, 2012).

Fundamental to such laws is the notion that goods from specific areas are *sui generis*, meaning that there is proof of a direct connection between the product's place of origin and its ultimate quality (Herrmann & Teuber, 2010).

The Commission of the European Communities (2008) states that PDOs and PGIs aid in revitalizing rural areas; the production of agri-foodstuffs may help to maintain the diversity of native plants, promote social cohesion, and benefit the local populace (CEC, 2008).

Geographical İndication at the Food and Beverages İndustry

Classifying food or agricultural products under one of two designations – PGI or PDO – provides protection against the use of registered trade names that could allow for the exploitation of reputation, as well as the abusive use, imitation, or evocation of origin of any false or misleading information regarding the provenance, origin, nature, or essential qualities of the product. Furthermore, it will refrain from any further actions intended to deceive the customer about the natural source of the product (Todea, Oroian, Holonec, Arion, & Mocanu, 2009).

Environmental variables that can significantly alter the chemical feature composition of a plant include soil fertility and components, elevation, temperature, and precipitation. A traditional product needs to be connected to place and tradition, and consumers' desire for traditional products could originate from a certain resistance to the growing industrialization and globalization of food production (Jordana, 2000).

The EU Commission had registered 5567 items as PDOs, PGIs, or GIs as of May 2024 (1515 PGI, 1992 PDOs, and 2060 GI). They are separated into primary product categories: (1)fresh meat; (2) meat products; (3) cheeses; (4) other animal-derived products; (5) oils and fats; (6) fruits, vegetables, and cereals; (7) fresh fish; (8) beers; (9) chocolate and its derivatives; (10) bread, pastry, and confectionary items; (11) drinks derived from plant extracts; (12) essential oils; (13) pasta; and (14) salt (GIview, 2024).

The distribution of GI registration was not uniform among product categories or nations. For example, 886 products from Italy were registered, including wines, vinegar (especially balsamic vinegar), fresh fish, molluscs, and crustaceans, as well as products made from them; other products included spirit drinks, fresh or processed fruit, vegetables, and cereals, as well as oils and fats (butter, margarine, oil, etc.). On the other hand, 766 products were registered from France, 378 from Spain, and 37 from Belgium (GIview, 2024).

Turkey started the GI system later than it needed to and has accelerated its development. The December 2016 publication of the Law represents a significant milestone for Turkey's GI system (Nizam, 2019). Turkey's suitable climate and environmental conditions allow it to produce a wide range of items that can each have their own reputation, construct systems around them, and utilize them for local development.

To this date, Turkiye has applied for 72 products as PDO and PGI, and 21 of them are registered. Registered products belong to Turkiye are Antakya Künefesi, Antep Baklavası/Gaziantep Baklavası and Maraş Tarhanası for PGI and Araban Sarımsağı, Ayaş Domatesi, Aydın Kestanesi, Aydın Memecik Zeytinyağı, Aydın İnciri, Bayramiç Beyazı, Edremit Körfezi Yeşil Çizik Zeytini, Edremit Zeytinyağı, Ezine Peyniri, Gemlik Zeytini, Giresun Tombul Fındığı, Malatya Kayısısı, Milas Zeytini, Milas Zeytinyağı, Safranbolu Safranı, Suruç Narı, Taşköprü Sarımsağı, Çağlayancerit Cevizi for PDO. Moreover, Bingöl Balı, Bursa Siyah İnciri/Bursa Siyahı, Bursa Şeftalisi, Hüyük Çileği, Osmaniye Yer Fıstığı were published as PDO, and Kayseri pastırması and Söke Pamuğuj were published as PGI (GIview, 2024).

Gastronomy Tourism and Geographical İndication in Turkiye

The wide range of agricultural food items with distinct qualities and their customs and culture enable the emergence of diverse culinary and meal cultures, both of which are the main pillars of tourist products. Gastronomy tourism includes domestic and international trips to places where local food and beverages are available. Tourists travel not only to see the historical places but also to taste local foods to expand their travel experience. Tasting traditional foods at the destination could enhance the tourists' perceived destination image and intention to revisit. GI is important in registering, promoting and turning local foods into tourism products (Canbolat & Cakiroglu, 2020).

Products with geographic indications support product sustainability in remote locations and places that prioritize regional values. It protects the cultural heritage, values, and practices of the area. It actively supports crop production, which enhances the reputation and promotion of the region to which it belongs. When considered from the tourism perspective, it contributes significantly to the industry by drawing visitors to the location of the cultural products produced. It boosts the region's tourism value, thereby helping in regional development. In this regard, it is possible to argue that geographic indications and tourism are closely related (Akgöz, Varol, & Oksuz, 2023).

Gastronomy tours include visits to local restaurants, gastronomy museums, festivals and villages where food and beverages such as fish, wine, coffee, cheese, fruit and tea are served. Therefore, food and beverages are considered the most critical elements in creating gastronomy routes. Special gastronomy tours are organized for those who want to experience various foods and beverages in different countries. Netherlands- Wine and Cheese tour, Chinese- Tea tour, Spain- Wine and Tapas tour, France-Cheese tour, Thailand- Street Food tour, Switzerland-Chocolate tour, Italy-Wine tour, USA- Bourbon Whiskey tour are just a few of these gastronomy tours.

Many people travel to the Champagne area of France because of the Champagne sparkling wine produced there, which has a geographical designation. Similar to France, Italy and Spain use a variety of tourism-related activities to promote their registered products, such as wine and cheese, and make financial contributions at the local and national levels (Kan, Gulcubuk, & Kucukcongar, 2012).

In Turkey, gastronomy tours are organized to various provinces for different foods and beverages. Alaçatı Grass Festival Tour (İzmir), Thrace Vineyard Route Tours (Edirne Kırklareli Tekirdağ), Antakya & Antep & Adana Local Cuisine Gourmet Tour (Antakya Gaziantep Adana), Mushroom Hunting and Walks Tour (İstanbul-Kastamonu), Tea Picking Tour (Rize), Olive Harvest Tour (Balıkesir), Morel Mushroom Festival (Muğla), Hazelnut Concentration Camp (Trabzon) are a few of them (Akyürek, 2020).

There are many geographically indicated products in different regions of Turkey. The province with the most geographically indicated products is Gaziantep, followed by Konya. Various local festivals and events are organized for geographically indicated products in Turkey. White Cherry Festival is organized annually for Eregli Starks Gold, Geographical Indicated Products of Konya's Ereğli district. Although there are many geographically indicated products in Konya, various initiatives are needed to use these products. Travel agencies should include more geographically indicated products in their tour programs, restaurants on their menus and public institutions on their websites as well (Akgöz et al., 2023).

In terms of gastronomic geographically marked products, Samsun province in the Black Sea region has the same number of gastronomic geographically marked products as Afyonkarahisar province, which is included in the UNESCO gastronomy cities list, and more than Hatay province, but it is insufficient in the recognition and use of these products as touristic products (Canbolat & Cakiroglu, 2020). Nonetheless, products with geographical markings support the local economy in the area in which they are produced. The province of Gümüşhane is among the best evidence of this. The province of Gümüşhane's economy has significantly benefited from the geographically indicated 'Gümüşhane Kömesi' (Esen, 2016).

Additionally, Doğanlı (2020) concluded that the geographical designation may be viewed as the cornerstone of brand development and that rural tourism activities supplemented by items with a geographical indication will contribute to both tourism and economic growth.

Saatçi (2019) concluded in his study that although destinations have geographically indicated products, they are not sufficiently included in the promotion of the destination, and those producers and consumers do not have enough information about protecting the products.

The study examining the importance of the geographically marked products registered in Çanakkale province in regional tourism development, it was revealed that the producers earned good economic gains through the geographically marked products in the province, and the rural economy was revitalized. In this study, it was argued that the promotional activities of the products were insufficient, as in the case of Samsun province. Tourists go to places where geographically marked products are produced and are curious about the lifestyles and cultures of those in that region. Therefore, geographically marked products contribute to revitalizing tourism in that region. It was stated that the province of Çanakkale has not yet reached this stage and that this could be possible with the necessary promotion and advertising efforts (Mercan & Üzülmez, 2014).

According to Ertas-Sabanci and Girgin (2023), geographically indicated items are a significant component of gastronomic tourism. Still, they must be promoted effectively to be utilized as a destination attraction feature. Thus, promoting local cuisine and growing tourism in the nation and region is adventageous once regionally designated products are included in promotional materials. The foods and drinks listed on the Gaziantep Culture and Tourism Directorate, Gaziantep Governorship, Gaziantep Municipality, and social media accounts for the Gaziantep Municipality, Gaziantep Governorship, and Gaziantep Culture and Tourism Directorate that are labeled with geographic information were the subject of this study. The study's findings led to the understanding that, regionally indicated products were featured more in source photos than in source texts on Gaziantep's company websites and social media accounts. According to the study, there are three recommendations for using culinary elements—particularly locally indicated products—in destination marketing: (1) using more geographically indicated products in promotion, (2) updating social media accounts, and (3) supporting source photographs with source texts.

Conclusion and Future Aspects

One of the sources of motivation that pushes people to travel is gastronomy. Tourists who come for gastronomy tourism spend 50 % more than ordinary tourists. Türkiye has a very rich culinary culture (Canbolat & Cakiroglu, 2020). At the same time, the number of products with gastronomic geographical indications, one of the most essential resources for gastronomy tourism in Turkiye, is relatively high. However, various efforts need to be made when using geographically indicated products within the scope of gastronomy tourism. First and foremost, the local population in each province needs to become aware of the gastronomic regional products. For this purpose, the number of organizations, such as festivals specific to geographically indicated products, should be increased. In addition, geographically indicated products should be included on the websites of public institutions and organizations.

A label for gastronomic geographically indicated products must be placed in the facilities and product packages where they are marketed. Good promotion and marketing of products that have received geographical indications significantly contribute to the country's economy.

References

Akgöz, E., Varol, F., & Oksuz, M. (2023). The Role of Geographically Indicated Products in the Determination of Gastronomy Routes in Konya. *Yaşar Üniversitesi E-Dergisi, 18*(72), 547–568.

Akyürek, S. (2020). *Gastro Turistlerin Deneyim Bileşenleri: Gastronomi Turları Kapsamında Bir Araştırma.* (PhD), Muğla Sıtkı Koçman Üniversitesi Muğla.

Canbolat, E., & Cakiroglu, F. P. (2020). Gastronomy Tourism and Geographical Indicaition: An Evaluation of Samsun Cuisine. *Journal of Tourism & Gastronomy Studies, 8*(2), 937–957.

Dimitrakopoulou, M. E., & Vantarakis, A. (2023). Does Traceability Lead to Food Authentication? A Systematic Review from A European Perspective. *Food Reviews International, 39*(1), 537–559. doi:10.1080/87559129.2021.1923028

Doğanlı, B. (2020). Coğrafi İşaret, Markalaşma Ve Kırsal Turizm İlişkileri. *İnsan ve Sosyal Bilimler Dergisi, 3*(2), 525–541.

Ertas-Sabanci, A., & Girgin, G. K. (2023). Destinasyon Pazarlamasında Coğrafi İşaretli Ürünler; Gaziantep Örneği. *Boyabat İktisadi ve İdari Bilimler Fakültesi E-Dergisi, 3*(2), 237–264.

Esen, Ş. (2016). Bir Farklılaşma Stratejisi Olarak Coğrafi İşaretler Ve Türkiye İncelemesi. *Bartın Üniversitesi İktisadi ve İdari Bilimler Fakültesi Dergisi, 7*(14), 447–464.

EU (2012). European Parliament and of the Council of 21 November 2012 on quality schemes for agricultural products and foodstuffs.

GIview. (2024). Geographical indications for foods and drinks.

Commission of the European Communities (CEC) (2008). Green Paper on Agricultural Product Quality: Product Standards, Farming Requirements and Quality Schemes. Brussels, Belgium.

Herrmann, R., & Teuber, R. (2010). Geographical Differentiated Products. In J. Lusk, J. Rosen, & J. Shogren (Eds.), *Oxford Handbook on the Economics of Food Consumption and Policy*. Oxford, UK: Oxford University Press.

Jordana, J. (2000). Traditional foods: challenges facing the European food industry. *Food Research International, 33*(3), 147–152.

Kan, M., Gulcubuk, B., & Kucukcongar, M. (2012). Coğrafi İşaretlerin Kırsal Turizmde Kullanılma Olanakları. *KMÜ Sosyal ve Ekonomik Araştırmalar Dergisi, 14*(22), 93–101.

Lawrence, S., Elliott, C., Huisman, W., Dean, M., & Ruth, S. V. (2022). The 11 sins of seafood: Assessing a decade of food fraud reports in the global supply chain. *Comprehensive Reviews in Food Science and Food Safety, 21*(4), 3746–3769. doi:10.1111/1541-4337.12998

Mercan, Ş. O., & Üzülmez, M. (2014). Coğrafi işaretlerin bölgesel turizm gelişimindeki önemi: Çanakkale ili örneği. *Dokuz Eylül Üniversitesi İktisadi İdari Bilimler Fakültesi Dergisi, 29*(2), 67–94.

Nizam, D. (2019). How to use geographical indication for the democratization of agricultural production: A comparative analysis of geographical indication rent-seeking strategies in Turkey *Geographical indication and global agri-food* (pp. 87–99): Routledge.

Saatçi, G. (2019). Coğrafi İşaretli Yiyeceklerin Tanıtım Unsuru Olarak Yöresel Yemekler Kapsamında. *Journal of Tourism and Gastronomy Studies, 7*(1), 358–374.

Soon, J. M., & Wahab, I. R. A. (2022). A Bayesian Approach to Predict Food Fraud Type and Point of Adulteration. *Foods, 11*(3). doi:10.3390/foods11030328

Todea, A., Oroian, I., Holonec, L., Arion, F., & Mocanu, C. (2009). Legal protection for geographical indications and designations of origin for agricultural products and foodstuffs. *University of Agricultural Science And Veterinary Medicine, 66*(2), 463–466.

Sabahat Deniz[1], Günay Ahmadli[2] and Hakan Koç[3]

Chapter 31 Green Management Practices in Tourism Businesses

Introduction

The tourism industry has a significant role in the global economy. However, the negative impact of tourism activities on the environment is a critical issue that needs to be taken into consideration. In this context, the adoption of green management practices by tourism businesses is an essential requirement in terms of environmental, economic and social impacts. This study examines the scope of green management practices and the benefits of these management applications for tourism organisations in detail. These benefits are discussed within the framework of economic, environmental and social dimensions of sustainable tourism. In addition, waste management and recycling, environmentally sensitive purchasing, energy and water use, use of technology are listed as the basic green management strategies that tourism enterprises can implement. This study investigates the challenges faced by tourism companies during green management practices and aims to contribute to the literature and the tourism industry with sample green management studies.

The Concept of Green Management

In the modern world, environmental pollution is one of the side effects of economic development, and so far both the economic growth and development in all countries around the world are based mostly on the utilisation of natural resources (Rutkowska & Sulich, 2021: 1). Hence, the concept of Sustainable Development (SD) was not introduced until the 1980s. SD has been defined as 'a management model in which meeting the needs of the current generation

---

1 Hacı Bayram Veli University, Ph.D. Student, Tourism Management Programme, 0000-0002-6215-5433, deniz.sabahat@hbv.edu.tr
2 Hacı Bayram Veli University, Ph.D. Student, Tourism Management Programme, 0000-0002-6035-8552, ehmedli.gunay@hbv.edu.tr
3 Hacı Bayram Veli University, Prof. Dr., Tourism Management Programme, 0000-0003-2850-0472, hakan.koc@hvb.edu.tr

does not reduce the opportunity to satisfy the needs of future generations' (Brundlandt, 1987).

There are various descriptions of green management in the literature. Green management is identified as 'the process and practice that an organisations presents in reducing, eliminating or ideally preventing the negative environmental impacts caused by its activities (Lee, 2010). According to another study, green management is defined as 'practices that produce environmentally friendly products and which reduce the impact on the environment through green production, green research and development and green marketing' (Lee et al., 2010). Lee and Ball (2003) consider green management as the management of the interaction of the business with the natural environment and its effect on the natural environment. According to Hart (2005), green management is seen as going above regulatory requirements and adopting innovative methods such as pollution prevention, waste prevention, product management and corporate social responsibility. According to Dwyer (2009), green management is an understanding that provides the opportunity to reduce environmental pollution and to design and commercialise products that reduce the risks to human health and the environment minimise the risks to human health and the environment. Haden et al. (2009) describe green management as an organisation-wide process of implementing innovation to achieve sustainability, waste reduction and social responsibility.

After all these identifications, in general terms, the practice of including environmental factors in an business's operations and strategies is called 'green management' and has also been defined as 'environmental management' or 'sustainable management' (Molina-Azorín et al., 2009). The common characteristic of these concepts is that they contain the mission of business activities to minimise the negative effects on the environment and to promote a sustainability culture (Alfred & Adam, 2009). Green management practices can include a wide range of activities such as reduction of waste and emissions, energy and water conservation, sustainable product sourcing and design of recyclable or biodegradable products (Bernauer et al., 2006). Indeed, green management is the whole system management (Taylor, 1992: 670). Green management is the process of applying organisation-wide innovation by adopting environmental goals and strategies that are fully integrated with the organisation's goals and strategies through continuous learning and development to achieve sustainability, waste and pollution reduction, social responsibility and competitive advantage (Loknanth & Azeem, 2017). Definitions can be generally considered as practices that aim to reduce the effects of the business on the environmental impact to a minimised level. These practices are aimed at ensuring that the environmental impact of businesses, government agencies and society in general is minimised (Loknanth & Azeem, 2017: 691). Therefore, one of the main objectives of green management is to maintain social welfare while balancing economic development with sustainable environmental order (Gavrilović & Maksimović, 2018).

Green Management Strategies in Tourism Businesses

Considering the damage to nature and the environment in the tourism sector, businesses should take various steps to diminish their environmental effects. In this context, 'green management' requires the integration of corporate behaviour and environmental awareness, minimising the environmental effect of corporate activities (Cramer, 1998), and incorporating environmental considerations into every business process, including product development activities and strategic planning (Wu & Wu, 2014). The main impulse of green management is to promote environmentally friendly practices, encouraging commitment to a variety of ecologically sound practices such as water and energy conservation as well as solid waste reduction (Manaktola & Jauhari 2007: 365).

Green management strategies applied in tourism enterprises are management strategies that aim environmental sustainability and protection of natural resources. Waste management and recycling, environmentally sensitive purchasing, energy and water use and the use of technology are known as green management strategies (Carter, Ellram & Ready, 1998).

Waste Management and Recycling: Waste prevention means reducing the amount of waste generated, reducing the hazardous content of waste and reducing its effect on the environment. Businesses should aim to reduce the amount of waste in their production processes. Waste minimisation strategies and recovery methods should be used to reduce waste generation and reduce environmental impacts (Zorpas et al.,2015). Wastes that can be generated in a hotel business in the sector are divided into different classes such as paper, plastic, glass, metal, food (Owen, et al., 2013). If an effective waste management system is not in position for a popular international hotel chain, it could cause reputational damage due to the negative impact on both the environment and customers (Shaalan, 2005).

Businesses have long been aware of the negative impact of recyclable materials in landfill and in the natural environment. As a result, businesses should carry out a number of studies to improve recycling processes and methods and to make waste reusable. These studies include, for example, the establishment of recycling programmes that encourage the recycling of waste and ensuring the efficient use of resources (Little, 2017: 37).

Environmentally responsible purchasing: Environmentally responsible purchasing is purchasing that reduces waste sources, encourages recycling and increases customer satisfaction with the products purchased (Min & Galle, 2001: 1222). Environmentally responsible purchasing is so favoured as it provides long-term benefits for both individuals and entire nations (Singh & Pandey, 2018). Businesses can reduce environmental impacts in their supply chains by promoting the use of environmentally friendly products and services.

Energy and water use: Tourism businesses consume large amounts of energy, water and perishable goods due to the nature of their functions, characteristics and services. Businesses target to reduce environmental impacts by using energy and water resources efficiently. It strives to optimise the use of resources by taking measures such as energy efficiency projects, the use of energy-efficient equipment, water saving measures, etc. (Trung & Kumar, 2005). The tourism sector has faced increasing pressure to ensure that environmental issues are given due attention. A clean environment is an essential element of quality service and is therefore an important issue for the development of the travel, tourism and hotel industries (Erdogan & Baris, 2007).

Use of technology: The use of technology as a green management strategy can help to reduce carbon intensity, improve energy and resource efficiency and prevent serious environmental degradation. The right use of technology not only reduces the risk of resource constraints, but also improves the quality of the environment, human well-being and social equity, leading to progress towards sustainable development goals (Pan, Fan & Lin, 2019: 1). Businesses need to invest in technologies for a greener environment to be successful in the national and international arena (Bahçet & Baysal, 2022: 131). During the effective implementation of green management in businesses, very high investments are required. At this stage, the use of technology is one of the strategies to be considered. Sometimes businesses do not work efficiently and consequently the resources used are wasted. Tourism businesses can use a range of opportunities such as smart booking, online travel platforms, smart building, energy management, smart transportation and digital communication in order to minimise environmental impact and achieve sustainability goals (Hüseyn, 2018).

Benefits of Green Management for Tourism Businesses

Green management practices provide many advantages for tourism businesses. These benefits are considered as directly related to the environmental, economic and social dimensions of sustainable tourism. According to the environmental dimension, there are many different reasons for green management practices of tourism organisations. The first one is that tourism enterprises want to protect environmental resources and influence others in the implementation process. Secondly, tourism businesses want to be seen as environmentally friendly in order to obtain organisational advantage and this can be achieved through an improved image. Thirdly, tourism businesses strive to save money or increase revenues via environmental practices and a green image. This can range from incentives to higher selling prices and cost savings (Font, 2001). Furthermore, green management practices contribute to the conservation of natural resources and the sustainability of ecosystems. In particular, improvements in areas such as energy efficiency and waste management help to reduce environmental impacts (Chengcai et al., 2017).

According to the economic dimension; green management practices enhance economic efficiency by reducing the costs of enterprises. According to the economic dimension; green management practices enhance economic efficiency by reducing the costs of businesses (Chengcai et al., 2017; Middleton & Hawkins, 1998; González & González,2006; Christmann, 2000).

On the social side, green management practices can be classified as; increase in employee motivation (Morrow & Rondinelli, 2002; Christmann, 2000), increase in commitment (Middleton & Hawkins, 1998), increase in productivity and job satisfaction (Christmann, 2000), promotion of environmental awareness (Christmann, 2000; González & González, 2006), positive developments in community relations and relations with local authorities, decrease in social habits that cause environmental damage, increased attractiveness for consumers (Middleton & Hawkins, 1998).

Challenges in Green Management Practices

The fact that tourism enterprises need to innovate in terms of management to keep up with the existing conditions is considered as a motivating factor in the adoption and implementation of green management (Zhou et al., 2018). Although the development and implementation of green governance is supported by positive factors, there are some challenges in implementing the practices. The first of these challenges is related to the high start-up costs of green management practices (Yazıcıoğlu & Aydın, 2018). Because the installation of technologies or systems for green practices can require a substantial investment (Bohdanowicz, 2006: 189). Tourism enterprises may not have the budget to absorb these costs. It shows that the cost of installing new systems can be compensated in a few years by saving water and energy and reducing waste (Iwanowski & Rushmore, 1994: 194). The tourism sector, which always gives priority to environmentally sensitive customers, can be seen as an opportunity for the tourism sector, although some challenges emerge for green management practices.. Within this context, it is important for tourism enterprises to determine their activities related to green management practices and hotel enterprises to determine the priority levels of these activities (Memiş, 2019). Another challenge is that customers may not believe in the effectiveness of green management practices in tourism enterprises. Therefore, enterprises must make sure that they persuade the customer about green management (Bhattacharjee, 2015:16).

Future perspectives on green governance are shaped by the rapid development of environmentally friendly technologies and digital transformation. Technological advancement and digital solutions such as data analytics enable more efficient green management and reduced environmental impact (McKinnon, 2015; Pacheco, 2020). Accordingly, in order to generalize green management practices in enterprises, it is important that their managers initially adopt an environmentally friendly perspective. Stakeholders need to develop long-term methods and implement them based on a roadmap. It is inevitable for local and central governments to initiate legal regulations and practices to encourage tourism enterprises to environmentally friendly practices (Seyhan & Yılmaz, 2010).

Exemplary Green Management Practices in Tourism Enterprises

Green management practices in the tourism industry have their origins in the concept of sustainable tourism and are broadly addressed (Chou, 2012). When viewed from this perspective, tourism, and especially the hospitality industry, can take responsibility for its impact on the environment and improve society by contributing to sustainable development (Gössling et al., 2009).

The concept of "green" in the hospitality industry is acknowledged as the reduction of environmental impact by way of environmentally friendly practices in hotel businesses (Han, 2020). Nowadays, hospitality businesses focus on environmental certifications, waste management, energy efficiency, consumer and staff awareness as "green management practices" (Gökdeniz, 2017: 70). There are also environmentally sensitive practices such as protection of water resources and air quality (Kahraman &Türkay, 2017: 147).

The number of hotels in Turkey that have obtained green stars over the years is shown in Figure 31.1. As of the end of 2022, the number of accommodation facilities with Tourism Management Certificates was 4,830, of which 441 (9.13 %) were certified with an environmentally friendly accommodation facility certificate (green star) (www.ktb.gov.tr, 2024). These data can be assessed as data showing that green management practices are being implemented in our country.

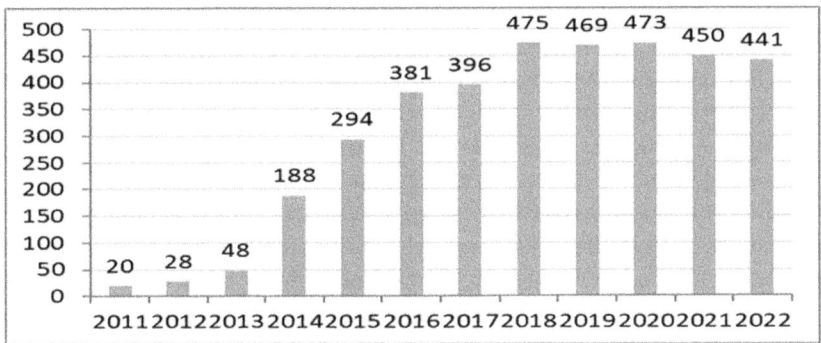

Figure 31.1. Turkey's Green Star Awarded Accommodation Establishments by Year.
Source: (Ministry of Culture and Tourism, 2023; Ministry of Environment and Urbanization, 2023)

Calista Luxury Resort is an exemplary tourism enterprise that successfully implements green management practices as Turkey's first 'Green Star' award-winning accommodation business with international status. This hotel publishes sustainability reports on a regular basis every year. The green management practices included in this report are energy management, water management, waste management, environmental education and information, social responsibility and contribution to society. Among these practices, in energy management; with the consciousness of energy efficiency, it uses lighting with sensors to save energy in rooms and general areas. In water management, information cards on water saving are placed in guest rooms. Photocell faucets are used in public places and aerators are used in shower heads. At the same time, all kinds of wastewater arising from the facility are discharged in accordance with the Water Pollution Control Regulation. The hotel controls and destructs packaging and solid wastes within the context of waste management. It encourages the use of recyclable materials. Chemicals are used with minimum damage to nature and environmentally friendly products are preferred. The human factor, which has an important place in sustainability, has been taken into account both internal and external customers, and a green management approach has been embraced. In this scope, training programs are organized for guests and employees to enhance environmental awareness. Finally, joint commercial processes with local people and businesses and social responsibility projects are carried out. In line with the target of gender equality, which is one of the sustainable development goals, supporting women's employment, making positive discrimination and improving working conditions by respecting the human rights and personal rights of employees are included among the basic policies of this accommodation facility (www.calista.com, 2024). When all these practices are taken into consideration, Calista Luxury Resort both ensures environmental sustainability and operates as an exemplary business in the tourism sector.

Restaurant businesses, which have an important place among tourism enterprises, also have an impact on the environment (Kahraman & Türkay, 2017: 185-186). Thus, the adoption of green management practices by restaurants provides significant benefits to their businesses when evaluated in terms of cost management, market competition and environmental protection activities (Yazıcıoğlu ve Aydın, 2018: 56). In the related literature, restaurants practicing green management are described as "green restaurants, enterprises that are environmentally friendly and prioritize the preservation of energy efficiency" (Yazıcıoğlu & Aydın, 2018: 56).

The Green Restaurant Association of America (GRA), a non-profit organization, has been encouraging the adoption of green practices in restaurants since 1990 (Chou, 2012). According to GRA, green management practices in restaurants can be listed as energy efficiency, water conservation, pollution prevention, environmental health, reuse and recycling programs, green purchasing, green materials, sustainable foods, and green design of buildings and spaces. Alongside these practices, GRA encourages restaurants to adopt environmental policies and train their employees on green practices. In contrast to green practices in other industries, the GRA emphasizes sustainable food sourcing, especially in restaurants, and aims to ensure that consumers continue to contribute to the environment while eating at restaurants can also be considered as a notable green practice (Chou, 2012).

The Michelin Star, which is of great importance for restaurants, has also created a new category called "Green Star" due to the importance it attaches to sustainable gastronomy and green practices. The main objective of this star is to offer restaurants that offer eco-responsible dining experiences the opportunity to become true role models that design a vision of gastronomy that is not only delicious, but also responsive and alternative (www.guide.michelin.com, 2024), As of 2024, more than 350 restaurants have received a Michelin Green Star out of the 15,000 restaurants recommended by the Michelin Guide (www.guide.michelin.com, 2024). The restaurant Neolokal, which also received the Green Star in Turkey in 2023, has succeeded in keeping its Green Star this year. The key practices behind this success are the collaboration with producers, artisans and farmers who are committed to clean and fair production using ancestral methods in order to preserve Turkish culinary traditions, and the inclusion of the garden, which has been adapted to the menus by valuing the land. In addition, it can be listed as giving importance to seasonality in menu planning, using a system where restaurant guests choose their menus in advance in order to ensure minimum waste generation and using the waste generated in the kitchen as compost in the garden within the restaurant (www.guide.michelin.com, 2024).

Businesses located at airports are also exposed to various regulations due to negative environmental impacts such as air and water pollution, waste, disruption of natural life. These negative environmental impacts pose significant problems for regions (Canöz & Ertek, 2020). As an airline flying to many countries around the world, THY strongly promotes sustainable development. Develops projects based on the Zero Waste principle in waste management practices. It supports sustainable products and services, taking into account the environmental impact of products and services, and prefers to choose these products and services as much as possible. Environmental Awareness and Greenhouse Gas Awareness Trainings were prepared in 2016 with the aim of ensuring more active participation of employees in preventing environmental problems, which is one of the requirements of the Environmental Management System, and to create environmental awareness and sensitivity among employees (Airlines, 2016: 5). Under the environmental management section of THY's sustainability reports, it is mentioned that THY, which works to minimize the impact of its operations on the environment to the lowest possible level and to combat climate change, holds the TSE ISO EN 14001 Environmental Management System certificate (Karakuş, 2023: 99).

References

Airlines, T. (2016). Türk Hava Yolları Çevre Performans Raporu.

Alfred, A. M., & Adam, R. F. (2009). Green Management Matters Regardless. *Academy Of Management Perspectives*, 23(3), 17–26.

Behçet, Ş., & Baysal, H. T. (2022). Çevrecilik Bağlamında Yeşil Yönetim Uygulamaları: Küresel Boyutta Faaliyet Gösteren Uluslararası İşletmelerden Yeşil Yönetim Uygulama Örnekleri. Uluslararası İşletme Bilimi ve Uygulamaları Dergisi, 2(2), 131–146.

Bernauer, T., Engel, S., Kammerer, D., & Sejas Nogareda, J. (2007). Explaining Green İnnovation: Ten Years After Porter's Win-Win Proposition: How

To Study The Effects Of Regulation On Corporate Environmental İnnovation?. *Politische Vierteljahresschrift, 39,* 323–341.

Bhattacharjee, K. (2015). Green Supply Chain Management-Challenges And Opportunities. Asian Journal Of Technology & Management Research, 5(01).

Bohdanowicz, P. (2005). European Hoteliers' Environmental Attitudes: Greening The Business, Cornell Hotel And Restaurant Administration Quarterly, 46, 188–204.

Brundlandt, G. (1987). Report Of The World Commission On Environment And Development: Our Common Future, Oslo.

Canöz, N., & Ertek, A. (2020). Yeşil Kuruluş Sertifikasının Yeşil İmaj Oluşumuna Katkısı: Türk Sivil Havacılığı Üzerine Bir Araştırma. International Journal Of Aeronautics And Astronautics, 1(1), 23–32.

Carter, C. R. Ellram, L. M. & Ready, K.J. (1998). Enviromental Purchasing: Benchmarking Our German Counterparts, International Jounal Of Purchasing And Materials Management, 34(4), 28–37.

Chengcai, T., Qianqian, Z., Nana, Q., Yan, S., Shushu, W., & Ling, F. (2017). A Review Of Green Development İn The Tourism İndustry. *Journal Of Resources And Ecology, 8*(5), 449–459.

Chou, C. J., Chen, K. S., & Wang, Y. Y. (2012). Green Practices İn The Restaurant İndustry From An İnnovation Adoption Perspective: Evidence From Taiwan. *International Journal Of Hospitality Management, 31*(3), 703–711.

Cramer, J. (1998). Environmental Management: From 'Fit' to 'Stretch'. Business Strategy And The Environment, 7(3), 162–172.

Dwyer, R. J. (2009). "Keen To Be Green" Organizations: A Focused Rules Approach To Accountability. *Management Decision, 47*(7), 1200–1216.

Font, X. & Tribe, J. (2001) Promoting Green Tourism: The Future Of Environmental Awards. International Journal Of Tourism Research 3(1), 1–13.

González-Benito, J., & González-Benito, Ó. (2006). A Review Of Determinant Factors Of Environmental Proactivity. Business Strategy And The Environment, 15(2), 87–102.

Gökdeniz, A. (2017). Konaklama Sektöründe Yeşil Yönetim Kavramı, Eko Etiket Ve Yeşil Yönetim Sertifikaları Ve Otellerde Yeşil Yönetim Uygulama Örnekleri. International Journal Of Social And Economic Sciences, 7(2), 70–77.

Gössling, S., Hall, C. M., & Weaver, D. B. (2009). Sustainable Tourism Futures: Perspectives On Systems, Restructuring And İnnovations. In *Sustainable Tourism Futures* (Pp. 1–16). Routledge.

Haden, S. S. P., Oyler, J. D., & Humphreys, J. H. (2009). Historical, Practical, And Theoretical Perspectives On Green Management: An Exploratory Analysis. *Management Decision, 47*(7), 1041–1055.

Han, H. (2020). Theory Of Green Purchase Behavior (Tgpb): A New Theory For Sustainable Consumption Of Green Hotel And Green Restaurant Products. *Business Strategy And The Environment, 29*(6), 2815–2828.

Hart, S. L. (2005). Innovation, Creative Destruction And Sustainability. *Research-Technology Management, 48*(5), 21–27.

Hüseyin, C. (2018). Endüstriyel Üretimde Döngüsel Çevre Politikaları. Nevşehir Bilim Ve Teknoloji Dergisi, 7(2), 111–122.

Iwanowski, K., & Rushmore, C. (1994). Introducing the Eco-Friendly Hotel. Cornell Hotel and Restaurant Administration Quarterly, 35(1), 34–38.

Karakuş, G. (2023). Türk Hava Yolları Sürdürülebilirlik Raporları Üzerine Bir Araştırma, Aerospace Research Letters (Asrel) Dergisi, 2(2), 86–113.

Lee, J. S., Hsu, L. T., Han, H., & Kim, Y. (2010). Understanding How Consumers View Green Hotels: How A Hotel's Green Image Can Influence Behavioural Intentions. *Journal Of Sustainable Tourism, 18*(7), 901–914.

Lee, K. H., & Ball, R. (2003). Achieving Sustainable Corporate Competitiveness: Strategic Link Between Top Management's (Green) Commitment And Corporate Environmental Strategy. *Greener Management International,* (44), 89–104.

Little, M. E. (2017). Innovative Recycling Solutions To Waste Management Challenges İn Costa Rican Tourism Communities. Jeta: Journal Of Environmental & Tourism Analyses, 5(1).

Loknath, Y., & Azeem, B. A. (2017). Green Management–Concept And Strategies. In *National Conference On Marketing And Sustainable Development,* 13(14), 688–702.

Mckinnon, A., Cullinane, S., & Browne, M. (2015). Green Logistics: Improving The Environmental Sustainability Of Logistics. Kogan Page Publishers.

Memiş, S. (2019). Konaklama İşletmelerinde Yeşil Yönetim Uygulamalarının Entropi Yöntemi ile Ağırlıklandırılması: Giresun İli Örneği. İşletme Araştırmaları Dergisi, 11(1), 653–665.

Middleton, V. T., & Hawkins, R. (1998). *Sustainable Tourism: A Marketing Perspective.* Routledge.

Molina-Azorín, J. F., Claver-Cortés, E., López-Gamero, M. D., & Tarí, J. J. (2009). Green Management And Financial Performance: A Literature Review. *Management Decision, 47*(7), 1080–1100.

Morrow, D. & Rondinelli, D. (2002). Adopting Corporate Environmental Management Systms: Motivations And Results Of Iso 14001 And Emas Certification, European Journal, (20) 2, 159–171.

Erdogan, N., & Baris E. (2007). Environmental Protection Programs And Conservation Practices Of Hotels İn Ankara, Turkey, Tourism Management, 28, 604–614.

Owen, N., Widdowson, S & Shields, L. (2013). Waste Mapping Guidance For Hotels İn Cyprus: Saving Money And İmproving The Environment. The Travel Foundation.

Pacheco, R. R., Borenstein, D., & Maletič, M. (2020). Sustainable Logistics And Supply Chain Management: Principles And Practices For Sustainable Operations And Management. Springer.

Pan, S. Y., Fan, C., & Lin, Y. P. (2019). Development And Deployment Of Green Technologies For Sustainable Environment. Environments, 6(11), 114.

Peng, Y. S., & Lin, S. S. (2008). Local Responsiveness Pressure, Subsidiary Resources, Green Management Adoption And Subsidiary's Performance: Evidence From Taiwanese Manufactures. *Journal Of Business Ethics*, 79, 199–212.

Rutkowska, M & Sulich, A. (2021). The Green Management Towards A Green Industrial Revolution, Ideas Working Paper Series From Repec, 1–13.

Seyhan, G., & Yılmaz, B. S. (2010), "Sürdürülebilir Turizm Kapsamında Konaklama İşletmelerinde Yeşil Pazarlama; Calısta Luxury Resort Hotel", İşletme Fa-Kültesi Dergisi, 11(1), 51–74.

Shaalan, I.M. 2005. Sustainable Tourism Development İn The Red Sea Of Egypt Threats And Opportunities, 13 (2), 83–87.

Singh, G., Pandey, N. (2018) The Determinants Of Green Packaging That Influence Buyers' Willingness To Pay A Price Premium. Austr. Market. J. (Amj), 26 (3), 221–230.

Taylor, S. R. (1992). Green Management: The Next Competitive Weapon. Futures, 24 (7), 669–680.

Trung, D.N., & Kumar, S. (2005). Resource Ese And Waste Management İn Vietnam Hotel Industry, J. Clean. Prod, 13, 109–116.

Wu, S. I., & Wu, Y. C. (2014). The Influence Of Enterprisers' Green Management Awareness On Green Management Strategy And Organizational Performance. International Journal Of Quality & Reliability Management, 31(4), 455–476.

Yazıcıoğlu, İ., & Aydın, A. (2018). Yeşil Restoran Uygulamaları Üzerine Nitel Bir Araştırma: İstanbul Örneği. Gazi Üniversitesi Turizm Fakültesi Dergisi, (1), 55–79.

Zorpas, A. A., Voukkali, I., & Loizia, P. (2015). The İmpact Of Tourist Sector İn The Waste Management Plans. Desalination And Water Treatment, 56 (5), 1141–1149.

https://calista.com.tr/tr/kurumsal/surdurulebilirlik / Date of access: 27.05.2024

https://cevreselgostergeler.csb.gov.tr/cevreye-duyarli-konaklama-tesisi-sayisi-i-85845#_ednref1 / Date of access: 27.05.2024

https://guide.michelin.com/tr/tr/article/features/the-michelin-green-star-tr / Date of access: 27.05.2024

https://guide.michelin.com/tr/tr/istanbulprovince/istanbul/restaurant/neolokal / Date of access: 27.05.2024

https://yigm.ktb.gov.tr/TR-277167/cevreye-duyarli-turizm-isletme-belgeli--konaklama-tesisi-istatistikleri.html / Date of access: 27.05.2024

Erdem Şimşek[1]

Chapter 32 The Future of Travel Assistance: SWOT Insights into Chatbot Usage in Tourism Industry

Introduction

Tourism plays an important role in the promotion and preservation of natural and cultural heritage, and in the current digital age, the industry is undergoing a profound transformation (Yaralı and Baloğlu, 2023). During this period, digital platforms such as websites and social media applications have become powerful tools that allow both the promotion of attractions and the sharing of experiences (Buhalis and Foerste, 2015). One of the AI-supported applications that has been increasingly used in the tourism industry in recent times is chatbots (Bozic et al., 2019; Dash and Bakshi, 2019; Thazhathethil et al., 2021; AlHumoud et al., 2022). Artificial Intelligence (AI), also known as machine intelligence, is the intelligence demonstrated by machines rather than humans (Russell, Norvig & Davis, 2010). It is generally used to describe machines that mimic "cognitive" functions associated with the human mind, such as "learning" and "problem-solving". Artificial intelligence can be divided into three different types of systems (Kaplan & Haenlein, 2019):

- "Analytical Artificial Intelligence" possesses characteristics consistent with cognitive intelligence, creates a cognitive representation of the world, and uses experience-based learning to inform future decisions.
- "Human-Inspired Artificial Intelligence" includes elements of both cognitive and emotional intelligence. In addition to cognitive elements, it understands human emotions and takes them into account in decision-making processes.
- "Humanized Artificial Intelligence" can be self-aware and possess self-consciousness in interactions, as it demonstrates characteristics of all types of competencies, including cognitive, emotional and social intelligence.

Nowadays, people use different travel applications for activities such as finding and booking flights, renting cars, and finding hotels. Although travel applications

---

[1] Ph.D., Ankara Hacı Bayram Veli University, Faculty of Tourism, Department of Recreation Management, erdem.simsek@hbv.edu.tr

are beneficial for tourists, they have some limitations in terms of efficiency. This is because each application focuses on different niche areas (for example, one application provides information about tourist attractions in a specific city, while another is used for hotel reservations), which means they occupy a significant amount of storage space on phones or tablets and often become unnecessary after the trip (Farkash, 2018). Examples of these travel applications include Skyscanner, Airbnb, Tripadvisor, TripIt, and Roadtrippers (Georgios, 2019). Travel chatbots offer a more practical alternative to travel applications since they do not require installing software on your device. Instead of installing an app, they allow receiving notifications, updates, and promotional offers using existing platforms such as browsers or messaging apps, which almost everyone already has on their phones (Farkash, 2018).

The Chatbots

Chatbots are specialized software programs that converse with users via natural language processing, a subfield of artificial intelligence that aims to make computers understand human language. The question "Can machines think?" posed by renowned mathematician Alan Turing gave rise to this idea initially (Al-Ghamdi et al., 2017). Virtual assistants, or chatbots, are gaining popularity as the basis for communications between people and systems. This increased interest can be attributed to two things: the natural communication technique that human languages offer and the methods and technology that allow computers to study and process these languages (Bozic et al., 2019). These software programs, also referred to as scripts, carry out automated online tasks. In most cases, bots are assigned simple and structurally repetitive tasks, which they perform at much higher speeds and accuracy rates than a human could alone (Georgios, 2019).

Even though chatbots are sophisticated apps, the process starts when a user messages or asks a question of the chatbot. Any platform's message interface can be used for this procedure. The chatbot comprehends and analyzes human input with "Natural Language Processing" (NLP). The chatbot ascertains the user's intent from the input they provide. Stated differently, it pinpoints the intention or aim underlying the user's inquiry. The chatbot recognizes entities in the user's input after determining the purpose. These entities contain specifics, including dates, locations, names, or other parameters, that are associated with the user's request. The chatbot maintains its context throughout the interaction by remembering previous interactions and answers. This helps provide a more consistent and personalized experience to the user as the chatbot can reference previous parts of the conversation. Once the purpose and entities are defined, the chatbot processes the user's query. It then determines the appropriate response or action based on predefined rules, algorithms, or machine learning models.

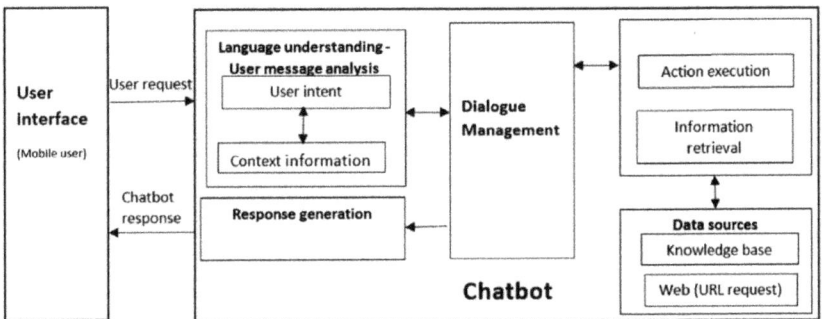

Figure 32.1. General Chatbot Architecture.
Source: Adamopoulou and Moussiades, 2020

Some chatbots use decision trees, while others use machine learning techniques to improve over time. If the chatbot needs to retrieve specific information from a database or external source, it retrieves the relevant data. This includes accessing details regarding flights, hotel availability or other information needed to fulfill the user's request. The chatbot generates responses in the form of text, images, links, or other media, depending on its capabilities and the nature of the interaction. A lot of contemporary chatbots are built to grow and learn over time. As part of this learning process, data is updated, algorithms are improved, and user feedback is included to help the systems comprehend and react to user inquiries more precisely (Sharma, Goyal and Malik, 2017; Hiremath et al., 2018; Lalwani et al., 2018; Adamopoulou and Moussiades, 2020).

Chatbots are artificial intelligence (AI) driven text and/or voice virtual assistants that converse naturally in human language with the goal of assisting users in getting what they want more quickly and accurately. The goal of these interactions is to resemble human interactions as much as feasible (Bradeško and Mladenić, 2012; Khanna et al., 2015). They are widely utilized in areas like reservation services, banking and insurance, education, media, food ordering, health and medicine, e-commerce, and tourism – all of which depend heavily on the caliber and speed of client communication (Cui et al., 2017; Almansor and Hussain, 2019). Rule-based and AI-powered chatbots are the two primary categories of chatbots that can be distinguished according to client needs.

Rule Based Chatbots

Rule-based chatbots operate based on a predetermined set of basic or sophisticated rules. The bot can offer solutions and solve problems in accordance with these criteria. Rule-based bots aren't able to learn from past interactions or get smarter because they can't answer any questions that fall outside of this range. These chatbots can be favored since they are easier to install, need less setup costs, and can be produced and taught more quickly (Alotaibi, Alharthi, and Almehamdi, 2020).

Artificial intelligence (AI) Powered Chatbots

Chatbots with artificial intelligence (AI) use machine learning techniques to comprehend natural language and produce intelligent answers to intricate queries. With every encounter, these bots learn more and require more training in order to become more intelligent and productive when interacting with users. They are able to recognize patterns in behavior, make snap decisions, and have the ability to understand several languages. They require more time to train, but they can handle massive volumes of data, which will ultimately save a lot of time (Alotaibi, Alharthi, and Almehamdi, 2020). Due to their competitive benefits, which include automated services, 24/7 customer assistance, and cost savings, AI-based chatbots have revolutionized the travel and tourism industry (Sheehan, 2018; Van Doorn et al., 2017).

Tourism companies, airlines, and hotels are currently using AI-based chatbots more frequently to communicate with customers in a variety of front-line activities like bookings, customer care, and recommendations (Li et al., 2021). More than 42 % of airports are choosing to employ AI-based chatbots to provide services, according Ghosh and Chakravarty (2018). Major tourism firms frequently employ AI-based chatbots in China, one of the top nations for chatbot-enabled applications, in order to better serve their clients' needs (Li et al., 2021).

Even though AI-based chatbots are being used more and more in the travel and tourism industry, there are still two significant research gaps in the literature on their application. First, the majority of research on chatbot adoption in the tourism industry currently in existence is based on theories of technology acceptance, such as the Unified Theory of Acceptance and Use of Technology 2 (UTAUT2) and the Technology Acceptance Model (TAM). However, these theories do not comprehensively explain the use of chatbots, as they focus solely on technological features and overlook other determinants (Zhu, Wang, and Pu, 2022). Chatbots for tourism are typically used to accomplish business goals.

Consequently, an in-depth examination of the factors influencing users' intentions to keep using these chatbots should be conducted utilizing a study theory from the domains of psychology or business. Second, research examining the moderating effect of gender on chatbot adoption in the tourism industry is lacking. Such a research gap can become an obstacle to system design and personalized marketing strategies. According to McDonnell and Baxter (2019), awareness of the gender dimension in AI-based service delivery is increasing. Therefore, it is necessary to test whether users' perceptions of chatbots towards tourism are homogeneous between genders.

Uses of Chatbots in the Tourism Industry

The tourism industry has experienced a significant transformation in recent years, with the integration of advanced technologies such as chatbot. Powered by artificial intelligence (AI) and natural language processing, chatbots offering a wide range of benefits to both businesses and customers (Samala et al., 2022). The literature on chatbots in the field of tourism began to increase in the mid-2010s (Adamopoulou, and Moussiades, 2020) and began to develop even later. In the coming years, the use of chatbots in the tourism sector may develop in the following ways (Dash and Bakshi, 2019; Alotaibi, Alharthi and Almehamdi, 2020; Pillai and Sivathanu, 2020; Melián-González et al., 2021; Pereira et al., 2022):

- Design customized vacation itineraries,
- Respond to commonly asked inquiries,
- Assist customers in locating and reserving lodging according on their preferences, spending limits and departure dates.
- Provide 7/24 customer service,
- Assist users with issues such as flight delays, cancellations or changes in travel plans,
- Travelers can overcome language barriers with real-time language translation skills,
- Travelers can receive real-time travel alerts, weather updates, and other pertinent information while they are on the road,
- By providing extra services they can expand organizations' revenue potential,
- Provide individualized travel recommendations by analyzing user preferences and past behavior,
- Improve the user experience by integrating with AR technologies,
- Chatbots can guide tourists to safe regions and provide emergency information during unforeseen catastrophes or emergencies,

- To increase user engagement, some chatbots include gamification features such as challenges and quizzes.

Nowadays, the tourism industry is using chatbots more and more, as they are an emerging technology. Travel agencies must take into account that visitors must become accustomed to this procedure and form the habit of launching a chatbot rather than contacting the business via phone or email (Melián-González et al., 2021). They ought to encourage their visitors to make use of chatbots. One advantage of chatbots related to tourism is that they can be accessed from multiple platforms, including desktop computers, laptops, and mobile phones. Additionally, the real-time answers and rapid ideas that these chatbots provide can help tourists travel more efficiently (Pillai and Sivathanu, 2020).

SWOT Analysis for the Use of Chatbots in Tourism

The use of chatbots in the tourism industry has completely changed how companies communicate with visitors, offering a smooth and effective approach to handle reservations, customer inquiries, and tailored recommendations. Assessing the industry's potential and effects of chatbot usage, a SWOT analysis provides a thorough framework for understanding the strengths, weaknesses, opportunities, and threats.

Strengths

When the strengths of chatbots are examined, the prominent features can be expressed as follows: Chatbots, which work seven days a week and twenty-four hours a day, can respond instantly to inquiries from anywhere in the world and allow this to be done without the need for extra human resources. Since it requires a much smaller number of employees, it offers businesses an advantage in terms of costs. It plays an important role in ensuring and increasing customer satisfaction, as inquiries are responded to immediately. Because chatbots are powered by artificial intelligence and learn from past inquiries, they can offer more personalized recommendations. It facilitates communication as it supports many languages. It allows collecting a large amount of data and processing this data to provide personalized offers.

Weaknesses

The weaknesses caused by the use of chatbots in the tourism sector are as follows: problems in making sense of complex questions and queries negatively affect the usage experience. Another negative aspect can be expressed as the need for an intensive internet and information technology infrastructure. Chatbots that lack a sense of empathy may encounter some problems in the field of customer relations. Finally, the high installation costs of high-level chatbots can also be given as an example of the weaknesses in this regard.

Opportunities

The ability of chatbots to work integrated with technologies such as AR and VR is an important opportunity. Other opportunities can be listed as follows: Working together with social media platforms will provide significant advantages in the field of marketing and guest relations. As it will reduce dependence on printed materials, it will also contribute to businesses reaching sustainable environmental standards.

Threats

Finally, when the threats posed by chatbots are examined, it is possible to say the following: As all data is stored digitally, concerns about data security will come to the fore. In addition, some users cannot get used to chatbots and use old method communication channels, malfunctions that may occur in the technical infrastructure, and legal regulations regarding the processing of personal data constitute other threats to chatbots.

Conclusion

The tourism industry benefits greatly from chatbots as they improve customer experience and make time easier. Thanks to chat robots, customers can ask and receive answers about their reservations, travel itineraries and destinations 24 hours a day, seven days a week. This positively affects the guest experience at the end of the day. Chatbots, which can prepare personalized suggestions thanks to personal data collected as a result of inquiries, contribute to the formation or strengthening of customer loyalty. Businesses gain significant benefits from chatbots, as they enable the creation of a structure that requires less human resources. While chatbots deal with simple transactions, employees may be more interested in more complex and time-consuming issues. Finally; Unlike human employees, whose physical and mental states constantly change, chatbots always respond to customers' questions in the same manner. This will prevent information accuracy and service quality problems that may occur in the service provided by people and will increase guest satisfaction at the end of the process.

References

Adamopoulou, E., Moussiades, L. (2020). An Overview of Chatbot Technology. In: Maglogiannis, I., Iliadis, L., Pimenidis, E. (eds) Artificial Intelligence Applications and Innovations. AIAI 2020. IFIP Advances in Information and Communication Technology, vol 584. Springer, Cham. https://doi.org/10.1007/978-3-030-49186-4_31

Al-Ghamdi, S.A., Khabti, J. & Al-Khalifa, H.S. (2017). Exploring NLP web APIs for building Arabic systems. In 2017 Twelfth International Conference on Digital Information Management (ICDIM), Sep. 2017, pp. 175–178. doi: 10.1109/ICDIM.2017.8244649.

AlHumoud, S., Diab, A., Aldukhai, D., AlShalhoub, A., AlAbdullatif, R., AlQahtany, D., AlAlyani, M. & Bin-Aqeel, F. (2022), Rahhal: A Tourist Arabic Chatbot, 2022 2nd International Conference of Smart Systems and Emerging Technologies (SMARTTECH), Riyadh, Saudi Arabia, pp. 66–73, doi: 10.1109/SMARTTECH54121.2022.00028.

Almansor, E. H., & Hussain, F. K. (2019). Survey on intelligent chatbots: State-of-the-Art and future research directions. Advances in Intelligent Systems and Computing, 534–543. https://doi.org/10.1007/978-3-030-22354-0_47

Alotaibi, R., Ali, A., Alharthi, H., & Almehamdi, R. (2020). AI Chatbot for Tourist Recommendations: A Case Study in the City of Jeddah, Saudi Arabia. International Journal of Interactive Mobile Technologies (iJIM), 14(19), pp. 18–30. https://doi.org/10.3991/ijim.v14i19.17601

Bozic, J., Tazl, O. A., & Wotawa, F. (2019). Chatbot testing using AI planning. In 2019 IEEE International Conference On Artificial Intelligence Testing (AITest) (pp. 37–44). IEEE.

Bradeško, L. & Mladenić, D. (2012). "A Survey of Chatbot Systems through a Loebner Prize Competition," Proc. Slov. Lang. Technol. Soc. Eighth Conf. Lang. Technol., vol. 2, pp. 34–37.

Buhalis, D., & Foerste, M. (2015). SoCoMo marketing for travel and tourism: Empowering co-creation of value. Journal of Destination Marketing & Management, 4(3), 151–161.

Cui, L., Huang, S., Wei, F., Tan, C., Duan, C., & Zhou, M. (2017, July). Superagent: A customer service chatbot for e-commerce websites. In Proceedings of ACL 2017, system demonstrations (pp. 97–102).

Dash, M., & Bakshi, S. (2019). An exploratory study of customer perceptions of usage of chatbots in the hospitality industry. International Journal on Customer Relations, 7(2), 27–33.

Farkash, Z. (2018). Travel chatbot — How chatbots can help city tourism. Medium. https://chatbotsmagazine.com/travel-chatbot-how-chatbots-can-help-city-tourism-a2f122c0896d

Georgios, C. (2019). Designing a Chatbot for Tourism [Unpublished master's thesis]. International Hellenic University School of Science & Technology.

Ghosh, J., & Chakravarty, R. (2018). Expedition 3.0: Travel and hospitality gone digital, KPMG and FICCI. https://home.kpmg/in/en/home/insights/2018/03/ficci-expedition-travel-hospitality-technologyinnovation-india-digital.html

Hiremath, G., Hajare, A., Bhosale, P., Nanaware, R., & Wagh, K. S. (2018). Chatbot for education system. International Journal of Advance Research, Ideas and Innovations in Technology, 4(3), 37–43.

Kaplan, A., & Haenlein, M. (2019). Siri, Siri, in my hand: Who's the fairest in the land? On the interpretations, illustrations, and implications of artificial intelligence. Business Horizons, 62(1), 15–25. https://doi.org/10.1016/j.bushor.2018.08.004

Khanna, A., Pandey, B., Vashishta, K., Kalia, K., Pradeepkumar, B., & Das, T. (2015). A study of today's A.I. through chatbots and rediscovery of machine intelligence. International Journal of u- and e-Service, Science and Technology, 8(7), 277–284. https://doi.org/10.14257/ijunesst.2015.8.7.28

Lalwani, T., Bhalotia, S., Pal, A., Rathod, V., & Bisen, S. (2018). Implementation of a Chatbot System using AI and NLP. International Journal of Innovative Research in Computer Science & Technology (IJIRCST), 6(3), 26–30.

Li, L., Lee, Y. K., Emokpae, E., & Yang, S. B. (2021). What makes you continuously use chatbot services? Evidence from Chinese online travel agencies. Electronic Markets, 31(3), 575–599. https://doi.org/10.1007/s12525-020-00454-z

McDonnell, M., & Baxter, D. (2019). Chatbots and gender stereotyping. Interacting with Computers, 31(2), 116–121. https://doi.org/10.1093/iwc/iwz007

Melián-González, S., Gutiérrez-Taño, D., & Bulchand-Gidumal, J. (2021). Predicting the intentions to use chatbots for travel and tourism. Current Issues in Tourism, 24(2), 192–210. DOI: 10.1080/13683500.2019.1706457

Pereira, T., Limberger, P. F., Minasi, S. M., & Buhalis, D. (2022). New Insights into consumers' intention to continue using chatbots in the tourism context. Journal of Quality Assurance in Hospitality & Tourism, 1–27.

Pillai, R., & Sivathanu, B. (2020). Adoption of AI-based chatbots for hospitality and tourism. International Journal of Contemporary Hospitality Management, 32(10), 3199–3226.

Russell, S. J., Norvig, P., & Davis, E. (2010). Artificial intelligence: A modern approach. Prentice Hall.

Samala, N., Katkam, B.S., Bellamkonda, R.S. & Rodriguez, R.V. (2022), "Impact of AI and robotics in the tourism sector: a critical insight", Journal of Tourism Futures, Vol. 8 No. 1, pp. 73–87. https://doi.org/10.1108/JTF-07-2019-0065

Sharma, V., Goyal, M. & Malik, D. (2017). "An Intelligent Behaviour Shown by Chatbot System." International Journal of New Technology and Research, 3(4), 52–54.

Sheehan, B. T. (2018). Customer service chatbots: Anthropomorphism, adoption and word of mouth [Doctoral dissertation]. Queensland University of Technology. https://eprints.qut.edu.au/121188/

Thazhathethil, B. V., Balasubramaniam, U., & Abraham, S. T. (2021). Coimbatore Destination Chatbot: A Study on Customer Preference. In 2021 IoT Vertical and Topical Summit for Tourism (pp. 1–6). IEEE.

Van Doorn, J., Mende, M., Noble, S. M., Hulland, J., Ostrom, A. L., Grewal, D., & Petersen, J. A. (2017). Domo arigato Mr Roboto: Emergence of automated social presence in organizational frontlines and customers' service experiences. Journal of Service Research, 20(1), 43–58. https://doi.org/10.1177/1094670516679272

Yaralı, M. C. & Özçelik Baloğlu, Ö. (2023). Dijital süreçlerin doğal ve kültürel miras turizminin gelişimine etkisi. Nevşehir Hacı Bektaş Veli Üniversitesi SBE Dergisi, İhtisaslaşma Özel Sayısı, 245–264.

Zhu, Y., Wang, R., & Pu, C. (2022). "I am chatbot, your virtual mental health adviser." What drives citizens' satisfaction and continuance intention toward mental health chatbots during the COVID-19 pandemic? An empirical study in China. Digital Health, 8, 20552076221090031. https://doi.org/10.1177/20552076221090031

Kürşat Başkan[1]

Chapter 33 Integration of Technology in Small Tourism Enterprises

Introduction

Technological products and processes, which have entered every aspect of life for many years and are now indispensable, have become tools that are frequently used by businesses. Undoubtedly, there may be various difficulties in using these technological opportunities. These challenges may not be difficult to overcome for large tourism enterprises. However, it is not possible to talk about the same thing for small-scale tourism enterprises. This is because small businesses have more limited opportunities in terms of both process adaptation and financial resources. While small businesses are expected to turn to more budget-friendly technological products and processes, the difficulties they may experience in these processes are also a matter of curiosity. An in-depth understanding of technology integration in small tourism businesses and the areas it encompasses, such as information technology integration, social media use and knowledge management integration, is considered important.

In an age where the pressure of global competition is increasing day by day, there is a very rapid production of information and the information produced is consumed in a shorter time than it is produced, technology and information opportunities have undoubtedly penetrated every aspect of life (Altınöz, 2008: 51). The development of technology is adapted to the tourism sector as in many business sectors. It can be said that small businesses in the sector need technology as much as large businesses do. For example, entrepreneurs in small tourism businesses in developing countries such as Malaysia and Ecuador are developing new ways of adopting innovations, particularly in relation to the use of the internet (Karanasios and Burgess, 2008). Information technologies can be involved in all functions of strategic and operational management. While information is crucial for tourism businesses, IT presents both opportunities and challenges for the sector. Competing businesses and destinations will increasingly have to make calculations. In other words, the strategic incorporation of

---

[1] Assist. Prof. Recep Tayyip Erdogan University, Ardeşen Vocational School. Department of Hotel, Restaurant and Catering. kursat.baskan@erdogan.edu.tr

technology into small tourism businesses is essential to increase competitiveness (Buhalis, 1998). For example, e-commerce, which uses information technologies intensively, can provide a competitive advantage for small and medium-sized tourism businesses (Halawani et al., 2013).

The tourism industry relies heavily on ICT, not only to reach target markets and interact between businesses, visitors and networks, but also as a tool to manage the day-to-day activities of tourism organizations. These organizations need to embrace Technology to maintain their place in the international tourism market in an increasingly globalized and networked world (Ali, 2014:11). In their 2014 study (Law et al. 2014), they examined the studies investigating the use of Information Technologies in the tourism sector. In the study where the studies were evaluated in 5 categories;

- e-marketing
- e-strategic management
- web design and analysis
- guest services
- e-security

Many studies in the field can be categorized from different perspectives. (Law et al. 2014)'s categorization was found valuable. In this study, it is useful to make some evaluations about the applications of different aspects of information technologies and technological opportunities in the business areas of tourism such as accommodation, travel and food and beverage sectors.

Technology in the Hospitality Industry

Businesses operating in the tourism sector must quickly adapt to digital transformation processes in order to survive in the increasing competitive environment and provide better service to their customers (Azadaliyev and Demirkol, 2023). Computer systems are instruments that hotels need to use to achieve their goals. A fully integrated hotel computer system provides hotel management with an effective means to monitor front office and backoffice activities (Azaltun, 2003:161). Technological solutions can bring significant benefits to the hospitality industry, such as improved operational efficiency, enhanced customer experience and increased profitability. Indeed, many studies show that ICT is widely used in the hospitality sector. In sectoral applications, ICT brings various capabilities to businesses. However, the full utilization of these capabilities may be limited by managerial ambition, attitude and finances rather than technological constraints. Therefore, managers should adopt a more proactive approach by integrating newly developed technologies into their daily business functions and ultimately incorporate ICT into their business mission (Law et al., 2014). These technologies are primarily used to overcome operational challenges in hotel management. These technological capabilities greatly facilitate daily operations (Law and Jogaratnam, 2005).

Today, a concept called smart tourism is widely used in the literature. Essentially, smart tourism is a new buzzword used to describe the increasing dependence of tourism destinations, industries and tourists on new information and communication technologies that allow the transformation of large amounts of data into value propositions (Gretzel et. al., 2015). In other words, smart tourism can be defined as the process of incorporating ICTs into different areas of tourism to enhance visitor experiences, increase operational efficiency and promote sustainable tourism development (Hamid et al., 2023). Smart tourism investments are seen in many different countries. In addition, it is possible to say that these investments, which are among the international trends of tourism, are supported at the state level.

For example, countries such as China officially support smart tourism projects to develop the tourism sector, and this concept is becoming widespread and adopted globally (Mehraliyev et al., 2019). In addition, Korea has made significant progress in establishing smart tourism cities that offer technology-driven on-site tourism experiences (Um et al., 2022). Similarly, in Hangzhou, China, smart tourism has integrated tourism resources and ICTs to provide satisfying information and services and enhance tourist experiences (Tian et al., 2019).

Smart tourism represents a new approach in the industry, utilizing technology to enhance visitor experiences, increase operational efficiency, and promote sustainable development. By creating smart ecosystems, destinations can provide tourists with better travel experiences through the integration of physical and information resources for societal benefit.

It is also possible to say that there are various concerns about some aspects of the integration of technology into the tourism industry. The first thing that comes to mind is that the security of personal data may be a threat, or in another dimension, there may be a fear that developing technology will replace people. For example, (Vatan & Dogan, 2021) evaluated the perceptions of Turkish hotel employees towards service robots in their study. It was concluded that the word robot arouses negative emotions on hotel employees. In addition, it was stated that there is a fear of problems in customer communication and the fear of unemployment in the future for people.

Smart tourism as an effective and efficient use of technology is the most tangible argument for its value in the tourism industry. Although it is a sector where manpower is a major factor, the tourism sector, and especially the hospitality sector, closely follows technological developments and uses them effectively by integrating them into production, marketing and management processes, as in other sectors.

Technology in the Travel Industry

The integration of technology has brought significant transformation to various aspects of travel services and operations within the travel industry. Airlines have been leading the way in using technological innovations to enhance effectiveness and customer contentment (Ying et al., 2016). These advancements have allowed airlines to reduce expenses, increase seating availability, improve security measures, and offer more flexible products and services. Moreover, the inclusion of ICT has not only altered social, domestic, and professional practices but has also revolutionized traveling methods and experiences (Hannam et al., 2014).

Being ready for technology is crucial in impacting customer happiness with travel technologies globally (Ying et al., 2016). Today, technology is widely adopted by consumers and businesses. This has led to the efficient use of technological innovations in the travel industry. Increasing technological possibilities can affect the strategy and competitiveness of businesses and destinations (Buhalis, 2019). The shift to e-Tourism and smart tourism signals a broader shift towards ambient intelligence tourism, where technology seamlessly integrates into the travel experience.

It cannot be denied that there is a widespread growth of digital technologies and digital space across the globe. This enables a prompt and sufficient reaction to significant market fluctuations. It is also viewed as a significant step by industry players in the tourism market to enhance business growth and cater to the needs of tourists. Effective utilization of these chances enables organizations to maintain a secure footing in the market. The primary factor in advancing the operations of tourism agencies is the shift to digital platforms, merging with worldwide systems, and incorporating digital marketing channels. This enables the increase of service sales and reduces reliance on supplier terms. The availability of new digital solutions allows for broader market reach and the attraction of tourists who favor aggregator platforms for browsing through various travel packages (Shmarkov et al., 2019).

Customer awareness in tourism has evolved significantly thanks to consumer-centered marketing (CCM) strategies that leverage information and communication technologies (ICT), in short, focusing on the motivations, habits, attitudes and values of businesses' customers (Niininen et al., 2007). By placing the customer at the center of marketing efforts and effectively using ICT tools, travel industry players are able to create personalized and engaging experiences for their customers. Moreover, the adoption of e-business applications, especially mobile e-business applications, has become increasingly vital for the existence of travel agencies that want to stay competitive and meet the changing needs of modern travelers (Wang and Cheung, 2004).

As another technological enabler, the use of mobile technologies such as QR code payment systems has been proven to increase tourist satisfaction and transaction experiences in the tourism industry (Lou et al., 2017). These technologies not only facilitate payment processes but also contribute to overall traveler satisfaction, indicating their potential to advance the tourism industry. Moreover, customer co-creation and adoption of new services are facilitated by tourists' propensity towards technology and their willingness to interact with smart applications for travel-related tasks (Sarmah et al., 2017).

In conclusion, the travel industry, especially the airline industry, has been revolutionized by technology leading to increased efficiency, cost reductions, improved safety and expanded services. Smart tourism applications have integrated travel experiences with technology. This can be seen as an important capability for businesses and destinations to remain competitive. Customized marketing strategies that use ICT tools effectively can provide significant convenience in determining customer needs. In addition, mobile technologies such as QR code payments and smart applications positively affect customer satisfaction and industry development.

Technology in the Food and Beverage Industry

Food and beverage businesses can improve their production processes and customer experiences by adapting to technological developments. Food waste is an important issue for the F&B industry. Food waste can be significantly reduced with a seemingly simple innovative method such as product labeling (Schanes et al., 2018).

While businesses turn to such innovations with commercial concerns such as gaining more competitiveness, the pandemic in 2019 has forced businesses to various innovations. Food and beverage businesses have had to innovate in many areas such as production, safety and transportation. This situation has accelerated the technological integration processes of these businesses (Chowdhury et al., 2020; Sivaraman et al., 2023; Ku, 2023). From implementing contactless payment systems to enhancing digital presence, technology has played a crucial role in helping food and beverage establishments navigate the challenges posed by the pandemic and adapt to the new normal in the industry.

By adopting digital solutions, restaurants can increase efficiency, offer innovative services and optimize various aspects of their business operations. For example, the use of digital menus and QR code technology in restaurants has not only increased operational efficiency but also improved the overall customer experience by providing a seamless and contactless dining experience (Suputra, 2024).

Production speed, which can be considered the most important contribution of technology, is perhaps most prominent in fast-food companies in the food and beverage sector. At this point digital transformation capabilities and tools play a key role in the fast-food industry by enhancing customer value and optimizing customer experiences (Daradkeh et al., 2023). Although, fine-dining restaurants can also benefit from digitalization of services to position themselves as innovators and gain competitive advantage (Vo-Thanh et al., 2022).

New technologies such as labeling have helped reduce food waste. The Covid-19 pandemic has been an opportunity to improve safety, operations and business processes in the food sector, and to some extent, businesses have had to quickly integrate technology into their processes. Technology has also increased efficiency in restaurants, providing innovative services and streamlining operations efficiently. Overall, technology has played a key role in transforming the food and beverage sector, improving various aspects of businesses and customer experiences.

Technology in Small Tourism Enterprices

As mentioned in the previous sections, technology is used effectively in the tourism sector in the areas of management, production and marketing. However, it is obvious that this use may be more limited in small businesses. Access to technological opportunities in the tourism industry can indeed present varying challenges for businesses of different sizes. Large tourism businesses often have more resources and capabilities to leverage technological advancements compared to small tourism businesses.

Large tourism businesses can make effective use of technologies such as cloud computing made possible by their internet access (Gretzel et al., 2015). These facilities play an important role in facilitating smart tourism goals, enabling large businesses to improve their operations and customer experiences through innovative digital solutions. However, small tourism businesses face challenges in improving their business processes through the effective use of ICTs (Ashari et al., 2014).

It can be stated that small businesses face various challenges in technology integration processes. The following are the main challenges faced by small tourism enterprises when integrating technology, supported by relevant research.

- *Limited resources:* Small businesses, which usually survive with small budgets and limited financial resources, may have difficulty investing in new technologies (Ali, 2014).
- *Lack of information:* Many small tourism businesses may lack awareness of the latest technological trends and their potential benefits.
- *Environmental factors:* Factors such as national infrastructure, market size, and country-specific technical aspects can influence technological integration processes in small tourism businesses.
- *Organizational factors:* Internal factors such as organizational awareness, ICT resource capabilities and top management support also play an important role in the technology integration process (Lama et al., 2020).
- *Resistance to change:* Resistance to change can be a problem for small tourism businesses in technology integration (Lei et al., 2023).
- *Lack of infrastructure:* Lack of regional technological infrastructure also complicates the process of technology integration for small tourism enterprises (Karanasios and Burgess, 2008).
- *Lack of qualified staff:* In small tourism businesses, existing employees may not be qualified to use new technological products and processes (Peters, 2005).
- *Complexity of integration processes*: Integrating various technologies and systems in small tourism businesses can be complex and challenging (Reino et al., 2016).

Conclusions

Businesses operating in the tourism sector must rapidly adapt to digital transformation processes in order to survive in an increasingly competitive environment and provide better service to their customers (Azadaliyev and Demirkol, 2023). Information-based systems, in other words office automation systems, are systems that change the way of doing business in the increasingly busy and stressful environment of offices, motivating employees by positively affecting their office life and increasing their productivity (Altınöz, 2008).

Many factors such as economic and technological changes that occur with globalization, computer-aided production and design systems, changing expectations of consumers, and the struggle for fast production force all tourism businesses, large or small, to digital opportunities and technological innovations.

Technology, such as information technology, the use of social media and knowledge management, can bring opportunities but also challenges. As seen in developing countries such as Malaysia and Ecuador, small tourism businesses need to adopt affordable technological products to increase their competitiveness. For example, e-commerce can give small tourism businesses a competitive advantage. Hospitality businesses need to adapt to digital transformation processes to provide better services. Smart tourism can effectively use ICTs to improve visitor experiences, operational efficiency and promote sustainable tourism development globally.

Smart tourism is an important tool to improve service delivery, management practices and resource optimization. Machine learning and data mining can improve tourist experiences by providing personalized attraction recommendations. Despite concerns over data security, smart tourism is seen as a valuable tool to enhance visitor experiences, increase efficiency and ensure sustainable development in the sector through the integration of physical technologies and information technologies.

Technology has revolutionized the travel industry. The use of information and communication technologies can be considered as a major step forward in travel experiences and practices. For all enterprices, being prepared for technological innovations is crucial for customer satisfaction. On the other hand, the integration of technology in food and beverage businesses has helped to reduce food waste, improve food safety and optimize operations. Overall, technology plays an important role in the survival of each level and unit of the tourism sector and in increasing customer satisfaction.

While larger companies have more resources to leverage technology, smaller businesses in the tourism sector may face challenges in accessing these opportunities. Large tourism businesses are leveraging technologies such as cloud computing to achieve their smart tourism goals and improve operations and customer experiences. For example, hotel chains are able to aggregate customer data from their hotels in different destinations and analyze this data to develop marketing policies. In contrast, small tourism businesses struggle with technology and challenges such as limited resources, lack of information, environmental and organizational factors, resistance to change, lack of infrastructure, lack of skilled staff and complexity of integration processes. With limited resources, these businesses may turn to smaller solutions. In other words, they can benefit from technological developments within their own means. At this point, a number of suggestions can be offered. To list them;

- Companies that can produce software and hardware in the field of tourism can increase applications for small businesses.
- Budget support for software and hardware can be provided to small businesses at the government level.
- Small businesses can come together through various formations and agreements and make agreements with technology companies in line with their common needs for software or hardware.
- In order to keep abreast of the inevitable technological developments, the educational content of the institutions providing tourism education at various levels should be organized especially in the relevant software areas.
- It should be ensured that existing staff or operators do not fall behind technological developments by giving importance to in-service or out-of-service trainings on the subject.
- Importance should be given to the technological infrastructure needed by rural areas where small businesses are more common.

References

Ali, V. (2014). Factor affecting the adoption of information and communication technology in the tourism sector of the maldives.. https://doi.org/10.26686/wgtn.17009135

Altınöz, M. (2008). Ofis otomasyon sistemlerinin bireysel performans üzerine etkisi. Selçuk Üniversitesi Sosyal Bilimler Enstitüsü Dergisi, (20), 51–63.

Ashari, H. A., Heidari, M., & Parvaresh, S. (2014). Improving SMTEs' Business Performance through Strategic Use of Information Communication Technology: ICT and Tourism Challenges and Opportunities. International Journal of Academic Research in Accounting Finance and Management Sciences, 4(3), 1–26.

Azadaliyev, S., & Demirkol, S. (2023). Turizm Sektöründe Artırılmış Gerçeklik ve Dijital Dönüşümün Değerlendirilmesi. Turizm Çalışmaları Dergisi, 5(1), 11–26.

Azaltun, M. (2003). Bilgisayarli otel yönetim sistemlerinde muhasebe paket programinin yeri ve gelirler için uygulama örnegi. Anadolu Üniversitesi Iktisadi Ve Idari Bilimler Fakültesi Dergisi, 19(1), 161–174.

Buhalis, D. (1998). Strategic use of information technologies in the tourism industry. Tourism Management, 19(5), 409–421.

Buhalis, D. (2019). Technology in tourism-from information communication technologies to etourism and smart tourism towards ambient intelligence tourism: a perspective article. Tourism Review, 75(1), 267–272.

Chowdhury, M. T., Sarkar, A., Paul, S. K., & Moktadir, M. A. (2020). A case study on strategies to deal with the impacts of covid-19 pandemic in the food and beverage industry. Operations Management Research, 15(1–2), 166–178.

Daradkeh, F., Hassan, T., Palei, T., Helal, M., Mabrouk, S., Saleh, M., ... & Elshawarbi, N. (2023). Enhancing digital presence for maximizing customer value in fast-food restaurants. Sustainability, 15(7), 5690.

Gretzel, U., Sigala, M., Xiang, Z., & Koo, C. (2015). Smart tourism: foundations and developments. Electronic markets, 25, 179–188.

Halawani, F., Rahman, M., & Halawani, Y. (2013). A proposed framework for e-commerce usage and competitive advantage on small and medium tourism enterprises (smtes) in lebanon. Journal of Social and Development Sciences, 4(6), 258–267.

Hamid, M. A., Rahmat, N., & Azmadi, A. S. A. (2023). Stakeholders' perception of smart tourism technology for tourism destination. International Journal of Academic Research in Business and Social Sciences, 13(4).

Hannam, K., Butler, G., & Paris, C. M. (2014). Developments and key issues in tourism mobilities. Annals of Tourism Research, 44, 171–185.

Karanasios, S. and Burgess, S. (2008). Tourism and internet adoption: a developing world perspective. International Journal of Tourism Research, 10(2), 169–182.

Ku, E. (2023). Role of inter-organizational systems in driving tourism businesses forward in the post-covid-19 new normal. Journal of Business and Industrial Marketing, 38(11), 2471–2484.

Lama, S., Pradhan, S., & Shrestha, A. (2020). Exploration and implication of factors affecting e-tourism adoption in developing countries: a case of nepal. Information Technology & Tourism, 22(1), 5–32.

Law, R. and Jogaratnam, G. (2005). A study of hotel information technology applications. International Journal of Contemporary Hospitality Management, 17(2), 170–180.

Law, R., Buhalis, D., & Cobanoglu, C. (2014). Progress on information and communication technologies in hospitality and tourism. International journal of contemporary hospitality management, 26(5), 727–750.

Lei, J., Indiran, L., & Kohar, U. (2023). Barriers to digital transformation among msme in tourism industry: cases studies from bali. International Journal of Academic Research in Business and Social Sciences, 13(3).

Lou, L., Tian, Z., & Koh, J. (2017). Tourist satisfaction enhancement using mobile qr code payment: an empirical investigation. Sustainability, 9(7), 1186.

Mehraliyev, F., Choi, Y., & Köseoğlu, M. A. (2019). Progress on smart tourism research. Journal of Hospitality and Tourism Technology, 10(4), 522–538.

Niininen, O., Buhalis, D., & March, R. (2007). Customer empowerment in tourism through consumer centric marketing (CCM). Qualitative Market Research: An International Journal, 10(3), 265–281.

Peters, M. (2005). Entrepreneurial skills in leadership and human resource management evaluated by apprentices in small tourism businesses. Education + Training, 47(8/9), 575–591.

Reino, S., Alzua-Sorzabal, A., & Baggio, R. (2016). Adopting interoperability solutions for online tourism distribution. Journal of Hospitality and Tourism Technology, 7(1), 2–15.

Sarmah, B., Rahman, Z., & Kamboj, S. (2017). Customer co-creation and adoption intention towards newly developed services: an empirical study. International Journal of Culture, Tourism and Hospitality Research, 11(3), 372–391.

Schanes, K., Dobernig, K., & Gözet, B. (2018). Food waste matters - a systematic review of household food waste practices and their policy implications. Journal of Cleaner Production, 182, 978–991.

Shmarkov, M. S., Shmarkova, L. I., & Shmarkova, E. A. (2019). Digital technologies in the organization and management of tourist organizations. In 1st International Scientific Conference" Modern Management Trends and the Digital Economy: from Regional Development to Global Economic Growth"(MTDE 2019) (pp. 98–101). Atlantis Press.

Sivaraman, T., Zahrin, M., Roni, M., & Kanapathipillai, K. (2023). A study on the factors that impact the future trends of marketing: evidence from the hospitality industry in klang valley, malaysia. European Journal of Management and Marketing Studies, 8(1).

Suputra, I. (2024). Pendampingan digitalisasi teknologi melalui menu qr code di mango taru restaurant & bar. Jurnal Sewaka Bhakti, 10(1), 60–68.

Tian, Z., Zheng, J., Hou, Q., Zhao, W., Shen, H., & Ye, S. (2019). Enhancing tourist experience with smart tourism— a case study of hangzhou. DEStech Transactions on Economics, Business and Management, (icaem).

Um, T., Kim, H., Kim, H., Lee, J., Koo, C., & Chung, N. (2022). Travel incheon as a metaverse: smart tourism cities development case in korea. Information and Communication Technologies in Tourism 2022, 226–231.

Vatan, A., & Dogan, S. (2021). What do hotel employees think about service robots? A qualitative study in Turkey. Tourism Management Perspectives, 37, 100775.

Vo-Thanh, T., Zaman, M., Hasan, R., Akter, S., & Dang, V. (2022). The service digitalization in fine-dining restaurants: a cost-benefit perspective. International Journal of Contemporary Hospitality Management, 34(9), 3502–3524.

Wang, S. and Cheung, W. (2004). E-business adoption by travel agencies: prime candidates for mobile e-business. International Journal of Electronic Commerce, 8(3), 43–63.

Ying, W., So, K. K. F., & Sparks, B. (2016). Technology readiness and customer satisfaction with travel technologies: a cross-country investigation. Journal of Travel Research, 56(5), 563–577.

Murat Hacimurtazaoğlu[1] and Gülsün Yildirim[2]

Chapter 34 Evaluation of the Use of Fuzzy Logic in Decision-Making Processes in Tourism

Introduction

Fuzzy logic allows for modelling real-world situations using values between two concepts ([0,1]), unlike traditional logic systems that use a binary (yes/no-0/1) decision system. Like human behavior, fuzzy logic is a computer logic revolution that assists computers in logical applications (Hacımurtazaoğlu, 2013). Decision-making processes in tourism are directly influenced by many factors such as customer expectations, preferences, satisfaction levels and environmental factors which may vary from person to person. The use of fuzzy logic in decision-making processes in tourism enables modelling uncertain and ambiguous situations with fuzzy logic systems to develop more accurate and flexible decision-making mechanisms.

The tourism industry has become increasingly reliant on information systems to support decision-making processes (Kutsenko et al., 2020; Chang & Chang, 2015). In recent years, researchers have explored the application of fuzzy logic to develop decision support systems for the tourism sector (Rochman et al., 2020; (Shpolianskaya et al., 2020).

In the tourism sector, various expert systems are being developed based on fuzzy logic-based decision support systems such as selecting accommodation facilities, pricing strategies, customer satisfaction analysis, and recommendations for activities based on weather forecasts. These developed expert systems provide advantages to tourism businesses in gaining deeper insights into customer behaviors and preferences, personalizing their services and improving their market positions. Additionally, expert systems developed with fuzzy logic help businesses better understand the industry dynamics and customer expectations, manage uncertainties they face in adapting to them, and enhance their capacity to provide customer-focused services. This study provides a comprehensive

---

1 Ph.D., Recep Tayyip Erdoğan University, Ardeşen Vocational School, Department of Computer Technology, murat.hacimurtazaoglu@erdogan.edu.tr
2 Ph.D., Recep Tayyip Erdoğan University, Ardeşen Tourism Faculty, Department of Gastronomy and Cuisine Arts, gulsun.yildirim@erdogan.edu.tr

examination of the applications of expert systems developed with fuzzy logic in decision-making processes within the field of tourism.

Decision-making and performance-determination processes involve the use of quantitative data. However, expressing some situations verbally in mathematical terms is quite challenging. At this stage, fuzzy logic allows for formulating information that cannot be described mathematically and includes uncertain information when making decisions (Ablyazov et al., 2020).

When making decisions on many issues, the Aristotelian logic is used, allowing us to qualify events and facts as true-false, hot-cold, and black-white. However, many expressions used in daily life have a blurry structure and Aristotelian logic falls short in classifying these uncertain areas. In this case, fuzzy logic plays an important role in measuring values that are considered not sensitive enough with classical logic. Fuzzy logic expresses that phenomena, events and decision-making mechanisms cannot be separated by clear lines like black-white or true-false (Özdemir & Kalınkara, 2020). There are hundreds of intervals and discontinuities between zero and one (Işıklı, 2004). Fuzzy logic theory holds an important place in resolving situations involving uncertainty.

Fuzzy Logic and Expert Systems

The fuzzy logic concept was first introduced by Lotfi Zadeh in 1965 (Hacımurtazaoğlu, 2013). Fuzzy logic is an artificial intelligence method used for modelling and solving problems that involve uncertainty (Zhu & Zhang, 2012). According to fuzzy logic, everything is a matter of degree and allows the representation of data on a continuum. Zadeh introduced concepts such as fuzzy sets, membership functions, fuzzy rules, and fuzzy reasoning to express situations with indefinite boundaries (Ben-Ari & Mondada, 2017). Fuzzy logic aims to imitate the human decision-making process which often involves subjective and imprecise information. By using fuzzy rules and membership functions, fuzzy logic systems can handle uncertain and vague information, providing more flexible and accurate solutions compared to traditional binary logic. Fuzzy logic can be described as a tool used to deal with complex real-world situations involving uncertainty. Expert systems are computer-based systems programmed with expert knowledge that can make decisions resembling human expertise. Expert systems typically store expert knowledge in a database, process data, and provide solutions for specific problems. The convergence of fuzzy logic and expert systems has expanded their applications in decision-making processes within the tourism sector (Priyanta, 2022). These methods help users make more informed decisions leading to dealing with complex or uncertain situations while facilitating effective solutions.

Fuzzy logic and expert systems are applied in various fields including decision-making and pattern recognition. Fuzzy logic has a wide range of applications in different disciplines such as engineering, computer-internet technologies, cybernetics, facial recognition systems, space vehicles, electronic technologies, robot technologies, warfare technologies etc. The theoretical basis of fuzzy logic includes the theory of dynamic systems and soft computing theories. One area of application for fuzzy logic is fuzzy control systems which have made classical control systems more efficient. Since the 1980s, the number of countries using fuzzy logic has been increasing. Japan leads these countries followed by France, Germany, Russia, Danish, and China (Işıklı, 2004).

Decision Making Processes and Management in Tourism

According to the needs of the era, the tourism sector is in a constant state of change and development, and decision-making processes have a complex structure that involves diverse information. Therefore, decision-making processes and management are of great importance in tourism (Uçak, 2008). Decision-making processes and management in tourism are significant factors that involve the analysis of situations affecting the operations of a tourism enterprise. In the decision-making process, it is necessary to analyze existing data first to evaluate the current situation (Bal et al., 2020). Thus, identifying the strengths and weaknesses, opportunities and threats of the business is possible. This allows for the development of strategies and action plans to meet customer expectations, gain competitive advantage, increase market share, and improve business performance. Businesses must provide a competitive advantage to effectively meet changing tourist demand and expectations (Sirakaya & Woodside, 2005). This situation makes decision-making processes and management in tourism important for businesses to achieve their sustainability and growth goals (Bal et al., 2020).

To optimize decision-making processes in tourism and minimize the error rate in decisions, the use of expert systems developed with fuzzy logic can create flexible movement space for businesses. These systems, based on a wide database, are capable of complex calculations and evaluating various decision alternatives to generate results. Additionally, they can provide quick solutions to problems commonly encountered in the tourism sector such as ambiguity and uncertainty. Expert systems developed with fuzzy logic may provide competitive advantages to businesses in the tourism sector and enhance customer satisfaction. These systems optimize crucial processes such as data analysis, making predictions, and managing risks during decision-making processes. The flexible capabilities of these systems ensure effective management of decision-making supported by expert systems utilizing fuzzy logic in the tourism sector (Taş & Çakır, 2022). The use of these systems is crucial for optimizing decision-making processes and enhancing customer satisfaction.

Fuzzy logic-based expert systems will rapidly increase their impact on the tourism sector. These systems can provide businesses with highly effective contributions, especially in areas such as data analytics, personalized services, and future prediction. The tourism sector should closely monitor these innovative technologies and utilize these systems to ensure the future success of businesses, enhance efficiency, and improve customer satisfaction.

The Role of Expert Systems in Tourism

The use of expert systems in the tourism sector not only influences the management processes of businesses but also transforms the tourism experience. These systems enable easy and fast travel planning, finding the most suitable prices, and making holiday experiences more comfortable and enjoyable. In the tourism sector, tour operators and travel agencies can enhance service quality by customizing tour packages and vacation plans with the information and analytic capabilities provided by expert systems.

Another important impact of expert systems in the tourism sector is related to destination and tourism area planning. These systems are utilized by destination managers in the areas of planning tourism activities, resource management, and developing sustainability strategies. Through advanced technology and analytical capabilities, these systems provide a competitive advantage to tourism businesses and can enhance the sustainability of the industry.

Expert systems play a crucial role in the tourism sector, particularly in automating and improving decision-making processes. They can assist tourism businesses in making better decisions in vital areas such as planning, marketing, reservation management, service quality assessment, risk analysis, and security management. Tourism businesses using expert systems can manage their operations more effectively and develop action plans based on understanding customer demands and expectations.

Advantages of Expert Systems Enhanced with Fuzzy Logic

Fuzzy logic technology can evaluate complex problems in the field of tourism. Expert systems developed with fuzzy logic can effectively process data containing uncertainty and complexity, leading to accurate results. Fuzzy logic does not require a large amount of data to produce results. This saves time and resources in the tourism sector. As a result, the use of expert systems developed with fuzzy logic in the tourism industry improves decision-making processes, facilitates solving complex problems, reduces time loss, and provides businesses with an advantage in making strategic decisions.

Areas of Application for Expert Systems Developed in Tourism Using Fuzzy Logic

Accommodation Reservation Management with Fuzzy Logic

The optimization of the reservation management process is crucial for the success of businesses. Expert systems used to evaluate reservation requests in accommodation facilities can provide the most suitable options by considering customer preferences. By recommending accommodation options that meet customers' needs and expectations, these systems can enhance customer satisfaction. Additionally, they can contribute to increasing occupancy rates by making demand predictions (Jayachandran et al., 2005). This enables businesses to optimize their operations and improve customer experiences. The use of these systems provides businesses with the opportunity to build better relationships with their customers and meet their expectations, ultimately leading to sustainable business models through enhanced customer relations.

Evaluation of Service Quality in Tourism with Fuzzy Logic

Specialist systems developed to improve customer satisfaction and business performance in the tourism sector can help tourism businesses analyze customer feedback more effectively and provide better service. Customer feedback helps businesses identify shortcomings and improve their service quality. As a result, businesses can create action plans based on the recommendations of specialist systems. With these action plans, businesses can take strategic steps to meet customer expectations and gain a competitive advantage.

Evaluation of reservation and staff performance statistics is among the significant factors affecting service quality. These statistics can aid businesses in measuring the effectiveness of the reservation process and customer satisfaction, as well as assessing staff skills, accuracy, and service quality. Expert systems can utilize data sources to analyze staff performance and indicate areas where businesses need to focus on personnel training or service improvement. Consequently, businesses can enhance staff quality and provide better service to ensure customer satisfaction. Furthermore, by analyzing industry practices and evaluating customer expectations, they can guide businesses in establishing service quality standards and continuously improving them to meet customer expectations effectively.

Tourism Promotion and Marketing Strategy Planning with Fuzzy Logic

One of the primary goals of businesses in the tourism sector is to enhance customer satisfaction and stand out in a competitive environment. To achieve these objectives, tourism enterprises need to effectively execute their advertising and promotional plans. During this period, expert systems can provide significant support to tourism businesses.

Expert systems can assist tourism businesses in analyzing their target audiences, creating personalized advertising strategies, and planning promotional campaigns. The use of these systems facilitates the acquisition of target audiences, thereby enabling tourism businesses to enhance their success and customer satisfaction in a competitive environment. Consequently, the utilization of expert systems developed with fuzzy logic holds great potential for the tourism sector (Putra et al., 2022)

Tourism enterprises need to first develop the right strategy and planning to effectively utilize expert systems. It is crucial for businesses to identify their needs and determine the areas where they will use expert systems. Therefore, proper training and guidance become essential for enterprises to use expert systems effectively (Liebowitz, 2007). Furthermore, for businesses to fully benefit from the potential of expert systems and gain a competitive advantage, they need to continuously update and improve expert systems.

Demand Forecasting and Capacity Planning in Tourism Using Fuzzy Logic

In the highly competitive tourism sector, it is crucial for businesses to make future and develop corresponding actions to be successful. Demand forecasting and capacity planning play an exceptionally important role in this process. Expert systems can provide accurate forecasts by considering factors such as seasonal fluctuations and special events, with the aim of improving demand predictions in the industry. Therefore, utilizing expert systems developed with fuzzy logic in the tourism sector can offer businesses the opportunity to make more precise forecasts and consequently make more effective decisions. With expert systems, businesses can enhance their ability to manage the complex dynamics of the industry, provide better service to customers, make smarter and more informed decisions, plan for a better future, and ultimately improve business success.

Hotels or tourism enterprises can effectively use expert systems in capacity planning. This enables businesses to determine the optimal capacity and enhance their ability to meet demand successfully, thus improving efficiency and maintaining customer satisfaction.

The continued use of expert systems in businesses can contribute to demand forecasting and capacity planning within the industry, effective management of operations, and enhancement of business competitiveness.

Target Market Determination in Tourism with Fuzzy Logic

In the tourism sector, the use of expert systems developed with fuzzy logic plays a significant role in the process of target market identification. These systems enable a detailed analysis of travel habits and can guide business managers in determining target markets based on preferences. Focusing on the right target audience provides businesses with a strategic advantage and increases their chances of success in today's competitive tourism sector. Advanced systems enable more effective analyses, market segmentation, and the development of accurate strategies targeting the desired audience. This in turn can facilitate businesses to increase their market shares and pursue a sustainable growth strategy.

Expert systems have a wide database on the analysis of travel habits, and through detailed examination of this data, they assist businesses in better understanding their target markets. In this way, businesses can develop accurate strategies and gain a significant competitive advantage.

Meeting customer expectations is a crucial part of increasing customer satisfaction and fostering loyal clientele. Expert systems assist businesses in better understanding customer expectations and devising strategies to provide superior service. The analyses and guidance provided by these systems enable businesses to plan more effectively and enhance their strategies.

Businesses in the tourism sector can benefit from using fuzzy logic and expert systems to better guide their marketing strategies and achieve greater success in their target markets. Additionally, these systems can provide substantial support to businesses in delivering customer-centric services.

Planning and Organizing Tourism Activities using Fuzzy Logic

Event planning and organization is a crucial aspect of the tourism sector, aiming to enhance travelers' experiences, increase attendance at events, and satisfy participants. The uncertain nature of tourist behaviors can make it challenging to plan and coordinate tourism events. A fuzzy logic system presents an appropriate tool for modelling these uncertain and fuzzy situations. By using fuzzy logic-based approaches, businesses can enhance customer satisfaction, organize events that meet expectations, and provide tourists with the best experience possible. These approaches have the capability to analyze natural uncertainties associated with decision-making processes in tourism-related activities, enabling personalized recommendations for tourism businesses and tourists alike. Consequently, event planning and organization in the tourism industry can be optimized by considering participant preferences and satisfaction.

The widespread use of developed expert systems can further advance the tourism sector. These systems can be utilized for planning and budgeting various activities such as tourism fairs, festivals, conferences, animations, and determining travel routes, as well as optimizing every stage of an event including program scheduling, budget allocation, location selection, and estimated number of participants.

In the increasingly popular tourism applications, the creation of e-tourism destinations has emerged as a social phenomenon that involves using information and communication technologies in addition to a destination. Fuzzy logic techniques can be utilized to pre-determine tourist preferences and facilitate the planning of activities offered to tourists about creating e-tourism destinations (Yapıcı & Yıldırım, 2021).

In previous studies, fuzzy logic-based applications have been developed to determine the selection of suitable destinations for tourists based on criteria such as budget, travel group size, and preferred types of places (Putra et al., 2022).

Investment Evaluation in Tourism with Fuzzy Logic

Given the growth potential of the tourism sector and technological advancements, identifying new destinations, innovative services, accommodation facilities, and technological innovations offers new opportunities for investors in the tourism sector.

In evaluating tourism investments, considerations of profitability, market share, competitive strength, and environmental impacts are essential. It is necessary to consider the uncertainties of these factors. Investment analysis in the tourism sector is a crucial process for determining the potential return and risks of an investment. Processing data with expert systems developed using fuzzy logic that involves uncertainty can help businesses conducting investment analysis in tourism make more informed investment decisions.

Investment analysis processes consider many factors such as market trends, competitive analysis, financial data, and destination selection. Fuzzy logic enables tourism investors to make more accurate, detailed, and information-based investment decisions. This can help investors make successful investments and gain a competitive advantage. The use of expert systems developed with fuzzy logic is important for the development of the tourism sector's investors.

These systems can serve as effective tools for investors and play a guiding role in monitoring and analyzing industry innovations. They can be used as a guide for investors to make strategic plans, identify business opportunities, and determine risk management strategies. Through these systems, industry investors can better assess potential risks, make efficient investments, and optimize profit margins.

Performance Evaluation of Tourism Enterprises with Fuzzy Logic

Measuring and evaluating business performance is crucial for the sustainability of businesses. In this context, it is important to objectively measure and evaluate the performance of businesses operating in the tourism sector (Eroğluer & Mert, 2020). Assessment of the financial and operational performance of tourism enterprises is important for both managers and investors (Özçelik & Kandemir, 2015). However, it is known that the nature of the tourism sector necessitates that traditional financial measurements alone are not sufficient. Therefore, in fields such as the tourism sector where there is high variability and uncertainty, using Fuzzy Logic models may be a more appropriate approach (Çanakçıoğlu & Ersan, 2020).

Expert systems developed for evaluating the performance of businesses in the tourism sector can analyze financial data, customer satisfaction, operational efficiency, and marketing strategies to thoroughly assess business performance. Businesses that utilize these technological solutions can enhance their productivity and gain a competitive advantage.

References

Ablyazov, V I., Драчев, О И., & Tisenko, V N. (2020). Fuzzy Control Under Uncertainty in Machine Building. https://doi.org/10.1088/1757-899x/971/2/022026

Bal, H., Dner, M M., Arslan, M., & Battal, A. (2020). Development of Organizational Entrepreneurship Scale in Turkey: Application in Manufacturing and

Wholesale/Retail Trade Sector. https://www.isarder.org/2020/vol.12_issue.3_article62.pdf

Ben-Ari, M., & Mondada, F. (2017). Fuzzy Logic Control. https://doi.org/10.1007/978-3-319-62533-1_11

Chang, J., & Chang, B. (2015). The Development of a Tourism Attraction Model by Using Fuzzy Theory. Hindawi Publishing Corporation, 2015, 1–10. https://doi.org/10.1155/2015/643842

Çanakçıoğlu, M., & Ersan, O. (2020). The Effect of Working Capital Management on Firm Performance: A Study on Cement Companies. Isarder, 12(3), 2749–2763. https://doi.org/10.20491/isarder.2020.1005

Eroğluer, K., & Mert, İ S. (2020). A Qualitative Exploratory Factor Analysis to Identify Effective Factors in the Emergence of Creative Ideas in Turkey. Isarder, 12(2), 1722–1738. https://doi.org/10.20491/isarder.2020.940

Hacımurtazaoğlu, M., (2013). Bulanık Mantık ile Manyetik Kilit Uygulaması. Akademik Bilişm (pp. 642–648). Antalya, Turkey

Isikli, Ş. (2004). Fuzzy Mantık ve Sibernetik'in Siber Toplum ve Yapay Zeka Üzerindeki Etkileri, Atatürk Üniversitesi, Yüksek Lisans Tezi.

Jayachandran, S A., Sharma, S., Kaufman, P A., & Raman, P. (2005). The Role of Relational Information Processes and Technology Use in Customer Relationship Management. https://doi.org/10.1509/jmkg.2005.69.4.177

Kutsenko, A A., Kovalenko, S., Kovalenko, S., Kashcheiev, L., Mikhnova, O., & Chaly, I. (2020). A Fuzzy Logic Based Approach to E-tourist Attractiveness Assessment. https://doi.org/10.1109/saic51296.2020.9239184

Liebowitz, J. (2007). Expert Systems for Business Applications. https://www.tandfonline.com/doi/abs/10.1080/08839518708927977

Özçelik, H., & Kandemir, B. (2015). Bist'de İşlem Gören Turizm İşletmelerinin Topsis Yöntemi İle Finansal Performanslarinin Değerlendirilmesi. 18(33), 97–114. https://doi.org/10.31795/baunsobed.645449

Özdemir, O., & Kalınkara, Y. (2020). Bulanık Mantık: 2000–2020 Yılları Arası Tez ve Makale Çalışmalarına Yönelik Bir İçerik Analizi. Acta Infologica, 4(2), 155–174.

Özışık Yapıcı, O., & Yıldırım, G. (2021). Endüstri 4.0'ın Turizm Alanındaki Kavramları Üzerine Bir Araştırma. IBAD Sosyal Bilimler Dergisi(11), 394–412. https://doi.org/10.21733/ibad.956298

Priyanta, D P P P. (2022). Tegal Tourism Object Selection Decision Support System Using Fuzzy Logic. https://jurnal.ugm.ac.id/ijccs/article/viewFile/70226/33167

Putra, D P., Priyanta, S., & Priyanta, S. (2022). Tegal Tourism Object Selection Decision Support System Using Fuzzy Logic. Gadjah Mada University, 16(1), 101–101. https://doi.org/10.22146/ijccs.70226

Rochman, E M S., Pratama, I., Husni, H., & Rachmad, A. (2020). Implementation of Fuzzy Mamdani for Recommended Tourist Locations In Madura - Indonesia. IOP Publishing, 1477(2), 022033–022033. https://doi.org/10.1088/1742-6596/1477/2/022033

Shpolianskaya, I., Dolzhenko, A., Степанов, Н Н., & Потапов, Л. (2020). A Fuzzy Model for Selecting Travel Services to Produce Personalized Offers to Internet Users. IOP Publishing, 1703(1), 012014–012014. https://doi.org/10.1088/1742-6596/1703/1/012014

Sirakaya, E., & Woodside, A G. (2005). Building and Testing Theories of Decision Making by Travellers. Elsevier BV, 26(6), 815–832. https://doi.org/10.1016/j.tourman.2004.05.004

Taş, M., & Çakır, E. (2022). A Hybrid Fuzzy MCDM Approach for Sustainable Health Tourism Sites Evaluation. https://doi.org/10.4018/978-1-7998-7979-4.ch004

Uçak, N Ö. (2008). Kütüphanecilik ve Bilgi Yönetimi Literatüründe Kullanıcı. 9(1), 20–40. https://doi.org/10.15612/bd.2008.326

Zhu, X., & Zhang, W. (2012). Research on Fuzzy Control System. https://doi.org/10.1109/icicee.2012.258

Seval Kurt[1] and Cansu Uzun Güripek[2]

Chapter 35 Ecological Approach to Tourism System

Introduction

Throughout the historical process, the impact of humans on the world has evolved, but has lagged behind the magnitude of the damage they have caused. Understanding the dynamics of the economy and the environment, and bringing damage control for evaluating the cost of damage, is a very late stage in human history. Understanding the dynamics of the economy and the environment, and bringing damage control for assesing the cost of damage, is a very late stage in human history. The fact that the intensity and speed of world and human relations have exceeded the renewal capacity of natural resources (Wackernagel vd., 2002) has brought environmental concerns into focus.

The economic structure, which evolved into the Industrial Revolution in the 18th century, technological inventions and colonialism in the 19th century, and space travel, technology and information systems in the 20th century, has entered a process that may result in environmental victory or disaster in the 21st century. In parallel with the economic, social and political developments and socio-economic pressures such as population growth, urbanization and pollution (Zhou, Yabar & Higano, 2016) that mark each century (Şahinöz, 2022), concerns about climate change, global warming, loss of biodiversity, scarcity of natural resources and extinction of cultural heritage have become increasingly common in the 21st century (Frank, Dalle Molle, Gerstlberger, Bernardi & Pedrini, 2016). These situations have confronted humanity with the metaphor of *sustainability*, which is defined as *"the process of meeting the needs of the present without compromising the ability of future generations to meet their own needs"*(UNWTO & UNEP, 2005: 8). Sustainable tourism is defined as *"tourism that meets the needs of visitors, industry, the environment and host communities, taking full account*

---

1 Asst. Prof., Tokat Gaziosmanpaşa University, Zile Dinçerler School of Tourism and Hotel Management, Department of Gastronomy and Culinary Arts, seval.kurt@gop.edu.tr
2 Ph.D., Tokat Gaziosmanpaşa University, Zile Dinçerler School of Tourism and Hotel Management, Department of Recreation Management, cansu.guripek@gop.edu.tr

of current and future economic, social and environmental impacts"(UNWTO & UNEP, 2005: 12).

The human population is an important part of the ecosystem. The lack of a clear explanation of the limits of the Earth's carrying capacity in the context of the human population is a fundamental problem (Costanza, Cumberland, Daly, Goodland & Norgaard, 1997: 107). Considering the increasing world population, the use and depletion levels of human and natural resources are very important in terms of delivering ecosystem resources to future generations. Per capita consumption, especially in developed economies, is higher than the population in developing countries. This situation creates excessive pollution and is therefore by far the biggest contributor to the approaching carrying capacity limits. Developing countries, on the other hand, are experiencing population growth much faster than their economies can afford and account for a large share of the world's population (Costanza, Cumberland, Daly, Goodland & Norgaard, 1997: 10). These economic developments have also deeply affected the tourism sector.

When we look at the development of tourism in the world, starting from European countries with a high level of development, it has spread to the Mediterranean basin and then to other regions with its climate, cultural and natural beauties and historical values. Over time, it crossed borders and continents and gained an international dimension (Jenkins, 1995:269-277). When analyzing tourism development trends around the world, three main conclusions emerge in connection with the potential of tourism (UNWTO, 2013: 14):

1. Tourism is a dynamic sector with upward and downward effects on other economic activities, whether in developed or developing countries.
2. It plays a key role economically by providing tourist flows to developing and underdeveloped countries with attractions such as culture, art, landscape, climate, wildlife, etc.
3. It reduces the level of poverty and increases the chances of competition with its foreign exchange earning effect on both developing and underdeveloped countries.

Given the rapidly increasing international tourism flows, the uncontrolled growth of tourism supply and demand leads to overuse of natural ecological resources by destinations and destruction of cultural attractions, social and economic environmental degradation (Mihalič, 2016, 2020). Another major problem is that developing countries tend to imitate culturally as they are exposed to tourist flows from developed economies (Capocchi, Vallone, Pierotti & Amaduzzi, 2019).

Tourism is generally recognized as an opportunity to promote economic and social development. However, tourism also represents a destructive force that can have a major negative impact on ecosystem structures and processes and degrade non-renewable natural resources. Exponentially increase in the number of tourists day by day and the spread of tourist activity to remote parts of the world reveals the paradoxical character of tourism based on the destruction of natural and cultural attractions (Hillery, Nancarrow, Griffin & Syme, 2001; Lacitignola, Petrosillo, Cataldi, & Zurlini, 2007; Lynn, & Brown, 2003; Mihalič, 2000).

The development of a responsible tourism approach, with a focus on reducing the negative and destructive impacts of tourism, is considered in three dimensions, focusing on *economic, social* and *environmental* consequences (Aguiñaga, Henriques, Scheel, & Scheel, 2018; Buckley & Pannell, 1990; Getz, 983; Elkington, 1998; Nyaupane & Thapa, 2006). At this point, the solution to these problems comes to life with sustainability.

When the tourism system is evaluated in terms of sustainable tourism, it is aimed to contribute to the local economy, the protection of natural and cultural heritage, and the improvement of the quality of life of local communities and visitors by minimizing the negative impacts on the economic, environmental and social consequences of the tourist sending regions and tourism destination.

The principles of sustainable tourism consist of 12 principles, each of equal importance, developed by the United Nations Environment Programme (UNEP) and the United Nations World Tourism Organization (UNWTO). These principles are based on the pillars of sustainability: economic, social and environmental perspectives (UNWTO & UNEP, 2005: 18–19):

Table 35.1. Twelve Aims for Sustainable Tourism

#	Principles	Explanation
1	Economic Viability	To ensure the implementation of policies that will ensure the long-term success and benefit of destinations and tourism enterprises and their competitiveness at national and international level
2	Local Prosperity	To strengthen and promote relationships with local producers and service providers in order to increase the number of overnight stays and expenditures of tourists
3	Employment Quality	To increase the number and quality of local jobs created in tourism in a way that encourages professional specialization without discrimination on the basis of race, gender, disability, etc.

(continued)

Table 35.1. Continued

#	Principles	Explanation
4	Social Equity	To align the economic and social benefits of tourism to include disadvantaged groups by improving incomes, opportunities and services
5	Visitor Fulfilment	To promote responsible tourism by offering tourists a safe, sustainable, innovative experience without discrimination on the basis of race, gender, disability, age or any other basis
6	Local Control	To strengthen cooperation between tourism stakeholders and local governments, and to ensure regional development of tourism with the support of local communities and civil society
7	Community Wellbeing	To create resources and opportunities to protect and improve the quality of life of local people in order to prevent social degradation
8	Cultural Richness	To protect and respect cultural heritage with sensitivity to authenticity and local distinctiveness
9	Physical Integrity	To contribute to the development of urban and rural areas in order to prevent physical and visual deterioration of the environment
10	Biological Diversity	To support natural areas, habitats, wildlife, species and endemism (endemic species) by observing a balance of conservation and utilization and to take measures to minimize damages
11	Resource Efficiency	To raise awareness of natural, cultural and historical values at national and international level in order to increase the efficiency of resource utilization in tourism enterprises and to observe the balance between conservation and utilization in resource utilization
12	Environmental Purity	To inform all stakeholders about the negative environmental impacts of tourism in order to minimize air, water and soil pollution and waste from tourism and to promote the concepts of responsible business and responsible tourists

Source: (UNWTO & UNEP, 2005: Making tourism more sustainable: a guide for policy makers, 18–19)

The Relationship between Economy and Ecology

If we look at the conceptual background, both words come from Greek. The term ecology is derived from *oikos* (house/household) and *logos* (science). The term economy is adapted from the word *oikonomia*. This word is composed of the ancient Greek *oikos* (house/household) and *nómou* (administration, law). Both concepts have the same origin: ecology extends from the home to all life on earth, while economics extends from the home to the production and consumption activities of all humanity on earth. Contrary to popular belief, both fields of science have very strong ties with each other. As in economic crises, ecological crises also lead to a decline in the welfare of humans and other living beings. Natural sciences, especially physics, have an important role in the development of economic thought. In fact, both fields of science have rules that have become fundamental laws. These rules can be characterized as follows (Kanbir, 2023):

Table 35.2. Common Key Issues in Ecology and Economics

Ecology (Nature)	Economy (Market)
Nature is a whole.	The economic structure is a whole.
Nature is finite.	Resources are scarce.
Nature has self-control.	The free market is self-regulating.
There is diversity in ecology.	The market is a system of many buyers and sellers.
In nature everything transforms.	The economy is based on income/spending and production/consumption relations.
There is no benefit without a price.	Everything has a price.
Nature finds the most appropriate solution.	The market finds the optimal solution.
Interfering with nature causes harm.	Government intervention in the market creates a crisis.
Nature finds its own balance.	The market stabilizes by itself.
Competition between species benefits the ecosystem.	Competition between firms is in the interest of the market.

Source: Kanbir (2023). Ekonomik gelişme ve ekoloji.

The Relationship between Tourism and Ecology

In line with the systems approach, the tourism system, which emerges from the interaction of different main elements (Govers, Van Hecke & Cabus, 2008; Hall 2005; Hall and Müller, 2004; Leiper, 1979), consists of the transit region that provides the connection between the generating region and the destination (Leiper, 1979; Cooper and Hall, 2008: 6). During this process, tourists receive services from tourism enterprises in many different areas (such as accommodation, food and beverage, entertainment, shopping, visits to ruins and museums) and interact with these enterprises. At the same time, the tourist, who temporarily leaves his/her place of residence, interacts with both the employees and the local people in the destination he/she reaches, thus paving the way for the emergence of economic and sociological consequences. On the other hand, natural, historical and cultural beauties and climate are very important as attraction factors in the preference of the destination. In this context, economic, sociological and environmental foundations, which are expressed as the pillars of sustainability, are among the basic criteria for tourism to exist in a region.

National legislators and local governments formulate plans and policies for the development of the region based on tourism. Within the scope of these policies, the effects of the social and environmental factors of the tourism event are tried to be determined and economic rules are defined. In this respect, tourism enterprises are guided in guiding, supervising and systemizing the practices of tourism enterprises, sustainability and protection of ecological balance. Businesses within the tourism system can be described as follows (Cooper & Hall, 2008: 8):

Table 35.3. Basic Elements of Tourism Production within the Tourism System

Generating Region	**Transit Region and Route**	**Destination**
Travel agencies Tour operators Online retailers Online distributors	Aviation services Bus and train services Cruise and ferry services Rental car services	Accommodation Meetings and exhibitions Theme parks Casinos Visitor centers National parks Restaurants Events Facilities Local transport

Source: Cooper & Hall, (2008). Contemporary tourism: an international approach.

A set of adaptable, destination-specific sustainable tourism indicators are being worked on at each stage of the policy process, collecting important data from stakeholders, designing and monitoring targets and indicators, and analyzing results in a continuous iterative approach. In the light of these indicators, a learning process is created to determine capacities, formulate plans and policies and mitigate the negative impacts of tourism. The relationships of these indicators are as follows (Twining-Ward & Butler, 2002):

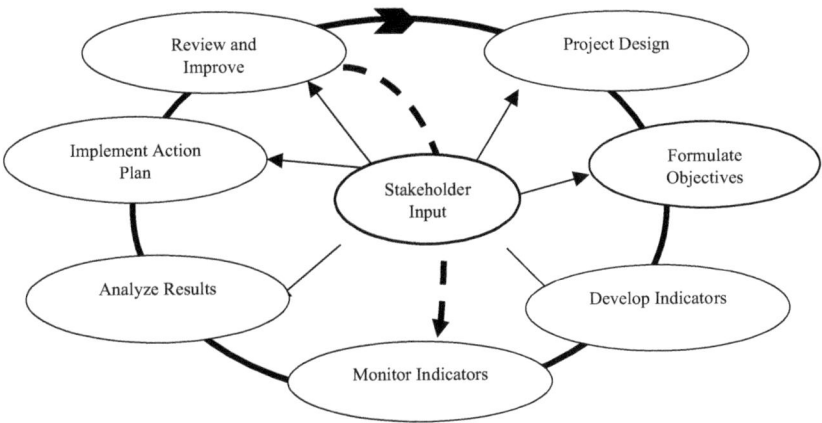

Image: Sustainability Indıcators

Source: Farrell & Twining-Ward, 2004; Twining-Ward & Butler, 2002.

In terms of sustainability, this process is a continuum with the impact of feedback. If there is a gap between the results and the achievement or deviation from the targets, it is important to take the necessary measures, eliminate the deficiencies and do better. Planning for a competitive/sustainable destination should go beyond economic and social benefits to include community and nature-based benefits. At this stage, the full integration and mutual support of planning and policy processes is vital for destination success. Plans and policies that are free from excessive commercialization, that contribute to the development of natural and cultural beauties and, moreover, that can prevent the negative damage that local people and tourists may cause to each other, and that both protect and improve the environment should be established (Goeldner & Ritchie, 2022/2012). With the awareness that the natural environment is an important factor in tourist mobility and destination choice, it is necessary to act with long-term and sustainable goals in order to develop the tourism system and move it forward (Güripek & Usta, 2018).

Conclusion

The positive contribution of the tourism system to the world economy can only be sustained with an ecological approach. Because the most important resource of the tourism system, which is essentially a mobility event, is the natural habitat. Considering that oxygen and water are the two most important elements necessary for the survival of living things, the tourism system is in direct need of these two elements in order to survive. Tourism can play a critical role in ensuring water access and safety, as well as hygiene and sanitation for all. Efficient use of water in the tourism sector, combined with appropriate safety measures, wastewater management, pollution control and technology efficiency, can be the key to protecting our most precious resource (UNWTO, 2015).

Ecology, which examines the relationship between living things and their environment with a human and living-centered, transformative, universal and integrated approach, should be integrated into the tourism system in the same way. Many negative actions of tourism enterprises and the people and vehicles involved in this system, such as harmful wastes, carbon footprints, destruction of the natural environment, fauna and flora, air, water and soil pollution, excessive pressure on the destination, architectural pollution, unnecessary energy consumption, should be replaced by a protected life cycle.

Ecological balance should be taken into account in each stakeholder of the basic elements of tourism production, which Cooper & Hall (2008) consider as tourist sending, transit and destination region. Tourists, who are the subject of tourism, should be made aware of developing an environmentally sensitive understanding and should be responsible for the waste and use of resources. Just as they avoid excessive consumption at home or in their own countries, they should also have the sensitivity to observe this during their vacations. This tourism system, which is used by tourists from the moment they leave their homes, requires stakeholders to reduce waste, save energy, minimize carbon monoxide emissions, take measures and precautions against environmental and nature destruction, and reduce excessive use of resources in terms of ecological balance.

The use of clean energy and renewable energy sources, environmentally friendly energy production, green marketing tactics, development of good agriculture, raising people's awareness of tourism and travel, careful architectural projects, and actions to increase natural resources will bring about a more sustainable ecological tourism approach. Therefore, calculating and executing the plans separately for each stakeholder in the tourism system (for example, starting from transportation vehicles to the facility where tourists shop) and transforming them into policies will support a sustainable world and subsequently a sustainable tourism system. However, at this point, all stages of sustainability indicators should be implemented in a controlled manner, feedback should be taken into account and any errors should be reviewed. In this context, steps should be taken with sustainability awareness at national and local level, sustainability indicators should be put into planning, environmental awareness should be created in the society and ecological balance should be given importance.

References

Aguiñaga, E., Henriques, I., Scheel, C., & Scheel, A. (2018). Building resilience: A self-sustainable community approach to the triple bottom line. *Journal of Cleaner Production*, 173, 186–196.

Buckley, R. and Pannell, J. (1990). Environmental impacts of tourism and recreation in national parks and conservation reserves. *Journal of Tourism Studies*, 1 (1), 24–32.

Capocchi A, Vallone C, Pierotti M, Amaduzzi A. (2019). Overtourism: a literature review to assess implications and future perspectives. *Sustainability*. 11(12):3303. 1–18.

Cooper, C. and Hall, C., M. (2008). *Contemporary tourism: an international approach*. Oxford: Butterworth-Heinemann.

Costanza, R., Cumberland, J. H., Daly, H., Goodland, R., & Norgaard, R. B. (1997). *An introduction to ecological economics* (1st Edition.). Florida: CRC Press.

Elkington, J. (1998). Partnerships fromcannibals with forks: The triple bottom line of 21st-century business. *Environmental Quality Management*, 8(1), 37–51.

Farrell, B. H., & Twining-Ward, L. (2004). Reconceptualizing tourism. *Annals of Tourism Research*, 31(2), 274–295.

Frank, A. G., Gerstlberger, W., Paslauski, C. A., Lerman, L. V., & Ayala, N. F. (2018). The contribution of innovation policy criteria to the development of local renewable energy systems. *Energy Policy*, 115, 353–365.

Getz, D. (1983). Capacity to absorb tourism: Concepts and implications for strategic planning. *Annals of Tourism Research*, 10(2), 239–263.

Goeldner C. R. & Ritchie, J. R. B. (2011). Tourism: principles, practices and philosophies. (12th Edition).

Goeldner C. R. & Ritchie, J. R. B. (2022). Tourism: principles, practices, philosophies. (Kurt, S., Trans.) in *Tourism: principles, practices and philosophies*. (12th Edition). (348–370). New Jersey: John Wiley & Son. (Original work published 2012)

Govers, R., Van Hecke, E. and Cabus, P. (2008). Delineating tourism: defining the usual environment. *Annals of Tourism Research*. 35(4). 1053–1073.

Güripek, E, & Usta, Ö. (2018) Turizm destinasyonlarının rekabet gücünün artırılmasında stratejik destinasyon yönetimi: Çeşme Alaçatı destinasyonu üzerine bir uygulama. *Journal of Tourism & Gastronomy Studies*, 6(4), 496–523.

Hall, C. M. (2005). *Tourism: rethinking the social science of mobility*. Harlow: Prentice-Hall.

Hall, C. M. and Müller, D. K. (Eds.) (2004). *Tourism, mobility and second homes: between elite landscape and common ground*. Clevedon: Channelview Publications.

Hillery, M., Nancarrow, B., Griffin, G., & Syme, G. (2001). Tourist perception of environmental impact. *Annals of Tourism Research*, 28(4), 853–867.

Jenkins, C. L., (1995). Tourism policies in developing countries, in Medlik, S. (ed.) *Managing tourism*, Oxford: Butterworth-Heinemann Ltd. 269–277.

Kanbir, Ö. (2023). Ekonomik gelişme ve ekoloji. *Liberal Düşünce Dergisi*, (109), 79–103.

Lacitignola, D., Petrosillo, I., Cataldi, M., & Zurlini, G. (2007). Modelling socio-ecological tourism-based systems for sustainability. *Ecological Modelling*, 206(1–2), 191–204.

Leiper, N. (1979). The framework of tourism: towards a definition of tourism, tourist, and the tourist industry. *Annals of Tourism Research*. VI(4). 390–407.

Lynn, N. A., & Brown, R. D. (2003). Effects of recreational use impacts on hiking experiences in natural areas. *Landscape and Urban Planning*, 64(1–2), 77–87.

Mihalič, T. (2000). Environmental management of a tourist destination: A factor of tourism competitiveness. *Tourism Management*, 21(1), 65–78.

Mihalič, T. (2016). Sustainable-responsible tourism discourse – Towards 'responsustable' tourism. *Journal of Cleaner Production*, 111(Part B), 461–470.

Mihalič, T. (2020). Conceptualising overtourism: A sustainability approach. *Annals of Tourism Research*, 84 (2020) 103025, 1–12.

Nyaupane, G. P., & Thapa, B. (2006). Perceptions of environmental impacts of tourism: A case study at ACAP, Nepal. *The International Journal of Sustainable Development and World Ecology*, 13(1), 51–61.

Şahinöz, A. (2022). Çevre sorunlarına "ekolojik ekonomi" açısından yaklaşım. *Efil Journal of Economic Research*, 5(2), 67–91.

Twining-Ward, L., & R. Butler (2002). Implementing sustainable tourism development on a small island: development and the use of sustainable tourism development indicators in samoa. *Journal of Sustainable Tourism*, 10(5):363–387.

UNWTO (World Tourism Organization), & UNEP (UN Environment Programme) (2005) *Making tourism more sustainable: a guide for policy makers*, UNWTO, Madrid and UNEP, Paris.

UNWTO (World Tourism Organization). (2013). *Sustainable tourism for development guidebook*. (1th Edition). Institutional and Corporate Relations Programme: Madrid.

UNWTO (World Tourism Organization). (2015). *Tourism and the sustainable development goals.* Madrid, Spain.

Wackernagel, M., Schulz, N. B., Deumling, D., Linares, A. C., Jenkins, M., Kapos, V., ... & Randers, J. (2002). Tracking the ecological overshoot of the human economy. Proceedings of *the National Academy of Sciences*, 99(14), 9266–9271.

Zhou, Q., Yabar, H., Mizunoya, T., & Higano, Y. (2016). Exploring the potential of introducing technology innovation and regulations in the energy sector in China: A regional dynamic evaluation model. *Journal of Cleaner Production*, 112, 1537–1548.

www.ingramcontent.com/pod-product-compliance
Ingram Content Group UK Ltd.
Pitfield, Milton Keynes, MK11 3LW, UK
UKHW021842210426
5322IPUK00022B/424